T0254654

The Anthropocene: Politik—Economics—Society—Science

Volume 25

Series editor

Hans Günter Brauch, Mosbach, Germany

More information about this series at http://www.springer.com/series/15232
http://www.afes-press-books.de/html/APESS.htm
http://afes-press-books.de/html/APESS_25.htm

Hans Günter Brauch
Úrsula Oswald Spring
Andrew E. Collins
Serena Eréndira Serrano Oswald
Editors

Climate Change, Disasters, Sustainability Transition and Peace in the Anthropocene

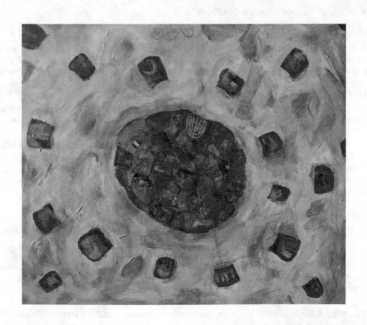

 Springer

Editors
Hans Günter Brauch
Peace Research and European Security
  Studies (AFES-PRESS)
Mosbach, Baden-Württemberg, Germany

Úrsula Oswald Spring
Centre for Regional Multidisciplinary
  Research (CRIM)
National Autonomous University
  of Mexico (UNAM)
Cuernavaca, Morelos, Mexico

Andrew E. Collins
Department of Geography
  and Environmental Sciences
Northumbria University
Newcastle upon Tyne, UK

Serena Eréndira Serrano Oswald
Centre for Regional Multidisciplinary
  Research (CRIM)
National Autonomous University
  of Mexico (UNAM)
Cuernavaca, Morelos, Mexico

More on this book is at: http://afes-press-books.de/html/APESS_25.htm.

ISSN 2367-4024   ISSN 2367-4032   (electronic)
The Anthropocene: Politik—Economics—Society—Science
ISBN 978-3-319-97561-0   ISBN 978-3-319-97562-7   (eBook)
https://doi.org/10.1007/978-3-319-97562-7

Library of Congress Control Number: 2018949876

© Springer Nature Switzerland AG 2019
This work is subject to copyright. All rights are reserved by the Publisher, whether the whole or part of the material is concerned, specifically the rights of translation, reprinting, reuse of illustrations, recitation, broadcasting, reproduction on microfilms or in any other physical way, and transmission or information storage and retrieval, electronic adaptation, computer software, or by similar or dissimilar methodology now known or hereafter developed.
The use of general descriptive names, registered names, trademarks, service marks, etc. in this publication does not imply, even in the absence of a specific statement, that such names are exempt from the relevant protective laws and regulations and therefore free for general use.
The publisher, the authors and the editors are safe to assume that the advice and information in this book are believed to be true and accurate at the date of publication. Neither the publisher nor the authors or the editors give a warranty, express or implied, with respect to the material contained herein or for any errors or omissions that may have been made. The publisher remains neutral with regard to jurisdictional claims in published maps and institutional affiliations.

Editor: PD Dr. Hans Günter Brauch, AFES-PRESS e.V., Mosbach, Germany.
English Language Editors: Margaret Gamberton, Harrogate, England, UK and Mike Headon, Colwyn Bay, Wales, UK.

The cover illustration is based on the artist's photograph of his painting "Power of Silence", for which the artist retains the copyright: © Narongrit Vannarat, Bangkok, Thailand. For Nop Vannarat's work as a painter, see at: https://onarto.com/artists/narongrit-vannarat/; https://www.facebook.com/narongrit.n.vannarat.
The illustration on the title page is based on the artist's photograph of his painting "Peace Project", for which the artist retains the copyright: © Peter Langguth, Mosbach, Germany, who also granted permission for its use. For details of Peter Langguth's work as a painter and musician, see at: http://www.badener-penwith.com/indexpeter2.html.

This Springer imprint is published by the registered company Springer Nature Switzerland AG
The registered company address is: Gewerbestrasse 11, 6330 Cham, Switzerland

*We dedicate this book to
all small children worldwide
who will face the destruction
we have left behind for the
present and future generations.*

Acknowledgements

This volume on *Climate Change, Disasters, Sustainability Transition and Peace in the Anthropocene* emerged from written papers that were presented to the *Ecology and Peace Commission* (EPC) during the 26th Conference of the *International Peace Research Association* (IPRA) on *Agenda for Peace and Development—Conflict Prevention, Post-Conflict Transformation, and the Conflict, Disaster Risk and Sustainable Development Debate* in Freetown, Sierra Leone, from 27 November to 1 December 2016.

The editors are grateful to Dr. Ibrahim Shaw, Co-Secretary General of IPRA (2012–2016), who organised the IPRA conference in Sierra Leone (2016) with the support of the Peace Research Centre of the University of Sierra Leone in Freetown.[1] We also thank all sponsors, especially the IPRA Foundation, who supported the participation of several colleagues from developing and low-income countries who had submitted written papers and who participated in the conference.

The co-editors of this book are grateful to all the authors who passed the double-blind anonymous peer review process and subsequently revised their papers to take into account the many critical comments and suggestions from these reviewers. Each chapter was assessed by at least three external reviewers from different countries.

We would like to thank all reviewers who spent much time reading and commenting on the submitted texts and who made detailed perceptive and critical remarks and suggestions for improvement—even for texts that could not be included here. The goal of the editors has been to enhance the quality of all submitted texts. The editors were bound by the reviewers' reports. Eleven texts were submitted, one manuscript was withdrawn, three drafts were rejected, and seven chapters were accepted with minor or major revisions.

The following colleagues (in alphabetical order) contributed anonymous reviews:

---

[1]See at: http://www.afes-press.de/html/pdf/2016/HGB/2016_04_ipra-news-letter6-vol-4.pdf (14 May 2018).

- Abubakar, Dr. Ayeshah, University of Malaysia in Sabah (UMS) (from The Philippines)
- Akuma, Joseph Misati, Ph.D., Senior Lecturer, Department of Social Studies, Maasai Mara University, Narok, Kenya
- Angom, Dr. Sidonia, Conflict Resolution and Peace Studies, Uganda
- Bang, Dr. Henry, Bournemouth University, UK
- Boersma, Associate Prof. Dr. Kees, Programme Director, Organisation Sciences, Project Leader, Amsterdam Research on Emergency Administration, Free University of Amsterdam, The Netherlands
- Brauch, Adjunct Prof. PD Dr. Hans Günter, Free University of Berlin (ret.) and AFES-PRESS, Chairman, Mosbach, Germany
- Bwalya, Dr. Martin, NEPAD, South Africa
- Chapa, Dr. Ana Cecilia, Faculty of Psychology, UNAM, Mexico
- Collins, Prof. Dr. Andrew E., Leader, Disaster and Development Network, Northumbria University, Newcastle upon Tyne, UK
- Dalby, Prof. Dr. Simon, CIGI Chair in the Political Economy of Climate Change Balsillie School of International Affairs, Wilfrid Laurier University, Waterloo, Canada
- Fosado, Dr. Ericka, UNAM, Mexico
- Fynn, Dr. Veronica Bruey, Adjunct Professor, Seattle University, School of Law, Seattle, Washington, USA
- Gomez, Dr. Oscar, Research Fellow, JICA Research Institute, Japan International Cooperation Agency, Japan
- Harris, Prof. Dr. Geoffrey, Technical University of Durban, South Africa (from Australia)
- Hurtado Saa, Prof. Dr. Teodora, Professor and Researcher, Department of Social Studies, Division of Social Sciences and Humanities, University of Guanajuato, León, Mexico
- Jansen, Dr. Bram, Sociology of Development and Change (SDC), Wageningen University, The Netherlands
- Jaspars, Dr. Susanne, University of Bristol, UK
- Jerez Ramírez, Dr. Deysi, Faculty of Social and Political Sciences, UNAM and Colombia
- Mahove, Dr. Golden, Deputy Team Leader and Agriculture Development Facility Lead, South Africa
- Mesjasz, Associate Prof. Dr. Czeslaw, Cracow Economic University, Poland
- Nombewu, Dr. Chuma, South Africa
- Nwenga, Dr. Dumisanie, Bulawayo, Zimbabwe
- Okpara, Dr. Uche, University of Leeds, UK (from Nigeria)
- Olokesusi, Prof. Dr. Femi, University of Ibadan, Nigeria
- Olorunfemi, Dr. Felix Bayode, Senior Research Fellow, Earth System Governance Project, Lund University, Sweden, Visiting Scholar, Department of Urban Studies and Planning, Massachusetts Institute of Technology and of the Universiti Teknologi Malaysia (MSCP) (from Nigeria)

- O'Riordan, Prof. Dr. Linda, Professor of Business Studies and International Management, Researcher and Lecturer in Business Studies and Corporate Responsibility (CR), Scientific Research Director, FOM KompetenzCentrum for Corporate Social Responsibility University of Applied Sciences, Essen, Germany (from the UK)
- Oswald Spring, Prof. Dr. Úrsula, UNAM, CRIM, Cuernavaca, Mexico
- Othman, Prof. Dr. Zarina, Strategic Studies and International Relations Program, School of History, Politics and Strategic Studies, Faculty of Humanities and Social Sciences, National University of Malaysia (UKM), Bangi, Selangor, Malaysia
- Padilla Menendez, Amb. Dr. Luis Alberto, Guatemala City, Director, Diplomatic Academy and of Iripaz, Guatemala
- Porter, Professor Dr. Elisabeth, University of South Australia, School of Communication, International Studies and Languages, Australia
- Quarmout, Dr. Tamer, Canada
- Raju, Dr. Emmanuel, Assistant Professor, Centre for International Law, Conflict and Crisis, København S, Denmark
- Richards, Prof. Dr. Howard, Santiago de Chile (from the USA)
- Rojas, Prof. Dr. Francisco, Rector, UN-mandated University for Peace in San Jose, Costa Rica (from Chile)
- Salim, Prof. Dr. Emil, Jakarta, Indonesia
- Serrano Oswald, Dr. Serena Eréndira, Secretary General of the Latin American Peace Research Council, CRIM, UNAM, Cuernavaca, Mexico
- Synott, Adjunct Prof. Dr. John, Sydney University (ret.), Sydney, Australia (from Brisbane, Australia)
- Zan, Prof. Dr. Myant, Myanmar, formerly with Multimedia University, Malaysia.

This book is the result of international teamwork among the editors and con-venors of IPRA's EPC. As co-convenors, *Prof. Dr. Úrsula Oswald Spring* and *PD Dr. Hans Günter Brauch* organised several sessions of IPRA's EPC in Freetown. Hans Günter Brauch prepared this volume, managed the peer review process, and did the copy-editing. As a native English speaker, *Ms. Margaret Gamberton* (York University, UK) volunteered to language-edit seven contributions to this book. *Mr. Mike Headon*, Colwyn Bay (Wales) language-edited the Introduction, Chapter 7, and the text in the front- and backmatter. The support of the above reviewers and the two language editors who contributed their time free of any remuneration is highly appreciated by the four coeditors.

Mosbach, Germany   Hans Günter Brauch
Cuernavaca, Mexico   Úrsula Oswald Spring
Newcastle upon Tyne, UK   Andrew E. Collins
May 2018   Serena Eréndira Serrano Oswald

Contents

Abbreviations

AfD	Action for Germany, right-wing populist party in Germany
AFES-PRESS	AG Friedensforschung und Europäische Sicherheitspolitik—Peace Research and European Security Studies
AIKS	African Indigenous Knowledge Systems
AMECIDER	Mexican Regional Development Association
APESS	The Anthropocene: Politik—Economics—Society—Science (book series)
ARTs	Assisted reproductive technologies
ASEAN	Association of Southeast Asian Nations
AWG	Anthropocene Working Group
CBD	Convention on Biodiversity
CC	Climate change
CE	Common Era or Current Era
CIGI	Centre for International Governance Innovation, Wilfrid Laurier University, Waterloo, Canada
COFEPRIS	Federal Commission for Protection against Sanitary Risk
CoK	Constitution of Kenya 2010
CONACyT	National Council of Science and Technology (Mexico)
CONAPRED	National Council to Prevent Discrimination (Mexico)
COP 15	Conference of Parties of UNFCC in Copenhagen in 2009
COROIPAS	Conferences on Research on International Peace and Security (predecessor of IPRA)
CO_2	Carbon dioxide
CR	Corporate responsibility
CRIM	Centro Regional de Investigaciones Multidisciplinarias (Regional Multidisciplinary Research Centre)
CSAFS	Climate-sustainable agriculture with food sovereignty
CSCE	Conference on Security and Co-operation in Europe
CTBT	Comprehensive Test Ban Treaty (1996)
DCRR	Disaster and Conflict Risk Reduction

DDN	Disaster and Development Network
DPAE	Département Provincial pour l'Agriculture et l'Elevage (Provincial Department for Agriculture and Livestock)
DRR	Disaster Risk Reduction
DWD	Dealing with Disasters (conference)
ecoSERVICES	Future Earth Project on the impact of biodiversity change on ecosystem functioning and services, and human well-being
ECR-FMP	Ecology, Conflict Risks, Forced Migration & Peace (successor to EPC)
ELRHA	Enhanced Learning and Research for Humanitarian Assistance
EPA	US Environmental Protection Agency
EPC	IPRA's Ecology and Peace Commission
EU	European Union
ESA	Earth Systems Analysis
ESDP	Environment, Security, Development and Peace
ESS	Earth System Science
EUR	Erasmus University Rotterdam
FAO	Food and Agriculture Organization of the UN
GADRI	Global Alliance of Disaster Research Institutes
GATT	General Agreement on Tariffs and Trade
GDP	Gross Domestic Product
GEC	Global environmental change
GHG	Greenhouse Gases
GIRE	Grupo de información en reproducción elegida (The Information Group on Reproductive Choice)
GMO	Genetic modified organisms
GSSP	Global Stratigraphic Section and Point
G 7	Group of seven major industrialised countries (Canada, France, Germany, Italy, Japan, UK, and USA)
G 8	Group of eight major industrialised countries (Canada, France, Germany, Italy, Japan, Russia, UK, and USA)
HIC	High-Intensity (Social) Conflict
ICRC	International Committee of the Red Cross
ICS	International Commission on Stratigraphy
ICSU	International Council for Science (Scientific Unions)
IDP	Internally Displaced Person
IDRiM	International Society for Integrated Disaster Risk Management
IFAD	International Fund for Agricultural Development
IFRC	International Federation of Red Cross and Red Crescent Societies
IGBP	International Geophysical–Biological Programme

IGEBU	Institut Géographique du Burundi (Geographical Institute of Burundi)
INEGI	National Institute of Statistics (in Mexico)
INGO	International non-governmental organisation
IPCC	Intergovernmental Panel on Climate Change
IPRA	International Peace Research Association
IPRA's EPC	IPRA's Ecology and Peace Commission
IPRAF	International Peace Research Association Foundation
ISS	International Institute of Social Studies
ISSC	International Social Science Council
IUGS	International Union of Geological Sciences
JICA	Japan International Cooperation Agency
KSI	Dutch Knowledge Network on Systems Innovation and Transition
KU Leuven	Catholic University, Leuven, Belgium
LGBTTTIQ+	Lesbian, gay, bisexual, transvestite, transgender, transsexual, intersexual, and queer/questioning populations, etc.
LSE	London School of Economics and Political Science
LUC	Land use changes
MAD	Mutually assured destruction
MFT	Master in Systemic Family Therapy
Mha	Million hectares
MIT	UTM Malaysia Sustainable Cities Program
NAFTA	North American Free Trade Agreement
NASA	US National Aeronautics and Space Administration
NASA	Network of African Science Academies
NATO	North Atlantic Treaty Organization
NEPAD	New Partnership for Africa's Development
NGO	Non-governmental organisation
NIH	National Institutes of Health
NNGO	National non-governmental organisation
NOAA	U.S. National Oceanic and Atmospheric Administration
NSF	U.S. National Science Foundation
NWO	Netherlands Organisation for Scientific Research
OCHA	Office for the Coordination of Humanitarian Affairs
ODI	Overseas Development Institute
OECD-DAC	Organisation for Economic Co-operation and Development—Development Assistance Committee
OSCE	Organization for Security and Co-operation in Europe
PD	Privatdozent (German academic title to be awarded after a habilitation)
PECS	Programme on Ecosystem Change and Society
PEISOR model	stimulus–response model on *P*ressure, *E*ffects, *I*mpacts, *S*ocietal *O*utcomes, *R*esponses
PGR	Mexican Attorney General's Office

PIK	Potsdam Institute for Climate Impact Research
PiS	Prawo i Sprawiedliwość (conservative Polish Party for Law and Justice)
PoC	Protection of Civilians sites
RedLara	Latin American Network of Assisted Reproduction
RETAC-CONACYT	National Council of Science and Technology, Network of Water Researchers in Mexico
SCJN	Supreme Court of Justice (Mexico)
SDC	Sociology of Development and Change, Wageningen University, Wageningen, The Netherlands
SDGs	Sustainable Development Goals
SFDRR	Sendai Framework for Disaster Risk Reduction
SIDA	Swedish International Development Agency
SNI	National Council of Researchers, Mexico
SOC	Soil organic carbon
SOM	Soil organic matter
SQS	Subcommission on Quaternary Stratigraphy
SRC	Stockholm Resilience Centre
SRT	Social Representations Theory
ST	Sustainability transition
STRN	Sustainability Transition Research Network
TKNs	Transformative Knowledge Networks
UCDP	Uppsala Conflict Data Program
UK	United Kingdom
UKADR	UK Alliance of Disaster Research
UKM	National University of Malaysia (Bangi, Selangor, Malaysia)
UMS	University of Malaysia in Sabah
UN	United Nations
UNAM	National Autonomous University of Mexico
UNCCD	United Nations Convention to Combat Desertification (1994)
UNCED	UN Conference on the Environment and Development (1992)
UNDP	United Nations Development Programme
UNEP	United Nations Environment Programme
UNFCCC	United Nations Framework Convention on Climate Change
UN GA	UN General Assembly
UN SC	UN Security Council
UNSDC	UN Sustainable Development Goals
UNU-EHS	United National University Institute for Environment and Human Security, Bonn, Germany
UNUIASTKI	United Nations University Institute for Advanced Studies' Traditional Knowledge Initiative
U.S.	United States

USA	United States of America
U.S. NSB	U.S. National Science Board
WBGU	German Advisory Council on Global Change
WFP	World Food Programme
WHS	World Humanitarian Summit
WMO	World Meteorological Organization
WSSD	World Summit on Sustainable Development in 2002 in Johannesburg
WTO	World Trade Organization

Chapter 1
Contextualising Climate Change, Disasters, Sustainability Transition and Peace in the Anthropocene

Hans Günter Brauch

This book emerged from peer-reviewed papers presented at the meetings of the *Ecology and Peace Commission* during the 26th General Conference of the *International Peace Research Association* in Freetown, Sierra Leone, from 27 November to 1 December 2016. Several of its authors are active in IPRA—as its Secretary General (Oswald Spring), as Secretary General of the Latin American Council for Peace Research (Serrano Oswald), in the IPRA Council (Brauch, Oswald Spring, Serrano Oswald) and as co-conveners of the EPC (2016–2018: Brauch, Collins, Mena).

The authors are established or retired senior scholars (Oswald Spring, Collins, Brauch), or active as a postdoctoral tenured researcher (Serrano Oswald) and as PhD candidates (Mena, Melis, Ombati), and they represent several social science disciplines: *political science* (Brauch, Mena, Melis), *anthropology* (Ombati, Serrano Oswald), *ecology* (Oswald Spring) and *geography* (Collins). They adopt different methodological and theoretical approaches. The chapters in this book address four broad themes in the Anthropocene, the new epoch of earth history that is replacing the Holocene, which has lasted from the end of the last glacial period about 11,700 years ago until the start of the nuclear era in 1945:

- *Climate change* (Oswald Spring, Ombati)
- *Disasters* (Collins, Mena, Melis)
- *Sustainability transition* (Brauch)
- *Sustainable peace* (Brauch, Oswald Spring),

as well as the theoretical approach of *sustainability transition* (Serrano Oswald).

Anthropogenic climate change and the accumulation of greenhouse gases, especially of carbon dioxide, in the atmosphere since the Industrial Revolution (1750) and

PD Dr. Hans Günter Brauch, Chairman, Peace Research and European Security Studies (AFES-PRESS), since 1987; co-convenor, IPRA's Ecology and Commission (2012–2016), Mosbach, Germany; Email: brauch@afes-press.de. He has been a co-convener of IPRA's EPC (2012–2018).

© Springer Nature Switzerland AG 2019   1
H. G. Brauch et al. (eds.), *Climate Change, Disasters, Sustainability Transition and Peace in the Anthropocene*, The Anthropocene: Politik—Economics— Society—Science 25, https://doi.org/10.1007/978-3-319-97562-7_1

more recently since the end of the Second World War and the start of the nuclear age (1945) have resulted in the first direct human interference into the composition of the atmosphere. In 2000, Nobel laureate Paul J. Crutzen (The Netherlands) coined the name 'Anthropocene' to describe this (Brauch, Chap. 8). This has provided a common context for four physical consequences: temperature change, variability of precipitation, rapid-onset environmental hydro-meteorological hazards and societal disasters, and slow-onset rise in sea level.

The chapter on climate change by *Úrsula Oswald Spring* (Mexico, Chap. 5) discusses "Climate-Smart Agriculture and a Sustainable Food System for a Sustainable-Engendered Peace" and that by *Mokua Ombati* (Kenya, Chap. 6) examines the "Ethnology of Select Indigenous Cultural Resources for Climate Change Adaptation: Responses of the Abagusii of Kenya". Of the three chapters on 'disasters', *Andrew E. Collins* (United Kingdom, Chap. 2) offers an *integrated Disaster and Conflict Risk Reduction* (DCRR) approach while of the two following literature reviews, the first by *Rodrigo Mena* (Chile/The Netherlands, Chap. 3) addresses "Responding to Socio-Environmental Disasters in High-Intensity Conflict Scenarios", while the second by *Samantha Melis* (The Netherlands, Chap. 4) discusses "The Fragile State of Disaster Response". The chapter by *Serena Eréndira Serrano Oswald* (Mexico, Chap. 7) applies the *Social Representations Theory* (SRT) approach to the "Family as a Social Institution in Transition in Mexico", while *Hans Günter Brauch* (Germany, Chap. 8) develops his argument on "Sustainable Peace through Sustainability Transition" by introducing this theme as an area for a "Transformative Science" as seen from a "Peace Ecology Perspective in the Anthropocene".

These chapters analyse the 'politics' and 'polity' dimensions but primarily address 'linkages' between 'policy' fields. They contribute to conceptual and empirical bridge-building, firstly between *disaster and conflict risk reduction*— these have only recently been framed together[1] by scholars and analysts, humanitarian organisations and sponsors of development assistance, secondly between *climate change impact research* and *adaptation* in the agricultural sector, and thirdly through a discussion that links research on *sustainability transition* in the social sciences with the normative goal of a *sustainable peace* in peace research.

---

[1]See the websites of the *United Nations International Strategy for Disaster Reduction* (UNISDR): on "Disaster and Conflict" at: https://www.unisdr.org/files/37777_plenaryintegratingdrrandccaforresil%5B1%5D.pdf; on "Conflict and disaster risk reduction" by UNESCO and the International Institute for Educational Planning (IIEP)" at: http://www.iiep.unesco.org/en/our-expertise/conflict-and-disaster-risk-reduction; the project description by the *Overseas Development Institute* (ODI, London, UK) at: https://www.odi.org/publications/10952-next-frontier-disaster-risk-reduction-tackling-disasters-fragile-and-conflict-affected-contexts. Several humanitarian organisations have analysed this linkage as "complex emergencies", for example, 'preventionweb' at: https://www.preventionweb.net/publications/view/34827 and reliefweb at: https://reliefweb.int/report/world/conflict-often-breeds-disaster-so-why-it-neglected-disaster-risk-reduction, and development agencies such as USAID at: https://www.usaid.gov/what-we-do/working-crises-and-conflict/disaster-risk-reduction/resources.

This author has put forward a special stimulus-response model (Brauch 2009), the PEISOR model (Fig. 8.2), that links the stages of *Pressure* (due to the burning of fossil fuels since the industrial revolution), *Effects* (through climate change) and their *Impacts* (e.g. environmental hazards and socio-environmental disasters), with *Societal Outcomes* (negotiations, crises, conflicts, wars) and *policy Responses*. From a peace research or 'peace ecology' perspective the implementation of the goal of 'sustainable development' through a process of a transition to sustainability (e.g. in the energy, transportation, production, and housing sectors) or 'sustainability transition' may reduce resource conflicts (e.g. by replacing fossil fuels with renewables) and prevent violent societal climate change impacts.

Andrew E. Collins (United Kingdom) argues in Chap. 2, "Advancing Disaster and Conflict Risk Reduction", that "enabling human survivability and improving quality of life for future generations requires reducing the risk of conflicts, destitution and environmental crises". He proposes a more "integrated *Disaster and Conflict Risk Reduction* (DCRR) framework" to provide "conceptual advances for better understanding, assessment, management and governance of risk and sustainability". He states that this integrated approach "emphasises early warning, rights based and resilience perspectives that build cross-cutting theoretical, policy and practice imperatives" for an expanded DCRR approach whose goal is to (i) build up earlier human well-being that can offset negative risk, (ii) live better with uncertainty and (iii) overcome barriers to mitigating and transcending disaster and conflict.

Rodrigo Mena (Chile / The Netherlands), in Chap. 3, "Responding to Socio-Environmental Disasters in High-Intensity Conflict Scenarios: Challenges and Legitimation Strategies", examines the process of "responding to socio-environmental disasters in places affected by high-intensity levels of conflict", and explores "the essential features and challenges that this type of conflict poses for disaster response." Using the concepts of "[the] humanitarian arena, legitimacy, and power relationships", Mena presents "the different strategies that aid and society actors ... use to respond in these complex settings, contributing to the study of the nexus between social conflicts and socio-environmental disasters such as earthquakes, droughts, or hurricanes." His chapter contributes to disaster response literature that uses "high-intensity conflict scenarios as an analytical category, to inform better policies and practices on disaster response in these specific types of conflict".

Samantha Melis (The Netherlands) claims in Chap. 4, "The Fragile State of Disaster Response: Understanding Aid–State–Society Relations in Post-Conflict Settings", that "natural hazards often strike in conflict-affected societies, where the devastation is further compounded by the fragility of these societies and a complex web of myriad actors". She argues that "to respond to disasters, aid, state, and societal actors enter the humanitarian arena, where they manoeuvre in the socio-political space to renegotiate power relations and gain legitimacy to achieve their goals by utilising authoritative and material resources". In conclusion she states that "post-conflict settings ... present a challenge for disaster response as actors are confronted with an uncertain transition period and the need to balance roles and capacity".

In Chap. 5, "Climate-Smart Agriculture and a Sustainable Food System for a Sustainable-Engendered Peace", *Úrsula Oswald Spring* (Mexico) argues that "in addition to extreme events due to climate change, losses of ecosystem services, soil depletion, water scarcity, and air pollution, in most emerging countries the importation of basic food items … has increased". She proposes "a *climate-sustainable agriculture with food sovereignty* (CSAFS) that combines … climate-smart agriculture … with the recovery of local food cultures, environmental diversity, and healthy food intake from a gendered perspective". In a case study the author illustrates "the nutritional impact on poor people of industrialised and imported food". In Mexico "the increase of food prices has forced many people to substitute nutritious fresh food with sugar and carbohydrates", and this has "increased obesity, diabetes, cardiovascular diseases, cancer and other chronic illnesses". Given this food crisis, the Mexican Government should develop a sustainable agricultural policy with "a healthy food culture, which may reverse environmental deterioration, increase the capture of greenhouse gases, mitigate climate change impacts, reduce the malnutrition of adults, and improve the chronic undernourishment of small children".

Mokua Ombati (Kenya) points in Chap. 6, "Ethnology of Select Indigenous Cultural Resources for Climate Change Adaptation: Responses of the Abagusii of Kenya", to the "consequences of climate change, and the need to adapt and spur livelihood challenges". During such periods "traditional African communities applied indigenous cultural resources to secure the agrarian sector which almost exclusively supported their livelihoods". In this chapter, Ombati aims to combine "insights from the theories of cultural functionalism and interaction rituals" to interpret "indigenous cultural resources [employed by] the Abagusii community" in response to climate change. He suggests utilising their "hitherto undervalued knowledge in partnership with contemporary climatological science" to support adaptation to climate change in Africa.

Serena Eréndira Serrano Oswald (Mexico) argues in Chap. 7, "Violent Gender Social Representations and the Family as a Social Institution in Transition in Mexico", that in the Anthropocene the social sciences are addressing "societal consequences of complex interrelations between global environmental change, lack of sustainable development, poor governance, inequality, social challenges, economic crises and risk society". She claims that in the Anthropocene "social relations, social dynamics and social institutions have … changed significantly" and that "the family [as] … the basic institution of society… has changed across time and space". In the last seven decades, the family "has experienced very visible changes which are redefining social knowledge, social relations and identities in multiple ways" because "broad societal changes reflected in the Anthropocene have also impacted on the family". The theory of Social Representations "is an epistemological, theoretical and methodological perspective evolving since the 1960s that deals with common sense knowledge, … understood as the link between knowledge and practice… in everyday life". She concludes that her chapter "develops a theoretical-conceptual framework to investigate the transitions, challenges and continuities of the family as an institution in the face the current *époque*, in the case of Mexico, especially following technological advances and legislative changes that have polarised the public".

Finally, in Chap. 8, "Sustainable Peace through Sustainability Transition as Transformative Science: A Peace Ecology Perspective in the Anthropocene", *Hans Günter Brauch* (Germany) contributes to "a conceptual discussion on the need for bridge-building between the natural and social sciences, among different social science disciplines, and the research programmes in political science focusing on peace, security, development and environment ('sustainable development'), by introducing the two new linkage concepts of 'political geo-ecology' and 'peace ecology'" that focus "on the policy goal of a 'sustainable peace' understood as 'peace with nature' in the … the Anthropocene". He argues that "this goal may be achieved by a process of 'sustainability transition' … where concerned individuals, families, local communities, states and nations as well as international governmental organisations and non-governmental bodies and social movements may contribute to the transition". His chapter addresses "fundamental conceptual, methodological, theoretical and action-oriented research needs … [that will contribute] to the realisation of a 'peace with nature' in the 'Anthropocene', where the societal outcomes of the physical effects of global environmental and climate change can be countered … by policies of adaptation, mitigation and an increase of resilience by the affected people." From a Hobbesian perspective and harnessing traditional scientific worldviews, "this goal may appear … to be utopian and … not achievable". It requires both a fundamental change in the dominant worldview and the neoliberal mindset aimed at a "'scientific revolution towards sustainability' … with a new scientific paradigm of a 'peace ecology'".

Contextualising the causes of anthropogenic climate change with increasing socio-environmental disasters, the process of sustainability transition and the goal of a 'sustainable peace' or 'peace with nature' in the new era of earth and human history remains a conceptual, theoretical, empirical but also practical challenge in the new epoch of earth and human history, the Anthropocene, an epoch that still has to be accepted by the geologist and where fundamental contributions are also needed from the social and policy sciences (Brauch 2020).

References

Brauch, Hans Günter, 2009: "Securitizing Global Environmental Change", in: Brauch, Hans Günter; Oswald Spring, Ursula; Grin, John; Mesjasz, Czeslaw; Kameri-Mbote, Patricia; Behera, Navnita Chadha; Chourou, Béchir; Krummenacher, Heinz (Eds.): *Facing Global Environmental Change: Environmental, Human, Energy, Food, Health and Water Security Concepts*. Hexagon Series on Human and Environmental Security and Peace, vol. 4 (Berlin–Heidelberg–New York: Springer): 65–102.
Brauch, Hans Günter (Ed.), 2020: *Politics, Policy and Polity in the Anthropocene: Tasks for the Social Sciences in the 21 st Century* (Cham: Springer International Publishing).
Crutzen, Paul J.; Stoermer, Eugene F., 2000: "The Anthropocene", in: *IGBP Newsletter*, 41: 17–18.

Chapter 2
Advancing Disaster and Conflict Risk Reduction

Andrew E. Collins

Abstract Enabling human survivability and improving quality of life for future generations requires reducing the risk of conflicts, destitution and environmental crises. A more *integrated Disaster and Conflict Risk Reduction* (DCRR) framework provides conceptual advances for better understanding, assessment, management and governance of risk and sustainability. This synthesis for sustainability and peace emphasises early warning, rights based and resilience perspectives that build cross-cutting theoretical, policy and practice imperatives in advancing DCRR. Derived DCRR systematics include (i) building up earlier human well-being that offsets negative risk, (ii) living better with uncertainty and (iii) overcoming political, behavioural and technical barriers in disaster and conflict risk transitioning.

Keywords Disaster risk reduction · Conflict risk · Imperatives Systematics

2.1 Introduction: Actuality of Disaster and Conflict Risk

Much of the debate about the nature of world disasters that has pervaded the human mind has been as to whether what is experienced as resultant human misery is a part of everyday life to be expected, is an abnormality induced by others or is self-inflicted. Understanding that no disaster scenario should be considered 'natural' meant that early scholarship on disaster management increasingly recognised the processes of human constructed events that lead to major crises (O'Keefe et al. 1976; Hartman/Squires 2006). The more recently updated disaster terminology indicates it is "a serious disruption of the functioning of a community or a society at any scale due to hazardous events interacting with conditions of exposure, vulnerability and capacity, leading to one or more of the following: human, material, economic and environmental losses and impacts" (UNISDR 2016). Should the

Prof. Dr. Andrew E. Collins, Department of Geography and Environmental Sciences / Disaster and Development Network (DDN), Northumbria University, Newcastle upon Tyne, UK, NE18ST. Email: andrew.collins@northumbria.ac.uk.

© Springer Nature Switzerland AG 2019
H. G. Brauch et al. (eds.), *Climate Change, Disasters, Sustainability Transition and Peace in the Anthropocene*, The Anthropocene: Politik—Economics—Society—Science 25, https://doi.org/10.1007/978-3-319-97562-7_2

human propensity for conflict be here considered to also be a hazardous event the definition already serves to describe conflict as disaster. Moreover, should disasters be considered simply as non-routine social events (Drabek 2010), a wide range of human conflict fit the definition. Hilhorst (2013) draws focus to the 'everyday politics of conflict and disaster and crisis response' that has implications for what needs to be studied as elements of disaster management beyond the 'technocratic point of view' (p. 2). This would also add legitimacy to the aid worker and recipient interface (Hilhorst et al. 2013). Typically, disaster risk assessment and management involves understanding the likelihood and expected losses caused (Smith 2001), or of exposure to hazards in terms of human vulnerability including compromised capacities (Blaikie et al. 1994; Cannon 1994; Cutter et al. 2003; Pelling 2003; Wisner et al. 2004; Bankhoff et al. 2004; Collins 2009a; Smith 2013). These basic formulae for understanding and addressing risk have been pervasive throughout the sector and in most instances sit comfortably with what emerged at the turn of the millennium as a more fundamental integration of disaster and development studies.[1]

The field of Disaster Risk Reduction (DRR) emerged through variously understanding a range of environmental hazards, human exposure to those events and the local and wider vulnerability contexts within which they are driven. The emphasis on varying human conditions and sustainability trajectories as the driver and recovery process by which disasters can be addressed at all levels is central to disaster and development studies. Further elaboration on a full range of hazards, such as for example including technological and pandemic health related, also helped build this field. Ultimately, a common thread conceptually is that the origins of disaster risk lie in complex and human managed systems of environmental, social and economic change. The transitioning of policy emphasis accompanying an ongoing scholarly investigation alongside the hard learnt realities from many disaster zones is now more clearly reflected in multiple global narratives. They include in particular the Sustainable Development Goals 2015–2030 (UN 2015a), the Sendai Framework for Disaster Risk Reduction (UN 2015b), the Framework Convention on Climate Change – Conferences of the Parties (UN 2015c) and the World Humanitarian Summit (UN 2016). Each promotes an agenda of *prevention* as an ultimate means to deal with multiple disaster types. However, whilst implicit to several of the SDGs, such as Goal 16 to promote just, peaceful and inclusive societies, and explicit in core commitments of the World Humanitarian Summit, conflict risk is not distributed across the frameworks conceptually or in terms of policy and practice. Moreover, perspectives and systematics to advance integrated disaster and conflict risk reduction, though implied or called for by some scholars, lacks advancement.

The human survivability and accompanying improvements in quality of life people would want to experience in the future requires meeting risk reduction and sustainability targets faster and more effectively, including with regard to both

[1]The world's first centre of international postgraduate studies in combined disaster management and sustainable development was developed at Northumbria University, United Kingdom in the late 1990s. Its first intake of students started during Millennium year 2000 and continues annually to date.

conflict and other disaster risk reduction, given an ongoing onslaught of both conflict and resources pressures through environmental degradation in many parts of the world (The Worldwatch Institute 2015). Further, few or any of the risks to life in the future can be effectively addressed as singular threats. Detailed knowledge of earthquake, flood, drought, cyclone, volcano, tornado, tsunami, wildfire, ice and so forth contribute part explanations of the impact resulting from the physical workings, timing, size, distributions and even rates of change in these phenomena. The human costs have been in part measurable though remaining often incalculable at a more personal level and in terms of longer-term human well-being. There is also little clarity and achievements in working out points of intervention in multiple types of hazards and disaster vulnerabilities together, whereby neglect is to go back on any objective of achieving greater human well-being. A further representation can be made that absence of mitigation and relief of disaster impact, where systems fail, is in effect a form of violence (Ray-Bennett 2018). The problem is definable in terms of human experiences of conflict disaster and mass displacement of lives and livelihoods through an inability to deal with complex systems within which politics, behaviour and technology ultimately increase or reduce disaster and conflict risk. It is within this context that this chapter aims to raise a call for progress in more integrated disaster and conflict risk reduction studies.

Conflict risk is in many ways already a part of what is meant by disaster and development studies in that both are fundamentally to do with an integrative understanding of process of change that provides opportunities for peace. Reducing disaster risk enables better development and the latter is what needs to be the focus of both disaster prevention and better recovery post emergency, the latter here being also represented by the so called 'build back better'[2] aspiration. The actuality is that all disaster and development aspirations are dependent on human cooperation as mutual support in coping with complex systemic risks to human survival. Whilst it will be no surprise to readers that human conflict is the antithesis to this survivability within nature in a rapidly transforming world, it is perhaps more concerning that progress in integrating disaster and conflict risk reduction as a combined interest is as yet a limited pursuit. As introduced above, whilst conflict risk can be seen within the SDGs it is not fundamental to their presentation, and was missed out altogether from the Sendai Framework for Action (SFDRR) (UN 2015b), crucially so for some regions such as Africa (Manyena 2016), and globally largely due to the political obstacles it would encounter. Benefits for conflict reduction that stem from the SFDRR, though acknowledged in WHS, even where indirectly, can however be clearly analysed as progressively positive for formulations of DRR (Glasser 2016; Stein/Walch 2017).

The need to build focus and advance DCRR is however not just an aberration of the policy world; it is with a sense of urgency that understanding conflict and other disaster and development risks riding together is key to prevention that can bridge

---

[2]Build back better dates back to writings such as Monday (2002) but is now also risen to prominence as the fourth of the priorities of the SFDRR (2015–2030).

the nexus. However, caution is needed in deriving simplistic cause and effect relationships here, as has become apparent in a recent debate on the role of climate in the origins of the Syrian conflict (Selby et al. 2017). Whilst a direct link to climate change effects in the region is highly contestable, environmental resource pressures as a part of a complex of socio-environmental, political-economic, regional and cultural linkages can be a factor in the overall weighting, and some instances triggering of conflict. Further, once conflict has become embedded, even when due to reasons other than physical environmental change, the use of environmental damage can become used as a weapon of war (Zwijnenburg 2016), historically involving a "scorched earth" approach during advances and retreats of combating forces.[3] The regional relationship whereby conflict is a context for other types of disasters is a well-established phenomenon and the more usual approach to the topic. For example, importantly the Overseas Development Institute estimated that 58% of disaster deaths occurred in the top 30 fragile states (Harris et al. 2013).

Another aspect of the actuality of conflict and other forms of disaster is that the aid and humanitarian industry, public and private emergency services required to respond to each type of crisis are often comprised of the same agencies, organisations and institutions. Some of the representational issues for these assemblages are also part to do with the longstanding nuance of understanding what is meant by 'disaster'. A well-trodden discussion about what is and what is not within different domains of human crisis, major incidents and disaster risk, and what terminology can be used to communicate this, persists over many years (Quarentelli 1998; Perry/Quarantelli 2005; Below et al. 2009; UNISDR 2016). Beyond terminology, however, most analysts, whether professionally recognised or not from either a disaster or conflict orientation, would recognise that 'conflict' as disaster dwarfs other categories of incident, even before it would be considered accentuated by environmental, social and economic stressors. Thus the great human crises of almost all regions of the world have been subject to conflict and depravation coexisting in time and place, from famines of South Asia into the 1970s, to those of many parts of sub-Saharan Africa until present. In Sierra Leone, where an early version of this chapter was presented at the end of 2016, the same communities of local people living in the vicinity of our conferencing hall[4] had lived through concurrent disaster events including 10 years of brutal civil war, followed by a brief recovery period before then living through the Ebola major incident of the 2000s and more recently an increase in catastrophic flash floods and landslides. Whilst these and many other

---

[3]"Scorched Earth" is an approach used in conflicts to hold up enemy approaches by destroying the resource base upon which their advances depend upon to survive – hence Napoleon's advance on Russia failed when food supplies and most of the infrastructure was destroyed by the retreating armies of imperial Russia, urban areas in modern day conflicts get land mined, and entire areas may be forcefully polluted by exploded oil wells.

[4]An early version of the chapter formed a part of the 2016 International Peace Research Association (IPRA) and Dealing with Disasters Conference (DwD) at Freetown, Sierra Leone Nov 27th–Dec 1st 2016.

examples of rapid onset mass fatalities through conflict, health and environmental risk are concentrated together and are not difficult to identify as concurrent, many are accompanied by slower onset and protracted poverty that pervades before, during and after these crises.

The argument of this chapter is in the first instance simply that a more integrated CDRR effort could open up much opportunity for progress in reducing the risks of ongoing and concatenated disasters. This can benefit from the current drive within disaster and development studies, otherwise referred to as DRR, to build cultures of smart prevention. It remains relatively easy to agree amongst those that see the evidence from the world's ongoing disaster zones. However, work needs to drive forward conceptual advances for better understanding conflict and disaster risk assessment, management and governance processes that accompany this challenge. Moreover systemic analysis of conflict and other types of disaster together is relatively scarce. This is a sentiment that has been expressed by King/Mutter (2015) who note that political and media figures make reference to it but scholarly work has remained somewhat separated. They go on to provide an explanation as to why the two, in actuality, ride together. The ODI also provided a review of when disasters and conflicts collide, beginning also to identify some of the links between disaster resilience and conflict prevention (Harris et al. 2013). However, despite these contributions and other similar recognition statements worldwide there is a need for integrated disaster and conflict students to begin to analyse more in depth the systematics that can take this forward. Indeed one of the recommendations of the ODI report is to develop and test conceptual frameworks and analytical tools by way of responding to the demand that such approaches would progress. This chapter continues by exploring a number of the hitherto neglected systematics that might be considered in advancing DCRR.

One starting point is to better unpack what viewing DCRR from different sustainability perspectives might look like. Table 2.1 is therefore a further basic adaptation from earlier analyses that emerged from a simple representation of environmental, social and economic emphases on sustainability begun in sustainability frameworks produced in the 1980s (Hatzius 1996). This was reworked by Collins (2009a) to demonstrate dimensions of sustainability in relation to disasters and is here expanded further to specifically identify a sustainable conflict risk reduction emphasis. Conflict in this context is not merely open warfare but also conflict over environmental resources whether through war or more subliminally, in everyday life or as presented by the mutually assured survival of nation states and economies. The table is by way of background framing and not meant to identify the range of perspectives and systematics that can be worked upon that are presented later in this chapter.

This understanding of different aspects of sustainability in relation to DCRR is nonetheless a suitable entry point to understanding the complex aspects of human systems that impinge on future survivability and in terms of underpinning an integrated DCRR. For example, it includes poverty and environmental degradation through population pressure, as accounted for in many a neo-Malthusian analysis.

Table 2.1 Viewing disaster and conflict risk from different sustainability perspectives. *Source* Adapted from Collins (2009a: 48)

	Environmental sustainability (Env.)	Sustainable development (Social)	Sustainable growth (Economic)
Purpose	• Ecological viability	• Social efficiency, justice	• Economic efficiency, sustainable production and reproduction
Policy rationale	• *Protect nature, educate people, equilibrium, holism, co- evolutionary ideas*	• *Empower people, build community, develop institutions and livelihoods*	• *Develop markets and internalise externalities*
Relationship of sustainability issues to disasters	• Environmental hazards, people and environmental change	• Vulnerability, human security and multiple dimensions of poverty	• Institutional security, infrastructure, economic control
Emphasising sustainable conflict risk reduction	• Environmental security • Environment access • Environmental justice	• Peace building • Trust and cooperation • Societal well-being	• Mutually assured survival of nation states and economies

The relationship can be represented as various forms of political ecology (Bryant/ Bailey 1997; Blaikie 1999; Forsyth 2002), ultimately unfolding as processes of poverty and environmental degradation combined with marginalisation, rapid onset environmental changes, humanitarian crises and all manner of outcomes or drivers of conflict. Reversing systems of deteriorating environments, societies and economies therefore requires inclusion and vulnerability reduction, environmental stability, well-being, disaster reduction and conflict mitigation. Whilst the starting point in bringing about change in any of these could be in any one part of the system, disaster and conflict risk reduction might better advance together, particularly in the case of extreme emergencies.

Moving from sustainability issues to those of DRR has conceptually not been difficult once commonalities are made clear. The integration agenda can be seen in that many initiatives now imply or directly address development as needing to be risk informed, that disaster and development are two sides of the same issue or simply that DRR to all intents and purposes is sustainable development (Collins 2009a, 2013). A number of approaches in the DRR lexicon have development explanations or are derivatives of this. Those addressed further here, as likely areas from which DCRR can increasingly draw, include early warning, rights based and resilience perspectives. Some explanation is introduced as follows and then expanded upon in new formulations of DCRR systematics in the later sections of the chapter.

2.2 Perspectives Common to Disaster Risk Reduction and Conflict Risk Reduction

2.2.1 Early Warning of Disaster and Conflict

Early warning as part of the prevention end of the disaster management cycle has developed significantly through a more analytical risk reduction approach. Rather than simply a system of sounding the alarm when danger appears to be approaching, this field is about the entirety of predictive models, early warning and risk assessments that variously help to identify the likelihood of disaster. The World Meteorological Organization (WMO 2016) indicates its core components as detection, monitoring and forecasting the hazards, analyses of risks involved, dissemination of timely warnings – which should carry the authority of government and activation of emergency plans to prepare and respond. However, in wider application for multiple types of hazards and emergencies early warning can include socio-economic, behavioural and physical research techniques with multiple institutions needed as part of an elaborate information system infrastructure.

Key challenges of early warning include the unpredictability in the information produced, the way it is communicated and how people react to warnings (Mileti/Sorensen 1990; Drabek 1999; Collins/Kapucu 2008). These are some of the processes already well analysed in contexts of emergency management. Understanding the early warning process also informs an expanded field of more subjective risk assessments and decision-making under conditions of uncertainty. Early warning within DRR therefore invokes the governance of wider preventative development processes that have potential benefits for safety both now and in the longer term future. Such an emphasis grew in response to the recognition that early warning for saving lives and assets is only as good as an accompanying early action (IFRC 2009). Effective intervention required individuals and communities that are engaged and motivated to respond and this comes through integrated early warning and development that engages from within those people and communities at risk (Collins 2009b). Ultimately, influences on early action are influenced by knowledge, power, culture and environment assemblages in affected areas (Collins 2009a, b).

Engaging community based early warning and action therefore requires combinations of investments in research and capacity alongside sustainable monitoring and management systems. The signals and communication necessary for this can include multiple techniques ranging from remote sensing, risk mapping, environmental assessment and market behaviour models through to participatory monitoring and evaluation, entitlement mapping, stakeholder and political analysis, ethnography. The early warning response process is complex since it includes all of hearing, understanding, believing, personalising, confirming, deciding and responding. This is clearly a broad field upon which DCRR can be built.

Conflict early warning, as a stand-alone field, is neither new. Many times during conflict related humanitarian crises questions as to what early warning was heard and what was ignored have been asked. For example, Adelman/Suhrke (1996: 79) identified that;

> although in the mid-nineties Rwanda genocide early warning was less critical than the willingness and ability to respond, the failure to respond adequately was in part influenced by the failure to collect and analyse the data that was available and to translate this information into strategic plans.

O'Brien (2010) reviews indices and modelling approaches that have been applied and queries the way forward for crisis early warning and decision support. Pham/Vinck (2012: 115) explain how the expansion of technology presents a likely positive role for greater inclusiveness of society in conflict early warning but that this is not really fully understood. They indicate that; 'conflict early warning systems still lack the ability to accurately predict violent events', and also the need for further study to understand; 'the usability of "big data" for the purpose of detecting trends and forecasting conflicts' and that 'the effectiveness of these technologies continues to be undermined by the lack of connection between warning and response'.

Considering the bringing together of early warning of conflict with that of other types of disaster risk in developing areas where data may be unavailable, where protection schemes have limitations and where there are few alternatives but to risk take presents common challenges of both early warning communication and overall capacity to prevent. Though significant gains in life saving has been made through preventative actions in Bangladesh over the last few decades, an array of literature is available documenting the struggle to implement early warning and action. The case of evacuation to safety in areas of wide scale flood and cyclone demonstrated the point that people will risk life and livelihood to protect their scarce assets, that evacuation to refuge points away from home can threaten cultural norms and security and that information is received in multiple ways (Alam/Collins 2010). The dilemmas faced are not entirely different to experiences of those facing approaching conflicts, whether to stay home, to abandon ones assets or be forcibly displaced in advance of a worsening crisis on the basis of which set of information available. It is axiomatic that conflict and disaster early warning requires security contingency planning in at risk locations, forward thinking for displacement alleviation and grounded data that is accurate, communicable and free of fake information.

2.2.2   Rights-based Approaches

Whereas rights based approaches to development became well-established as a field for analytical development and practise, there had been little significant advances in mainstreaming the approach for the case of reducing the risk of disasters in policy and practice environments. The achievement of human rights is both a means to securing sustainable development and an objective of development and therefore a rights based approach to people's protection from disaster risk would seem entirely feasible. However, the less tangible field of risk assessment struggles to engage enforceable legislations. There are no conventions in the world that currently aim to impose legislation on nation states to reduce disaster risk (Eburn et al. 2018). In relation to DRR the ILCs *Articles on the Protection of Persons in the event of Disasters* (2016) would be the nearest to a basis of any binding international convention. Nonetheless, the concept is strong in that the process of people trying to exercise their rights in the context of disaster risk is engaging and can be impactful. For example, when people know their land rights and collectively build food and livelihood from this resource they would be less prone to being shifted into situations of vulnerability by developers who would attempt to concentrate capital for smaller numbers of beneficiaries. The struggle to define the scope of the legality of rights based approaches to DRR could benefit from greater integration of more closely related conflict based analyses, for in the case of the latter some legislation to the Geneva Conventions are possible to apply.

In relation to conflict there can be recourse to international hard law as set out in the Geneva Conventions. This forms the basis of international humanitarian law with minimum obligations being applied during times of armed conflict including the rights of citizens to be protected and to be provided with aid. Breach of the conventions may amount to a war crime and offenders may be prosecuted before the International Criminal Court that was established by the *Rome Statute of the International Criminal Court* (1998). Though more in the responsive rather than preventative mode of operation, humanitarian work is also assisted by minimum standard of response as outlined in the Sphere Project (2000 and later editions), which is based on rights, refugee and humanitarian law. There is not yet this approach of minimum requirements for disaster or conflict risk reduction but it stands to reason that progress in framing this would be best if addressing the human side of disaster and conflict risk reduction together. This is because rights in disaster relief are supposed to in any event span the rights to survival, dignity, development, and security. Human rights in disaster response ultimately include all of the rights that people aspire to during normal periods of development.

The human rights approach means that people are able to better resist disaster events since, where a lack of rights increases, vulnerability risks are also increased. This increased vulnerability is usually reflected by the condition of *human security*, which includes food and livelihood, health, environmental and other forms of entitlement. In that future generations have rights too, including the right to live in a viable and peaceful environment in which disaster scenarios are not elevated beyond a reasonable level, suggests the complimentary integration of rights based risk reduction and basic principles of sustainable development. It echoes much of the issues within the struggle for human rights throughout the ages, relating to power, control, representation, freedom, values, and survival. Identifying the role of people's rights in contexts of DCRR therefore soon exposes the need for good governance of risk. Governance, the action, manner or system of governing, in

DRR emphasises that environmental crises can be natural hazards triggered but are more often the result of failed development and poor governance.

Disaster avoidance therefore comes with an imperative for better governance of risk. Ojha et al. (2009: 365), also quoted in Jones et al. (2016) indicate that what;

> hinders effective governance in most situations is the prevalence of complex interplay of power and knowledge among diverse groups of actors with unequal command over resources to influence mutual interactions that underpin governance actions.

Meanwhile good local risk governance can enable proactive engagement that reduces disasters. This is because people at the local level often already know best how to adapt to hazards, manage risks, demand rights, develop resilience and secure livelihood niches. Locally owned prevention and response has been witnessed to also strengthen community cohesion and to counteract moral/social downturns in society, with potential economic and environmental benefits as a result. The benefits extend not least to engaging knowledge, attitudes and practice, more sustainable systems with an effect of addressing multiple hazards and risks together. However, it is not necessarily clear, and can only be identified on a case by case approach, as to how a rights and responsibilities emphasis for risk reduction can be best addressed through organized local governance. It may otherwise be more reliant on adaptation based human agency already latent within communities. However, risk governance depends on rights based risk reduction to be effective. Good disaster risk governance in contexts of conflict and other disaster categories is a development process, depending on people with rights and other capabilities, locally grounded, with good risk communication and delineated responsibilities of states and citizens. In the context of DCRR, and as a basis upon which to advance, it is suggested it needs to be:

- Informed – by ongoing real or perceived threats of the governed
- Practitioner orientated – guided by a perpetual interpretation and review process
- Proactively engaged – including with hazards, vulnerability, and coping to facilitate resilience
- Guided by lessons learnt – through evaluation before, during and after risk reduction activities
- Related to localised knowledge – made relevant through grounded research
- People centred – driven and motivated disaster assessment that is multidisciplinary, integrated and perpetually reassessed
- Invested in – through political will, institutional and personal commitment to disaster reduction and sustainable development.

Ultimately good risk governance not only offsets DRR and DCRR but invests in local well-being that would be sustainable. This is what can be considered as building resilience to disasters.

2.2.3   Resilience to Disaster and Conflict

DCRR is considered here to represent a set of actions taken in contexts of conflict and other hazards that would lead to greater human resilience. It is not necessary here to examine anew some hundreds of resilience definitions, rather to highlight an example of how its theoretical stance supports DCRR, such as can be reflected in the definition by Turnbull et al. (2013: 160) who captured a relatively common working narrative as follows:

> Resilience refers to the capacity of an individual, household, population group or system to anticipate, absorb, and recover from hazards and/or effects of climate change and other shocks and stresses without compromising (and potentially enhancing) its long-term prospects.

Whilst this, and other definitions, relate resilience to hazards and climate change, definitions such as that of Turnbull et al. above, and many more, typically add reference to 'other shocks', implying that conflict impacts are included. Human resilience in the context of conflicts need not be entirely reflected by these definitions but can be observed in the day to day survival of many millions of people living in the ruins of war and failed states. Resilience for advancing the DCRR paradigm is therefore relative to its surrounding topology. As such, it is pertinent to ask; resilient to what? It could for example be any of environmental hazard, unsustainable development, socio-economic destitution, political exploitation, others stressors and combinations of all of these. The process of resilience building may be through achieving greater adaptation and flexibility or alternatively a form of hardening (Hyslop/Collins 2013). These characteristics are relevant here but there is no intention in this chapter to provide a further rethinking of the connotation of resilience since this has already been advanced and redeveloped in other literature (Manyena 2006; Davoudi 2012).

Being resilient as a foundation on which to advance DCRR relative to its nature within surrounding topologies can implicate for example, environmental, social and economic dimensions of everyday life that are resistant to drought and to conflict. The ability of communities, such as for example pastoralist groups in Southern Angola, to be resilient to both the extreme climate whilst also managing to avoid getting caught up in major military conflicts during the 1980s and 90s was due to migrating livelihoods between pasture and water flexibly through the seasons using techniques adapted over thousands of years. Resilient people have to reduce conflict and disaster risk by getting out of the way of oncoming threats. Further, there are many actions that people take in pursuit of making their circumstances more resilient that contributes to the rationale of DCRR as in part about emergency adaptation. The pedagogy of community resilience could be then considered to resemble problem based and adaptive responses from people centred learning. Partnership building and establishment of communities of practice are further ways in which resilience produces risk reduction.

Ultimately, cooperation between groups of people and their institutions, rather than an obsession with the prizes and devastation wrought by competition, can be a key driver of resilience. Building resilience as part of risk reduction may take many forms, through the strengthening of livelihood security and other forms of human security such as health. It can involve organising community action and motivated building of systems of good governance. Where activities build what is considered to be resilience to environmental hazards, examples from around the world indicate that these processes are generally low cost and can be sustainably monitored. They are adaptable to local knowledge and perception, felt as effective within the community and therefore adopted, serving as a stimulus for wider risk reduction. Reference can be made, for example, to local flood, drought or cyclone committees in Mozambique, or flood action groups in the UK; community based DRR to build local resilience is now more widespread. This represents a step towards grounded awareness and an energy that could be advanced in the interests of also reducing the risk of conflicts.

The process of building resilience knowingly or unknowingly in policy and practice cannot however be an end in itself since people desire for more than just to survive; people prefer to thrive. This sentiment is the basis of an orientation towards well-being interpretations and outcomes that can be analysed in terms of DCRR. At the heart of a well-being agenda is that it is possible to build up quality of life in a way that offsets disaster risk. The theme is returned to further in the next section.

2.3   Some Systematics for Advancing DCRR

The justification for advancing DCRR presented so far in this chapter has been based on the actuality of conflict and other types of disasters drawing from a rationale and perspectives that are common to both existing DRR approaches and mutually supportive for conflict risk reduction, either directly or implied. The early warning, rights based and resilience aspects of DCRR are however subject to limitations of understanding risk. A first priority of the SFDRR is to better understand risk and implies that in depth work will be required to do so, engaging an entire partnership of institutions that can help progress the sciences and technologies necessary for the framework to function with impact. Building on these themes it is therefore important to try to advance existing DRR scholarship, practice and policy in a DCRR agenda. The systematics that follow are what emerges as key challenges, albeit also opportunities, confronting DRR in terms of its making a significant impact towards achieving the goal of the SFDRR.[5] It is being asserted in this chapter how these are also aspects that may be best analysed as DCRR.

---

[5]Goal of Sendai Framework for Disaster Risk Reduction – "Prevent new and reduce existing disaster risk through the implementation of integrated and inclusive economic, structural, legal, social, health, cultural, educational, environmental, technological, political and institutional measures that prevent and reduce hazard exposure and vulnerability to disaster, increase preparedness for response and recovery, and thus strengthen resilience." (UNISDR 2015, Clause 17, p. 12).

Addressing challenges to DCRR, rather than to DRR or CRR on their own enables more opportunity to understand and advance each domain, assisting in expanding disaster and development concepts, policy and practice going forward. Whilst what follows are only some of the potential systematics, they are fundamental to further developing the risk reduction paradigm and in identifying opportunities of DCRR.

2.3.1   Build up Human Well-being Earlier to Offset Disaster and Conflict Risk

Building-up human well-being earlier to offset negative risk impact is a different concept to the current emphasis on building back better post disaster as promoted in the SFDRR. What this variant approach advocates is that it is conceptually and most probably for many instances a practice and policy strength to invest in well-being even prior to risks are confirmed and early warnings given. Human well-being in the context of DCRR is about a attaining a combined state of health, resilience and human security (Collins 2009a). Specifically health requires nutrition, water and sanitation, shelter and energy, health care and longevity to mention some aspects. Meanwhile, resilience requires ability to cope, capacity, adaptability and creativity, social, economic and cultural resources (or capital). Human security requires rights, access and resources, representation and empowerment to name a few aspects. Overall, the build up early approach referred to here is about strengthening society well in advance of the propensity for an increased vulnerability to hazards. In so doing a tendency for conflict is also reduced, whilst by recognising opportunity for conflict reduction all manner of well-being is increased, perpetuating the offsetting of negative risks. It is also related to the idea of building back better in that targeting of areas and people can be identified by both pre and post disaster reference points. A version of the approach was related to human health being also core to a study in Bangladesh on using health security for disaster resilience (Ray-Bennett et al. 2016). It was proposed that people's resistance to physical and mental stresses or shocks, reduction of poverty and ill-health and presence of basic rights were what was required to have health security. Part of this agenda was to gain household's opinions regarding what protected them from flood, cyclone and drought. Money and good health were cited the most times (Nahar et al. 2013).

It is clear from people living with the threat of flood, cyclone or drought, who are vulnerable to these and other hazards, that there needs to be a concerted effort by states and other development and disaster risk reduction advocates to find ways to strengthen the well-being of those exposed, so that they are able to resist, get out of the way or help their family or neighbour to avoid disaster. The effects of this targeting of well-being building would be to reduce conflict risk. It is not entirely proven, but is theoretically sound and practically desirable. This however requires a shift from response modes that are brief to longer term prevention investments that are able to address DCRR. It is not advocated that attention should be detracted from the build back better objective needed for sustainable recovery post disaster. It is suggested that with DCRR in mind in particular that investment in well-being pre-disaster in areas at risk of disruption, peacebuilding, even where small scale would offset significant higher levels of conflict whilst be addressing other forms of disaster risk. This needs to be considered as an approach linked to smart early warning and resilience building that is grounded in finding pathways to local well-being strengthening during non-risky periods as well.

2.3.2 Live with Uncertain Disaster and Conflict Risk

The second of the systematics presented here is better living with uncertainty. A weakness in risk reduction approaches has been the relativity of what is considered a risk, a known risk or an unknown risk and derivatives of these (Vasta 2004; Collins 2009a, b; Olson/Wu 2010). There are problems also with what is considered a norm of human suffering and therefore what is an "acceptable risk". For example if diarrhoeal disease epidemics kill thousands of young children in the economically poorest countries each year and the world gets used to this occurring, this somehow becomes a globally more acceptable risk of occurrence in comparison to only a few cases of the same diseases occurring more sporadically in wealthy locations. Further uncertainty challenges loom large from the re-emphasising of the human side of disaster risk. This is because whilst it might be possible to improve levels of certainty as to when a next flooding event will occur due to improved weather forecasting and hydrological river basin modelling, predicting human behaviour and reactions to risks in real world context is not a well-developed field. As there begins to be greater recognition of DCRR this aspect of the human condition becomes particularly in focus. Understanding social relations and systems of meaning in DRR requires awareness of cultures of conflict, peace, development and risk as an interconnected field. DCRR is a field concerned not only with major rapid onset threats but also the everyday underlying tendencies to generate negative types or risk behaviour. Social relations can be considered to include social capital, communication, accountabilities, responsibilities, dependencies, emotional and kinship ties, symbiosis and empathy. The systems of meaning are the lenses through which these are examined, such as the intrinsic value of natural systems or human life, mediation and cooperation, hope and expectation, trust, consciousness, rationality, justice and rights.

The evidence base for such an expansive agenda for unique places, times and people would present too vast an array of uncertainties to tackle. For DCRR there are alternative considerations so that advances could be more effective. For example, more could be made of approaches that accept the use of 'unknowing'. It needs to be applied with care as generally not knowing may simply increase the possibility of increasing exposure to risk. What is meant rather is that it is possible to move in the right direction of DCRR without knowing everything. Not knowing everything should not be an excuse for inaction. Some fields concerned with environmental conservation and management have already been aligned with this in part through the 'precautionary principle' (Harremoës et al. 2002). As a process of better living with uncertainty of conflict or other types of disasters forthcoming, the unknowing frame of analysis suggests the merits of dealing with a form of 'non-experiential learning' (Collins 2015b), of working between the gap of known and unknown risks. In this context is could be helpful to also consider unknowing along the lines of that engaged by many peoples through mysticism,[6] the use of the 'imaginary unit' or 'j operator' in mathematics (Peterson 2004; Nahin 2010) and other responses to complexity such as 'Info Gap' (Ben-Haim 2001) across multiple disciplines. The randomness that starts to form order within chaotic systems or cosmic patterns provides opportunities for living with gaps in knowledge. This suggests that future survivability in relation to climate or complex and largely unpredicted emergencies, such as occurred with the combined earthquake, tsunami and nuclear meltdown in northeast Japan in 2011, definitively requires greater predictability but also realism linked to better gap filling for the unknown.

Uncertainty relates to the three perspectives outlined earlier and the other systematics in that people's unknowing can also be a reason for their injection of hope into a situation including during professional relief and response operations. Projects facilitated by the Northumbria programmes and elsewhere working in contexts of mass displacement had noted that to have uncertainty about the future when a child in a refugee camp can breed hope and this counteracts fatalism and hopelessness to encourage opportunities for resilience, survivability and well-being.

A problem facing DRR programmes is the tendency for people to not engage, since many risks are not regular or as actual in everyday life as the need to have food on the table. However, a consideration that can be applied in this context is that the evidence of engaged risk reduction will be proportionate to the evidence of certainty multiplied by evidence of hope – further research would need to be routinely carried out to understand this as a DCRR systematic should hope also be inversely proportionate to conflict. This however is only one way of advancing the paradigm. It is important to note that effectiveness of building more hope is mediated by economic, cultural and bio-geophysical contexts. The sentiment is particularly appropriate to DCRR where people at risk of conflict and other disasters may be better equipped to live expectations in everyday life through transitioning hope into risk reduction actions. Hence, dealing with uncertainty better through a DCRR perspective could be key going forward. This is also because it is conducive to inclusion of all people and as such can be an aspect of what is meant by an all of society approach.

The all of society approach to living with uncertainty introduced here promotes the role of collective learning. Learning is a function of experiences, secondary

---

[6]For example, in the Cloud of Unknowing, a text written by a late 14th Century Christian Mystic it suggested to surrender one's mind and ego to the realm of unknowing, at which point, one may begin to glimpse the nature of God.

sources and of feelings and belief that occur individually, though in dealing with uncertainty are checked and complimented by processes of collective learning. An open learning environment demands cooperative structures and shared under-standing. This approach to learning is participatory and empowering, and impor-tantly can encourage smart coping with known and unknown threats within complex social, economic and environmental systems. The level of cooperation required and resulting then becomes conducive to community building, mutual support and consequent conflict risk reduction.

2.3.3   Overcome Political, Behavioural and Technical Barriers in Disaster and Conflict Risk Transitioning

The third systemic of DCRR involves understanding its boundaries and optimal means to transitioning, or transcending, these to achieve potentially wide ranging outcomes. This suggests a further complex array of factors unique in time and place and for which merely better identification of the barriers would begin a process of progressing opportunities for risk reduction and sustainability. It is suggested that understanding of transitioning processes, which may also be thought of as trans-formative or transcendental, is more realistic if considered in a DCRR domain rather for DRR without conflict included. If transitioning depends on early warning, rights and resilience the purpose of this approach would be to identify the nature of the barriers to early warning and action; those which are preventing people from living better with uncertainty and which are therefore in this analysis obstructing resilience and well-being.

It was introduced in Sect. 3.1 above that a desired outcome of DCRR is to transition to wellbeing from vulnerability and that this requires perspective that contribute to understanding how biological susceptibility, insecurity and mental impairment becomes health, human security and resilience. Processes of change that require recognition and removal of barriers are likely to be the most cost effective approach available, empowering people to be able to prevent disaster and conflict whilst being likely to open up pathways to greater sustainability. This contrasts with approaches that for example would seek to input additional projects, programmes and policies and that can then disrupt the propensity for people to advance DCRR for themselves. Some examples of boundaries or barriers in tran-sitioning relate to risk qualifiers or quantifiers, perceptions, security systems, communication, market forces, knowledge, trust, habitat, values and other factors driven by the nature of places, cultures and social economies. To transcend or transform the transitioning process might be more or less risky, finite, self-regulating, equalising, accelerating, entropic, mobile, learnt, commodified, diffusive and creative, to indicate a few. The outcomes therefore include people, systems and places that have stayed the same or changed, becoming more or less secure, vulnerable, fragile, complex homo or heterogeneous, capable, sensitive, placed, ethical, included and peaceful.

2.4   Conclusions

Disaster and conflict risks span many aspects of everyday life and critical situations. Knowledge that produces practical ways forward through strengthened risk reduction analyses involves addressing integrated, proximate and underlying disaster and conflict situations that are variably visible. Bringing the two types of risk context together will assist in developing the means to all of society engagements with DCRR since one element is usually dependent on, or informed by, the context of the other. This then helps identify pathways to DCRR awaiting further exploration. It is proposed that better understanding and engagement with systematics of DCRR presented here could help advance more effective strategies of DRR through sustainable development. This would be the case in contexts of climate change adaptation, mass migration, humanitarian intervention and poverty reduction. Longstanding and new systematics of building up earlier, living with uncertainty and barrier transitioning are required for both reducing conflict and other disaster risks. Building DCRR more into risk assessment through integrated common objectives and approaches to bridge gaps, address ethics, enhance human reactions to managing risk, motivate and engage good risk governance, rights and responsibilities for all of society leads to sustainability and peace. Whilst seemingly a vast vision to promote, there is opportunity and some hope in the process of the various global targets for 2030 to make advances in offsetting poverty and disaster and conflict risk with investments in well-being. Ultimately, an expansion of DCRR within the disaster and development paradigm and DRR would significantly contribute.

References

Adelman, Howard; Suhrke, Astri, 1996: *The International Response to Conflict and Genocide: Lessons from the Rwanda Experience*, Joint Evaluation of Emergency Assistance to Rwanda (Copenhagen).
Alam, Edris; Collins, Andrew E, 2010: "Cyclone Disaster Vulnerability and Response Experiences in Coastal Bangladesh", in: *Disasters*, 34, 4: 931–53.
Bankhoff, Greg; Frerks, George; Hilhorst, Dorothea, (Eds.), 2004: *Mapping Vulnerability: Disasters, Development and People* (London: Earthscan).
Below, Regina; Wirtz, A; Guah-Sapir, Deberata, 2009: Disaster Category Classification and Peril Terminology for Operational Purposes. CRED (Brussels: Munich RE).
Ben-Haim, Yakov, 2001: *Information-gap Decision Theory: Decisions Under Extreme Uncertainty* (San Diego: Academic Press).
Blaikie, Piers; Cannon, Terry; Davis, Ian; Wisner, Ben 1994: *At Risk*, 1st edition (London: Routledge).
Blaikie, Piers, 1999: "A Review of Political Ecology: Issues, Epistemology and Analytical Narratives." In: *Zeitschrift für Wirtschaftsgeographie*, 43: 131–47.
Bryant, Raymond L; Bailey, Sinéad, 1997: *Third World Political Ecology* (London: Routledge).

Cannon, Terry, 1994: "Vulnerability Analysis and the Explanation of "Natural" Disasters", in: Varley, Ann (Ed.): *Disasters, Development and the Environment* (Chichester: John Wiley): 13–30.

Collins, Andrew E., 2009a: *Disaster and Development* (London: Routledge Perspectives in Development Series).

Collins, Andrew E., 2009b: "The People Centred Approach to Early Warning Systems and the 'Last Mile'", in: International Federation of the Red Cross and Red Crescent Societies (IFRC): *Focus on Early Warning, Early Action* (Geneva: World Disaster Report): 39–68.

Collins, Andrew E., 2010: "Human Rights", in: *Encyclopaedia of Disaster Relief 1* (SAGE Publications Inc): 310–13.

Collins, Andrew E., 2013: "Editorial - Linking Disaster and Development: Further Challenges and Opportunities", in: Special Edition, *Environmental Hazards*, 12, 1: 1–4.

Collins, Andrew E., 2015a: "Risk Governance and Development", in: Paleo, Urbano F., (Ed.): *Risk Governance: The Articulation of Hazards, Politics and Ecology* (London: Springer): 477–80.

Collins, Andrew E., 2015b: "Beyond Experiential Learning in Disaster and Development Communication", in: Egner, Heike; Schorch, Marén; Voss, Martin (Eds.): *Learning and Calamities: Practices, Interpretations, Patterns* (London: Routledge) 56–76.

Collins, Matthew L.; Kapucu, Naim, 2008: "Early Warning Systems and Disaster Preparedness and Response in Local Government", in: *Disaster Prevention and Management*, 17, 5: 587–600.

Cutter, Susan L.; Boruff, Bryan J.; Shirley, Lynn W., 2003: "Social Vulnerability to Environmental Hazards", in; *Social Science Quarterly*, 84, 2: 242–61.

Davoudi, Simin, 2012: "Resilience: A Bridging Concept or a Dead End?", in: *Planning Theory and Practice*, 13, 2: 299–333.

Drabek, Thomas E., 1999: "Understanding Disaster Warning Response", in: *The Social Science Journal*, 36, 3: 515–23.

Drabek, Thomas E., 2010: *The Human Side of Disasters* (London: CRC Press).

Eburn, Michael; Collins, Andrew E.; Da Costa, Karen, 2018: "Recognising Limits of International Law in DRR as Problem and Solution", in: Samuel, Katja; Aronsson-Storrier, Marie; Bookmiller, Kirsten N. (Eds.): *The Cambridge Handbook of Disaster Risk Reduction and International Law* (Cambridge University Press).

Forsyth, Tim J., 2002: *Critical Political Ecology* (London: Routledge).

Glasser, Robert, 2016: "Tackling disaster reduces risk of conflict," UNISDR commen; https:// www.unisdr.org/archive/51734 (16 February 2018).

Harris, Katie; Keen, David; Mitchell, Tom, 2013: *When disasters and conflicts collide. Improving links between disaster resilience and conflict prevention* (London: ODI).

Hartman, Chester; Squires, Gregory D., (Eds.), 2006: *There is no Such Thing as a Natural Disaster: Race, class and Hurricane Katrina* (Oxon: Routledge).

Harremoës, Poul; Gee, David; MacGarwin, M.; Stirling, Andy; Keys, Jane; Wynne, Brian; Guedes Vaz. (Eds.), (2002): *The Precautionary Principle in the 20th Century: Late Lessons from Early Warnings* (London: Earthscan and European Environment Agency).

Hatzius, Thilo, 1996: "Sustainability and Institutions: Catchwords or New Agenda for Ecologically Sound Development?", in: *IDS Working Paper* 48.

Hilhorst, Dorothea, (Ed.), 2013: *Disaster, Conflict and Society in Crises: Everyday Politics of Crisis Response* (Oxon: Routledge Humanitarian Studies).

Hilhorst, Dorothea; Andriessen, Gemma; Kemkens, Lotte; Weijers, Loes. 2013: "Doing Good/ Being Nice? Aid Legitimacy and Mutual Imaging of Aid Workers and Aid Recipients", in Hilhorst, Dorothea (Ed.): *Disaster, Conflict and Society in Crises: Everyday Politics of Crisis Response* (Oxon: Routledge Humanitarian Studies): 258–74.

Hyslop, Maitland P.; Collins, Andrew E., 2013: "Hardened Institutions and Disaster Risk Reduction", Special Edition – Linking Disaster and Development: Further Challenges and Opportunities, in: *Environmental Hazards*, 12, 1: 19–31.

IFRC (International Federation of the Red Cross and Red Crescent Societies), 2009: *Focus on Early Warning, Early Action* (Geneva: World Disaster Report).

ILC (International Law Commission), 2016: *Draft articles on the protection of persons in the event of disasters* (Geneva: United Nations).

Jones, Samantha; Oven, Katie J.; Wisner, Ben, 2016: "A Comparison of the Governance Landscape of Earthquake Risk Reduction in Nepal and the Indian State of Bihar", in: *International Journal of Disaster Risk Reduction*, 15: 29–42.

King, Elizabeth; Mutter, John C., 2015: "Natural Disasters and Violent Conflicts", in: Collins, Andrew E; Jones, Samantha; Manyena, Bernard; Jayawickrama, Janaka, (Eds.): *Hazards, Risks and Disasters in Society* (Oxford: Hazards Risks and Disasters Series, Elsevier): 181–98.

Manyena, Bernard S., 2006: "The Concept of Resilience Revisited", in: *Disasters Journal*, 30, 4: 433–50.

Manyena, Bernard S., 2016: "After Sendai: Is Africa Bouncing Back or Bouncing Forward from Disasters?", in: *International Journal of Disaster Risk Science*, 7, 1: 41–53.

Mileti, Dennis S.; Sorensen, John H., 1990: *Communication of Emergency Public Warnings: A Social Science Perspective and State-of-the-Art Assessment* (Oak Ridge, TN: ORNL-6609, Oak Ridge National Laboratory, Department of Energy).

Monday, Jacqueline L., 2002: "Building Back Better: Creating a sustainable community after disaster", in: *Natural Hazards Informer*, 3: 1–11.

Nahar, Papreen; Collins, Andrew E.; Bhuiya, Abbas; Alamgir, Fariba; Ray Bennett, Nibedita S.; Edgeworth, Ross, 2013: "Indigenous Indicators of Health Security in Relation to Climatic Disasters in Bangladesh", in: Special Edition – Linking Disaster and Development: Further Challenges and Opportunities, *Environmental Hazards*, 12,1: 32–46.

Nahin, Paul J., 2010: *An Imaginary Tale: The Story of $\sqrt{-1}$* (Princeton and Oxford: Princeton University Press).

O'Brien, Sean, 2010: "Crisis Early Warning and Decision Support: Contemporary Approaches and Thoughts on Future Research", in: *International Studies Review*, 12: 87–104.

Ojha, Hemant R.; Cameron, J.; Kumar, C., 2009: "Deliberation or Symbolic Violence? The Governance of Community Forestry in Nepal", in: *Forest Policy and Economics*, 11, 5–6: 365–74.

O'Keefe, Phil; Westgate, Ken; Wisner, Ben, 1976: "Taking the Naturalness out of Natural Disasters", *in: Nature*, 260: 566–7.

Olson, David L.; Wu, Desheng D., 2010: "A Review of Enterprise Risk Management in Supply Chain", *Kybernete*, 39, 5: 694–706.

Pelling, Mark, 2003: *The Vulnerability of Cities: Natural disasters and social resilience* (London: Earthscan).

Perry, Ronald W.; Quarantelli, Enrico L., 2005: *What is a Disaster? New Answers to Old Questions* (USA: International Research Committee on Disasters).

Peterson, John C., 2004: *Technical Mathematics with Calculus*, Third Edition (New York: Thomson).

Pham, Phuong N.; Vinck, Patrick, 2012: "Technology, Conflict Early Warning Systems, Public Health, and Human Rights", in: *Health and Human Rights*, 14, 2: 106–17.

Quarantelli, Enrico L., 1998: *What is a Disaster? Perspectives on the Question* (London: Routledge).

Ray-Bennett, Nibedita S., 2018: *Avoidable Deaths: A Systems Failure Approach to Disaster Risk Management* (Cham: Springer International Publishing).

Ray-Bennett, Nibedita S.; Collins, Andrew E.; Edgeworth, Ross; Bhuiya, Abbas; Nahar, Papreen; Alamgi, Fariba, 2016: "Everyday Health Security Practices as Disaster Resilience in Rural Bangladesh", in: *Development in Practice*, 26, 2: 170–83.

Rome Statute of the International Criminal Court, 1998: United Nations Diplomatic Conference of Plenipotentiaries on the Establishment of an International Criminal Court (Rome: International Criminal Court).

Selby, Jan; Dahi, Omar S.; Fröhlich, Christiane; Hulme, Mike, 2017: "Climate Change and the Syrian Civil War Revisited", *Political Geography*, 60: 232–44.

A. E. Collins

Smith, Keith, 2001: *Environmental Hazards: Assessing Risk and Reducing Disaster*, Third Edition (London: Routledge).
Smith, Keith, 2013: *Environmental Hazards: Assessing Risk and Reducing Disaster*, Sixth Edition (London: Routledge).
Stein, Sabrina; Walch, Colin, 2017: *The Sendai Framework for Disaster Risk Reduction as a Tool for Conflict Prevention* (Brooklyn, NY: Social Sciences Research Council, 15 July).
The Sphere Project, 2000: *Humanitarian Charter and Minimum Standards in Disaster Response* (Geneva: Sphere Project).
The Worldwatch Institute, 2015: *Confronting Hidden Threats to Sustainability* (New York: Island Press).
Turnbull, Marilise; Sterrett, Charlotte L.; Hilleboe, Amy, 2013: *Toward Resilience: a Guide to Disaster Risk Reduction and Climate Change Adaptation* (Rugby, UK: Practical Action).
UN (United Nations), 2015a: *Transforming our World: The 2030 Agenda for Sustainable Development*, Seventieth Session, Agenda Items 15 and 116, A/Res/70/1 (New York: UN General Assembly).
UN (United Nations), 2015b: *The Sendai Framework for Disaster Risk Reduction 2015–2030* (Third UN World Conference in Sendai, Japan, March 18, 2015).
UN (United Nations), 2015c: *Report of the Conference of the Parties on its twenty-first session, held in Paris from 30 November to 13 December 2015*, FCCC/CP/2015/10 (Bonn: United Nations Climate Change Secretariat).
UN (United Nations), 2016: *Outcome of the World Humanitarian Summit (WHS) Report of the Secretary General*, 71st Session, Agenda Item 70, A/71/353 (New York: UN: General Assembly).
UNISDR (United Nations International Strategy for Disaster Reduction), 2016: "Report of the Open-ended Intergovernmental Expert Working Group on Indicators and Terminology Relating to Disaster Risk Reduction"; at: http://www.preventionweb.net/files/50683_oiewgreportenglish.pdf (12 April 2017).
Vasta, Krishna S., 2004: "Risk, Vulnerability, and Asset-based Approach to Disaster Risk Management", in: *International Journal of Sociology and Social Policy*, 24, 10/11: 1–48.
WMO (World Meteorological Organization), 2016: *Multi-hazard Early Warning Systems*, http://www.wmo.int/pages/prog/drr/projects/Thematic/MHEWS/MHEWS_en.html (16 February 2018).
Wisner, Ben; Blaikie, Piers; Cannon, Terry; Davis, Ian, 2004: *At Risk: Natural Hazards, People's Vulnerability and Disasters*, Second Edition (London: Routledge).
Zwijnenburg, Wim, 2016: Environmental Damage as a Weapon of War? Open Source Industrial Risk Analysis of the Mosul Battle, *Bellingcat online resource*, accessed https://www.bellingcat.com/news/mena/2016/10/25/environmental-damage-weapon-war-open-source-industrial-risk-analysis-mosul-battle/ (16 February 2018).

Chapter 3
Responding to Socio-environmental Disasters in High-Intensity Conflict Scenarios: Challenges and Legitimation Strategies

Rodrigo Mena

Abstract This chapter reviews the process of responding to socio-environmental disasters in places affected by high-intensity levels of conflict, and explores the essential features and challenges that this type of conflict poses for disaster response. Using the notions of humanitarian arena, legitimacy, and power relationships, the chapter presents the different strategies that aid and society actors (those for whom humanitarian aid action is part of their core function and those for whom is not) use to respond in these complex settings, contributing to the study of the nexus between social conflicts and socio-environmental disasters such as earthquakes, droughts, or hurricanes. This chapter makes an original contribution to the disaster response literature by reflecting on the utility of using high-intensity conflict scenarios as an analytical category, to inform better policies and practices on disaster response in these specific types of conflict.

Keywords Disaster response · High-intensity conflict · Aid-society actors
Legitimacy · Humanitarian arena

3.1 Introduction

The earthquake in Afghanistan in 2015, as well as the decade-long drought in Somalia, exemplify the challenges faced by multiple type of actors, including local and international ones, when responding to a socio-environmental disaster[1] such as

---

Mr. Rodrigo Mena is a research assistant on disasters response and humanitarian aid at the International Institute of Social Studies (ISS) of Erasmus University Rotterdam in The Hague; Email: mena@iss.nl. This chapter was made possible by a VICI grant of the Netherlands Organisation for Scientific Research NWO, grant number 453-14-013.

---

[1]The concept of *socio-environmental* disaster is addressed in more detail below, including an explanation of the relevance of stressing the social aspects of it. In this paper, the terms disaster and socio-environmental disaster will be used interchangeably.

© Springer Nature Switzerland AG 2019
H. G. Brauch et al. (eds.), *Climate Change, Disasters, Sustainability Transition and Peace in the Anthropocene*, The Anthropocene: Politik—Economics—Society—Science 25, https://doi.org/10.1007/978-3-319-97562-7_3

earthquakes, droughts or hurricanes, in places affected by high levels of social conflict. Access and security issues of all involved stakeholders contribute to the political and social strategies required to develop a comprehensive and effective disaster response. This chapter examines the process of disaster response in places affected by high-intensity levels of conflict. The purpose of this chapter is to contribute to disaster response policies and practice by understanding better the special features required in responding in places where, among other response challenges, wide-spread violent social conflict occurs.

The reasons for this approach are three-fold. First, multiple studies demonstrate that the occurrence of socio-environmental disasters may affect social conflict and, vice versa, social conflict affecting the response to and ocurrence of disasters (e.g. Harris et al. 2013; Nel/Righarts 2008; Spiegel et al. 2007; Wisner 2012). However, little political and academic attention has been given to the differences between multiple conflict scenarios and the unique challenges that each of them represents for disaster response. Disaster response models and international agreements do not incorporate scenarios where disasters occur in situations of conflict. For example, the Sendai Framework,[2] the most recent active and long-term international agreement on disaster risk reduction, does not mention the concept of conflict or crisis. Secondly, regardless of how unfortunate it might seem, the co-occurrence of conflict and disaster happens, especially in places with widespread violent conflict or facing a complex emergency. During the decade from 1995–2004, a total of 87% of complex emergency sites were affected by socio-environmental disasters (Spiegel et al. 2007). Despite this trend, the features of responding to disaster in places affected by violent social conflict are under-studied or addressed in overly narrow manner. Thirdly, various studies give an account of the common social base that disasters and conflicts share, stressing the need to deal with them in a coordinated manner (Bankoff et al. 2004; Hilhorst 2013b; Wisner 2012).

Exploring the multiple dynamics of the social and political aspects of the co-occurrence of disaster response and widespread violent conflict is a critical issue. Using the term *high-intensity conflict* (HIC) as an analytical category to understand disaster response, this chapter sets itself the following questions: 'what does it mean to respond to socio-environmental disasters in places affected by HIC' and 'how can actors respond?'

The chapter has four main sections. Following the introduction, the key elements of HIC and disaster response are described. Next, the challenges that this type of conflict poses for disaster response are explored, and the actors involved in the process are identified. With this discussion as a basis, the chapter then explores the

---

[2]This framework refers to an international document – the Sendai Framework for Disaster Risk Reduction (2015–2030) – adopted by the UN state members. It seeks to achieve in the next fifteen year the following outcome: "The substantial reduction of disaster risk and losses in lives, livelihoods and health and in the economic, physical, social, cultural and environmental assets of persons, businesses, communities and countries" (United Nations 2015b: 12).

strategies used by different actors enabling them to perform in a socio-environmental disaster. Finally, the conclusion offers reflections, including a critical assessment of the value of using high-intensity conflict scenarios as an analytical category to inform disaster response. A summary of the main results and a critical review of them is presented in this final section.

The chapter is also an attempt to map and document the available literature related to the question being addressed in an effort to fill the identified knowledge gap. The conceptualisation of high-intensity conflict is proposed and developed in order to add to the existing literature. The theoretical concepts of *aid-society, humanitarian arena, legitimacy, and power relationships* are introduced as a method of studying the *problematic* presented.

These four terms are crucial in addressing the issues in question. In order to understand the complex, socially-constructed nature of the response in HIC settings, it is necessary not only to know how aid agencies and all society (state and non-state) actors respond, but also to know how the response is affecting, and is affected by, their interactions. The notions of *aid-society relationships* and *humanitarian arena* offer an appropriate analytical framework to observe the complex fabric of processes and actors that each specific context presents. The basic premise of the chapter is that the response is essentially socially constructed and embedded in wider social (power) relationships and scenarios. An effective response to a disaster is enhanced when the response is legitimate in the eyes of the affected population and other stakeholders. Even under a state of emergency such as HIC in which the option of coercive power is more available, the legitimacy of aid is crucial as the access, distribution and allocation of aid, and the protection of all people involved, depends on many actors on the ground. At the same time, aid resources can also offer legitimacy to actors that seek power, including the government or contesting parties. A focus on legitimacy thus shifts attention to the everyday politics of aid delivery in which actors invest their meaning and seek to enhance their strategic interests by engaging, altering or disengaging from the terms of aid. Consequently, aspects of the legitimacy, negotiation, empowerment, and institutional change associated with the response are also reviewed to understand disaster response in HIC settings better.

Methodologically, the chapter is based on an extensive literature review on humanitarian aid, disaster response, violent social conflict, and on legitimacy and institutional power relationships. The review included books, journal articles, reports, policy documents, and protocols[3] published or released up to November

---

[3]'Policy documents and protocols' refer to documents written by United Nations, NGOs, donors, and other aid organisations describing procedures, norms and/or standards. E.g. The Sphere Handbook, the International Humanitarian Law, security guidelines of some NGOs.

Table 3.1 Description of interviews. *Source* The author

Code	Interviewed	Gender	Description
AC1	Academic	Male	Professor of humanitarian aid with vast experience in consultancies and evaluation
AC2	Academic	Male	Researcher on humanitarian aid with experience in projects management with international non-governmental organisations (INGOS) and the United Nations
AP1	Aid practitioner	Female	INGO project manager with more than 10 years of experience in emergency projects, some of them in HIC areas
AP2	Aid practitioner	Male	National NGO project manager with experience in emergency response and Water, Sanitation and Hygiene (WASH) programmes. NGO from a HIC country
C1	Consultant	Male	International consultant on disaster risk reduction and resilience with experience working with United Nation agencies, INGOs, Donors and developmental organisations. Experience in HIC countries
B1	Aid beneficiary	Female	Beneficiary of humanitarian aid, affected by extreme drought in a HIC affected country
B2	Aid beneficiary	Male	Person affected by extreme floods in areas of high intensity conflict, who then volunteered for rescue and humanitarian relief operations

2016. It also included grey literature and audiovisual material, including blog entries, websites and documentaries.[4] After this desk research, seven interviews were carried out with two academics, two aid practitioners, one consultant and two aid beneficiaries. Table 3.1 provides more details on each interviewed. The aim of the interviews was to present and discuss the results of the literature review with different actors and identify analytical blind spots. Finally, the chapter is also to some extent informed by the author's own experience conducting fieldwork in HIC countries like South Sudan or Afghanistan, although the interviews, participant observations, and other data gathered in those cases are not formally included in this chapter.

Regarding data analysis, a thematic content analysis was carried out by tabulating all the information obtained. Analytical codes consisted of 44 initial analytical categories and the construction of new emergent sub-categories. The codes, the sample, and further information are presented in the Appendix.

---

[4]Grey literature is commonly unpublished and less formal information, usually defined as a 'genre of literature [that] includes theses and dissertations, faculty research works, reports of meetings, conferences, seminars and workshops, students' projects, in-house publications of associations and organizations… [forming a] body of materials that cannot be found easily through conventional channels such as publishers, but which is frequently original and usually recent' (Okoroma 2011: 789). Every time that grey literature was used, the information was validated with peer-reviewed documents, official data and statistics, or via interviews and triangulation of the information presented.

3.2 Unwrapping High-Intensity Conflict Scenarios (HIC) and Disaster Response

As presented by Demmers (2012), it is important when studying violent conflict to be clear about the differences that exist with the concept of war and also to understand that there are multiple types of conflict, not all of them violent. For example, it is easy to find in the literature the notions of low-intensity conflict and post-conflict. However, there is very little discussion of high-intensity types of conflict. In this chapter, it is proposed that 'high-intensity' represents a valid type of conflict which allows situations or scenarios to be described that includes not only the presence of violent conflict but also of a particular set of governmental arrangements and social problems, without necessarily being a conflict which is called a war. Moreover, this scenario imposes specific challenges for disaster response, shaping the response itself.

To unwrap the notion of high-intensity conflict (HIC) scenarios, it is necessary first to understand better the role of violence and its relationship with conflict. Violent social conflict is generally depicted as a competition, clash, or contradiction between two or more social groups or actors over a specific goal, resource, or interest involving the use of manifest violence to pursue the objectives (Oberschall 1978; Homer-Dixon 1994; Galtung 1996; Demmers 2012; Estévez et al. 2015; Ide 2015). Manifest violence is here conceptualised as a "visible, instrumental and expressive action. It is this kind of violence that is generally defined as 'an act of physical hurt'" (Demmers 2012: 56). Sometimes it is also termed physical violence, when one person "is physically damaged or physically restricted without giving consent to the activity" (Cameron 1999 in Gasper 1999: 10). Although in HIC scenarios the manifest and direct forms of violence are more evident, structural and cultural forms of violence are also important. Structural violence is embedded in social structures or institutions, preventing people from meeting their basic needs or reducing their potential for realisation (Galtung 1996). Cultural violence is symbolic, lost-lasting, and present in many aspects of a culture that legitimises the other forms of violence (Galtung 1990). In other words, structural and manifest violence are 'legitimised and thus rendered acceptable in society' (Galtung 1996: 196).

Taking into consideration only the violent part of the conflict, it would be easy to conflate HIC and war; but HIC is broader. For example, war can be defined as a type of HIC where usually states are involved against each other or against non-state actors, and the casualty threshold reaches a thousand people through battle-related deaths per annum in international wars and per conflict in civil and intra-state wars (Collier/Hoeffler 2001; Demmers 2012). HIC scenarios, however, occur in more than those places where wide-spread social violent conflict involves over a thousand casualties. Other characteristics of HIC include places where, due to the level of conflict, local authorities and governments have minimal or no effective control over the country or regions, generating a high level of state fragility. The provision of goods and basic services is irregular or fragmented, causing, together with the levels of violence, high rates of migration of people looking for safety from their localities, regions, or countries (see: Demmers 2012; Grünewald 2012; Healy/Tiller 2014; HIIK 2016; Hilhorst/Pereboom 2016; HPN/OPM 2010; Keen 2008; Maxwell/Majid 2015). As a result of this displacement, conflict spreads over the territory and beyond, creating impacts on neighbouring countries and regions (Keen 2008; Maxwell/Majid 2015). The provision of aid and response is difficult and restricted due to a range of challenges (detailed below), with access and security being the most overt ones.

An important consideration is that HIC scenarios are not permanent, isolated, nor occurring once and then disappearing. Most of the time, they represent specific moments in a protracted crisis, developing out of or leading into low conflict or post conflict periods. Some examples of HIC scenarios can be observed in South Sudan, Afghanistan, Yemen, Syria, or Somalia. In all these countries, it is possible to observe all the characteristics mentioned above, even though stronger in some cases or weaker in others. In some of those countries, the government is stronger than in another, but in all of them there are regions where the control of the territory is in the hands of state-contesting parties. Over a thousand casualties have occurred in all the cases, a large number of people have fled, and the provision of goods is fragmented in parts of the territory. Moreover, even within the HIC category, there are a variety of possible different cases.

A concept that includes similar elements to HIC is 'complex emergencies' which is used to describe a humanitarian crisis resulting from the combination of large-scale violent conflict, political and economic instability, and/or disasters, usually requiring an external humanitarian response (Keen 2008; OCHA 1999). However, although helpful in understanding HIC scenarios, they differ in some important respects. The concept of 'complex emergencies' describes the *outcome* of a diverse range of factors and the process of responding to them, mostly by aid interventions (Keen 2008). The construct of complex emergencies emerges not only because of the complexity of the emergency itself, but also because of the complexity of the responses to these scenarios that must take into account numerous factors such as dangerous settings, political use of aid, or donor dependency, to name a few (Davey et al. 2013; Duffield 1994). HIC scenarios describe a range of social and political arrangements without describing them as an emergency and without questioning the need or ways to respond to them. Moreover, HIC seeks to contribute to understanding that particular moment when the conflict reaches the highest socially violent period resulting in producing the conditions listed above. It might be possible to say that, if complex emergencies are 'protracted political crises' (Duffield 1994: 4), HIC are moments within them, describing key features of the conflict. The idea of complex emergencies has important attributes for the understanding of HIC including the relevance of the relationship between humanitarian aid and the military, peacekeeping operations and other protection groups (Duffield 1994; RPN 1997; Stoddard et al. 2006). Another relevant distinction is that HIC enables an analytical distinction from other types of conflict, notably low-intensity conflict. Complex emergencies and the large number of studies about it are also useful in understanding other types of conflict scenarios, including post-conflict settings.

Similar to 'complex emergency', 'fragile state' is another concept regularly used to study scenarios similar to HIC situations. A state defined as fragile is 'unable to perform its core functions and displays vulnerability in the social, political, and economic domains' (Sekhar 2010: 1). These states are also framed as failing in their role of providing human security due to the concentration of poverty they generate (Duffield 2007). Conflict is mentioned sometimes as a cause of fragile states, as much as fragile states are the cause of conflicts. Fragile states can suffer HIC moments, but also experience low- and post-conflict scenarios. Moreover, due to the vulnerability of their population, fragile states present a higher risk of suffering a socio-environmental disaster (Shreya/Vivekananda 2015). The following section will discuss one of the biggest challenges for disaster response in HIC which is dealing with fragile states. Fragile states play a role as a causative factor for both conflict and disaster.

The presence of fragile states in HIC scenarios does not necessarily mean that their governments are not strong in many respects. In every case of HIC studied, the national government had a tight level of control over sections of the territory and over some, or all, borders with neighbouring countries; and they still performed some level of international activity. Moreover, in all cases reviewed, national governments are one of the parties involved in the conflict. These features can be seen in Afghanistan, Yemen, South Sudan, Syria, or Somalia – with some important differences among them, though. This situation plays into a dual complexity in terms of the governance and coordination of disaster response. On the one hand, the national government has the main role in coordinating disaster response while their fragility and involvement in the conflict might hinder their capacity to act and manage disaster response. In fact, HIC-affected countries rely heavily on international aid in their responses and the coordination of it. On the other hand, by being the official government part of the conflict, aid actors adopting the principle of neutrality and independence may be persuaded not to include government in the coordination as it would compromise their access to territories held by contesting parties. At the same time, the strength of the government can mean that, at some level, aid actors should inform, respect, and seek authorisation for their actions from the national authority. This paradox and the ways in which aid and society actors deal with it is a familiar situation for emergency and developmental aid programmes but has not yet been a feature of disaster response models. The legitimation strategies the aid and society actors have adopted to manoeuvre through this challenge are described later.

These ideas about the role of states and the vulnerabilities of the local population reinforce the proposition that studying disaster response in places affected by high-intensity levels of conflict is more than just knowing how an action (the disaster response) occurs in a specific context. It is about understanding the shared social factors explaining the conflict and the disaster, an exercise in revealing a dynamic process where each phenomenon plays a role with the other. Socio-environmental disasters, as well as conflicts, result from a complex combination of multiple factors. On the one hand, natural events have the potential to damage property, produce social and economic disruption, cause death or injury, and environmental degradation (UNISDR 2009: 4). On the other hand, vulnerable human populations lack the mechanisms, response institutions, resources, and knowledge to prevent being affected by, or to mitigate the impact of, socio-natural hazards (Aboagye 2012; Hewitt 2013; Todd/Todd 2011; Wisner et al. 2003). When a natural event affects people and their livelihoods significantly, the result is a socio-environmental disaster; the impact of natural forces or events that have severe consequences on vulnerable human populations and their possessions.[5] The use of the words *social* and *environmental* instead of the traditional phrase *natural disaster* seeks to stress the relevance and presence of social factors in these events, such as people's vulnerability, lack of preparedness, or poor environmental management, to name a few.

Natural events with the potential to cause damage are also termed *hazards*, defined as events that "may cause the loss of life or injury, property damage, social and economic disruption or environmental degradation" (UNISDR 2009: 4). Hazards also include latent conditions representing future threats but to produce direct social damage requires a particular set of conditions (leaving aside the effects on natural environments, e.g. the effects of volcanic eruption on an isolated island) (Parker 2006; Todd/Todd 2011; UNISDR 2009). Therefore, socio-environmental disasters are a social construction triggered by a natural hazard. These physical, social, economic, and environmental conditions which determine the susceptibility of a community to the impact of hazards are generally termed '*vulnerabilities*' (UNISDR 2009). Risk is another common term used, a function of hazards and vulnerability, establishing the likelihood of people being affected by hazards (Collins 2008; UNISDR 2009; Wisner et al. 2003). Risk can be reduced and managed by reducing people's exposure to hazards and/or reducing people's vulnerability (Todd/Todd 2011; UNISDR 2013).

Hazard also plays a role in the general classification of disasters. The speed of onset determines the time that it takes for a hazard to reach its peak manifestation or impact. Based on the speed of onset, disasters are usually classified into two categories: slow and rapid onset disaster. Slow onset emergencies, like disaster, is defined by the *United Nations Office for the Coordination of Humanitarian Affairs* (OCHA) as those "that do not emerge from a single, distinct event but one that emerges gradually over time, often based on a confluence of different events" (2011: 3). Rapid-onset disasters (sometimes also named sudden-onset disaster) develop, as the term implies, rapidly or almost immediately. The speed of onset must not be confused with the predictability of an event. Although there is no internationally agreed list classifying disasters or determining what is 'slow' or 'sudden', most disasters are classified as sudden-onset. In general terms, earthquakes, cyclones, typhoons or hurricanes, flash flooding, landslides, avalanches, and volcanic eruptions are seen as rapid-onset disasters. Some examples of slow-onset disaster are droughts, sea level rise, water salinisation, and erosion.

---

[5]In the present article, I am using the term disaster or socio-environmental disaster interchangeably.

Disasters, in brief, result from *vulnerable* populations being exposed to natural hazards (Bankoff et al. 2004; Cannon 1994; Harris et al. 2013). Conflict scenarios, on the other hand, play a key role in the development and maintenance of social vulnerabilities, resulting in disaster response in HIC scenarios becoming muddled with other relief and aid efforts related to the crisis (Hilhorst 2013a). Furthermore, people's lack of coping and responding mechanisms is also a result of conflict and other social situations, such as poverty (Bankoff 2001; Shreya/Vivekananda 2015). Vulnerability is, in this sense, a key concept working as a link between conflict and disaster. As defined by Bankoff (2001: 24), vulnerability "denotes much more than an area's, nation's or region's geographic or climatic predisposition to hazard and forms part of an ongoing debate about the nature of disasters and their causes".

To prevent, manage, and respond to disasters, disaster risk managers, specialized institutions, and aid agencies uses a multi-phase disaster management cycle. This cycle "includes [sic] sum total of all activities, programmes and measures which can be taken up before, during and after a disaster with the purpose to avoid a disaster, reduce its impact or recover from its losses" (Vasilescu et al. 2008: 44). The cycle has three main phases: The first is *pre-disaster*, including all prevention, mitigation, risk reduction, and preparedness activities and measures. This phase seeks to reduce human and property losses and vulnerability. The second phase is *disaster-response* including an initial damage and impact assessment and assistance to affected victims to ensure that needs and provisions are met and suffering is minimised (Todd/Todd 2011; UNISDR 2009). Media coverage and delivery of information are also part of this phase. Alongside and before this formal disaster response phase, a more spontaneous or less official response starts among the same people affected and local actors. The third phase is *post-disaster*, with a first sub-stage focused on providing continuity with the previous phase, initial infrastructure recovery, and rehabilitation of affected communities. In a second sub-stage, social and economic long-term recovery plans are implemented, together with risk reduction measures and activities focusing on enabling community self-protection (Parker 2006: 4–6; Vasilescu et al. 2008: 47).

The decision to focus the analysis on the disaster response phase is mainly because at that specific moment the opportunity exists to observe a larger number of actors, actions and procedures. During disaster response, all the other elements of the cycle are present in addition to the actors and actions that only occur at that precise moment of the emergency. Moreover, HIC are periods of a particularly protracted crisis and disaster responses are also periods in a longer continuum of the disaster management cycle. When both periods coincide, due to the nature of each of them, the impacts that the actions might have on the wider population are significant. Finally, as it will be shown, in HIC scenarios, disaster response occurs in ways not yet well understood thus providing the opportunity for a scholarly and political inquiry.

Studying disaster response in HIC entails multiple challenges. Firstly, disaster response is a complex process: alongside its technical and economic aspects, it is also highly political, social, and contextual-historical (Cannon 1994; Hilhorst 2013a). HIC scenarios never show clear distinctions between the conflict and the disaster. It is difficult to know if the response is tackling the effects of one, the other, or both. The response, therefore, may always address planned and unplanned sufferings, as termed by Gasper (1999), like manifest intentional violence (planned) or reduced local capacity to respond due to societal dysfunction (unplanned). Moreover, every place is exposed to different hazards, and each population has its own vulnerabilities (Wisner 2010), and every society has its own history at the base of their conflict.

Another challenge lies in the fact that several theoretical prerequisites of disaster response on the ground in HIC places may not be present. For instance, in theory, disaster response activities may be organised and executed by local or national authorities. The organisation of international aid and humanitarian agencies is, supposedly, also coordinated by states within known protocols (Todd/Todd 2011). In reality, the process usually begins with local people, including the ones affected, providing aid to each other. Later, aid agencies assume the task, relating to local actors and modifying the shape that the response takes. The collection of information about what happened, the number of people affected and meeting basic needs is neither linear nor fast (Comfort et al. 2004; Walle/Turoff 2008). The former also applies to slow-onset disasters such as droughts because the defining process to classify them as a disaster in need of response can also be a complex and lengthy one (Maxwell/Majid 2015; OCHA 2011). In cases where the disaster occurs in places affected by violent social conflict, as in HIC scenarios, extra layers of complexity are added to the response (Harris et al. 2013; Keen 2008).

Disaster response, moreover, is supposedly a short-term intervention in advance of a long-term and more permanent response by governments and other organisations. In other words, disaster response seeks to focus in saving life and assessing the damages, leaving long term intervention (like recovery or reconstruction actions) to following phases. However this is not always the case: protracted crises tend to produce protracted aid and responses (Harmer/Macrae 2004). The actions to save lives tend to prolong and perpetuate, entering a cycle of response or emergency, not transitioning in a timely sequence to the following phases. The challenge here is to recognise when disaster response is moving into the post-disaster phase. In HIC scenarios a similar dilemma is faced by the actors responding to the conflict. Reaffirming these observations, one of the practitioners interviewed (AP1) mentioned a question frequently raised in HIC environments: 'until when are we providing emergency aid for the conflict and when do we need to start moving or we are already developing development programmes?'.

This discussion reveals that HIC scenarios are dynamic. However, better understanding on how they change and what those changes might mean for disaster response are yet to be explored. HIC so far has been exemplified using countries as cases, but certainly some countries exhibit differences between cities and regions. Is it possible to have cases of environments with different conflict scenarios in play, and if so, how would the disaster response process be different? From the literature reviewed and the interviews it may be possible to hypothesise that the dynamics of the HIC scenario will dominate other types of conflict, for example, low- or post-conflict. During a protracted crisis, the HIC scenario tends to develop suddenly. From the time a conflict turns violent and the most overt challenges emerge, the response to disasters occurring (like drought or floods) or suddenly striking in a particular area changes immediately. However, once the level of conflict diminishes, most of the actors continue to respond in the same way for a while with a kind of inertia. It may be that aid and society actors decide to wait until they are sure the level of conflict has really changed. Another option is that the transition period from HIC scenarios to low-level or post-conflict scenarios is slow and with no clear demarcation. Although the violence and other characteristics of HIC may not be present, many other challenges are still in place requiring a response. In this regard, the discussion would be enriched with further studies on the escalation/de-escalation process between HIC and other conflict scenarios, and how they relate to disaster response.

Despite the challenges of studying disaster response in HIC (not just from a theoretical point of view, as doing fieldwork in those cases has also proved to be challenging), the analytical categories of HIC scenarios present an opportunity to study various aspects of disaster response. This section unwrapped disaster response and HIC and presented some challenges in studying responses to these scenarios. The following section will explore what challenges HIC scenarios present for disaster response, and for whom. In other words, it will examine who in HIC is responding to disasters and what this singular type of conflict means for their actions. A subsequent section unwraps how these actors overcome these challenges and are enabled to respond.

3.3 Actors and Challenges of Disaster Response in HIC

3.3.1 Humanitarian Arena and Aid-society Actors

In responding to a disaster, several actors are present. In HIC scenarios, the available literature suggests that most commonly present are the single-mandate organisations – those with a "strict focus on life-saving humanitarian assistance" (Hilhorst/Pereboom 2016: 85) – and diaspora groups, while in humanitarian aid multi-mandate organisations are the majority (OCHA 1999; Wood et al. 2001b; Keen 2008; Demmers 2012; Maxwell/Majid 2015; Hilhorst/Pereboom 2016). Although mentioned less in the literature, the presence of other actors must not be ignored, as also stressed in four interviews (AC2, AP2, B2). For instance, local people and the private sector together create a large group of respondents. As an example of the scale of these actions, medium or large humanitarian operations may include tens of NGOs, *United Nations* (UN) agencies, different components of the *International Federation of Red Cross and Red Crescent Societies* (IFRC), *International Committee of the Red Cross* (ICRC) and national societies, plus a dozen other private and corporate organisations as well as local people, institutions and governments (ALNAP 2015; Weiss 2007; Wood et al. 2001a).

Studying this large group of actors can be difficult. One way to facilitate the process is to find ways to organise or divide them into groups. It is easier to observe and analyse these sets of actors in aggregate mode, which also makes it possible to discover common patterns among groups. The sorting can be done via cluster analysis, the "art of finding groups in data" (Kaufman/Rousseeuw 2005: 1) but this requires studying all actors and then finding commonalities among them. Another option is to develop typologies (theoretical categories) and then, based on the attributes describing each group, to classify the actors (Babbie 2013). This section uses a typology analysis classifying the actor as part of the aid or society categories.

As an analytical concept, the aid-society construct is dynamic and represents the relationships between different actors of the aid and society spheres without always identifying to which specific sphere the actor belongs. Aid actors are those for whom humanitarian actions are part of their core function while they are usually part, or at least linked to, international institutions. Society actors play relevant roles in the response, but humanitarian aid is not part of their core function. Local state and non-state institutions and local people are some of these society actors. Aid actors, however, should not necessarily be seen as totally external to the realities of the places where they act: they 'add a layer to the complexity of governance in crisis-affected settings, creating an imprint on the institutional landscape as it unfolds' (Hilhorst 2016: 5). Conversely, society actors interact with aid in strategic ways to pursue their interests and agendas. As a result, all the actors involved in disaster response form an *aid-society arena* – an aid-society relationship that occurs within a humanitarian arena.

From an actor-oriented perspective the term 'humanitarian arena' seeks to represent "the outcome of the messy interaction of social actors struggling, negotiating and trying to further their interests" (Bakewell 2000: 108–9 in Hilhorst/Jansen 2010: 1120). However, the arena is not 'out there' but rather built by the multiple actors, institutions and stakeholders involved in the process, including those without exclusively humanitarian interests (Hilhorst/Jansen 2010; Hilhorst/Pereboom 2016). Humanitarian action is, in this sense, an arena where all actors related to the response, including recipients, negotiate and shape the outcomes of aid (Collinson/Duffield 2013; Hilhorst/Jansen 2010).

An aspect of the arena is that aid gets shaped in practice, in contrast with the concept of humanitarian space, as aid is not limited to the physical, working, and ideal spaces where it should be delivered following well-known humanitarian principles (Hilhorst/Jansen 2013). The notion of humanitarian space is also frequently used by many actors to legitimise their actions and interest, framing themselves as neutral, ethical, needed, or distant from local political contexts (DeChaine 2002; Hilhorst/Jansen 2010). The concept of arena, in contrast and as presented by Hilhorst and Jansen (2013), is empirical and built on people's practices, including all social-political strategies and negotiations, formal and informal actions, and everyday practices occurring in, and for the delivery of, aid. Therefore, this approach allows for observation of the ways in which it is possible for multiple actors to respond in HIC, recognising practices and the shape that the response takes as a result of the relationships amongst all involved players.

Table 3.2 Aid-society actors. *Source* The author

AID	SOCIETY
UN system and agencies	
Regional and inter-governmental humanitarian organisations	
International aid and humanitarian organisations	
IFRC – ICRC	
National relief organism	
ICRC national societies	
INGOs (International non-governmental organizations)	
Inter-regional or transnational organization	
International-multinational private and corporate organisms	
Donors	
Military and armed groups: Peacekeepers, blue helmet, national armies, armed rebel/opposition groups, mercenaries	
Media, journalist, photographers	
Evaluation teams (methodologist, evaluators, evaluation manager, facilitators)	
Volunteers	
Religious institutions	
Researchers	
NGOs	
Funding and financial institutions	
	Other national governments
	National government
	Ministers and national agencies
	Parallel states-governments and state-contesting parties
	Local governments and authorities
	Local institutions
	Local people

Table 3.2 is an initial attempt to map aid-society actors in these two categories, accounting for the diversity of players involved in HIC scenarios for disaster response and humanitarian aid.

Each group of institutional actors is constituted out of an important number of sub-actors playing a particular role. It must be noticed, though, that given the combined effect of the disaster and the conflict, it becomes impossible to differentiate accurately between actors responding primarily to the conflict, or to the disaster. As stated by Wood et al. (2001a: 3), "to determine who are the actors participating in the humanitarian system seems to be an impossible mission, as it usually includes thousands of individuals worldwide and uncountable organisations".

3.3.2 Challenges of Disaster Response in HIC Scenarios

In HIC scenarios such as the ones here described, the actors have to respond to what is termed 'dual' disasters, "where a humanitarian crisis with human-made political

roots overlaps with a humanitarian crisis induced by environmental disaster"
(Hyndman 2011: 1). These dual disasters present multiple challenges for the
response that encompass all activities, processes, and mechanisms associated with
affected victims and which ensures that their needs are met, suffering is minimised,
and an initial damage and impact assessment is carried out.

The challenges in HIC scenarios (to be reviewed in detail below) include issues
of security, access, reduced supply of services and goods, deficiency of information,
complex governance at the local or national level, economic problems, difficulties
of reaching people in need, challenges in the establishment of refugee camps and
settlements. As presented in a report from Médecins Sans Frontières, as a result of
these challenges "UN agencies and INGOs are increasingly absent from field
locations, especially when there are any kind of significant security or logistical
issues" (Healy/Tiller 2014: 4). These challenges on the ground affect not only
disaster response but humanitarian aid actions.

Among the challenges (Illustrated in Fig. 3.1), weak or complex governance
systems are an overarching challenge from which many others derive, such as
reduced access to information or economic crisis. The governance issue also plays a
significant role as a link between the response to disaster and to conflict. Complex
systems of governance can involve the complexity of multiple and parallel systems
of governance in one territory and can include different economic and political
systems in some parts of the territories. For example, a study of the Central African
Republic (CAR) mentioned the presence of "three parallel governance structures:

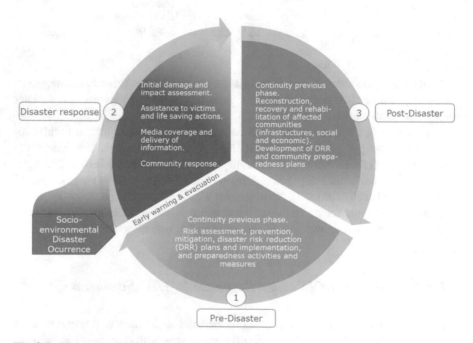

Fig. 3.1 Disaster management cycle. *Source* The author

local government or civic administration; the tribal administration for different tribal groups; and humanitarian governance structures which include the United Nations organizations, international, and national non-government organizations, and donor countries" (Young/Maxwell 2009: vii). Their complexity is not only based on the many (and sometimes unknown) governance systems in a place, but also from the lack of knowledge that could enable a way of manoeuvring through them. For example, the presence of parallel governance systems implies that the coordination of responses is not only fragmented but entails negotiation and coordination with multiple parties (Loeb 2013; Magone et al. 2011; Wood et al. 2001a). These multi-governed scenarios make the coordination, access to information and the whole process more complex and issues of legitimacy and power are intertwined with this challenge.

For humanitarian actors from the international aid community, the challenge of complex systems of governance includes, for example, having to negotiate with state-contesting parties that often fall under the political label of rebels or terrorists. Another name given to these actors is non-state armed groups. These negotiations are generally driven by political interest from both the armed groups controlling territories and from donors, national governments and the international community which may wish to have a say in allowing or participating in negotiations (Jackson/Davey 2014; Magone et al. 2011). For the UN system or donor countries, deals with these actors can be an opportunity to negotiate and/or pursue other agendas by imposing conditions on aid (Atmar 2001). Negotiating and engaging with those parties has also confronted many humanitarian actors with ethical, legal and political dilemmas, especially the international aid agencies (Jackson/Davey 2014; Loeb 2013). At the same time, local actors – both responders and aid beneficiaries – also pursue their agendas and interests in the negotiations with humanitarian players.

In addition, the contesting parties and the open social conflict affecting the territories may drive the development of norms, legal frameworks, and protocols that, although being developed most of the time within the framework of increasing the protection of people, many times might hinder disaster response. In HIC scenarios the use of drones is widely contested, but many disaster responders are using them to obtain data about affected areas. Introducing medicine or medical equipment to these places can be trapped in large multiple 'bureaucratic layers', as described by the two practitioners interviewed (AP1, AP2). The International Federation of the Red Cross and Red Cresent-United Nations Development Program (IFRC-UNDP) (2015) document 'The Checklist on Law and Disaster Risk Reduction' also presents more examples of these situations, especially with regard to the laws and legal frameworks required for appropriate disaster risk reduction and response. However, one academic interviewed (AC2) stated that in places with HIC levels of conflict, disaster response was not prioritised or facilitated because every action is read as a move in the conflict. The political reality of the conflict thus permeates into disaster response.

This politicisation of humanitarian aid is therefore another challenge for humanitarian response to disaster in conflict-ridden areas (Atmar 2001; Davey et al.

2013; Hilhorst 2013b; Kelman 2012). Humanitarian actors' decisions and actions unfold in a political arena (Hilhorst/Jansen 2013; Magone et al. 2011). These dynamics do not only occur at the local level, as the geopolitical use of aid and disasters reach regional and international arenas, too (Barnett 2011; Wood et al. 2001a). The political aspect of the disasters can also be seen as a window of opportunity, for example, in cases of disaster diplomacy, where disaster-related activities may reduce conflict by inducing cooperation, peaceful negotiation and diplomatic opportunities between the parties involved (Kelman 2006). For example, the case of the 1999 earthquake affecting Greece and Turkey explored by Ganapati et al. (2010) showed that, under specific conditions, disaster could lead to long-term collaboration between countries, including "disaster-related collaboration at non-governmental level" (Ganapati et al. 2010: 176). This politicisation also extends to the response funding processes in the HIC area. In some cases, as presented by Wood et al. (2001a), governments are cautious in support actions in these HIC scenarios so that, along with UN agencies grants and pool funds, NGOs and other responders working in this setting depend on funding coming from private sources, including bank loans. However, cases like South Sudan or Afghanistan showed that government-driven funds represent the majority of aid funding (Financial Tracking Service (FTS) 2016) (Fig. 3.2).

Another main challenge for disaster response in HIC is related to security. This challenge includes the protection and safety of multiple actors from different threats. One concern is the protection of affected people from the disaster itself and its related events or effects, for instance aftershocks, unstable terrains, or contaminated flood water (Healy/Tiller 2014; HPN/OPM 2010). In HIC scenarios, the protection of affected population and respondents from other people must be added to those concerns (Grünewald 2012; Healy/Tiller 2014; Maxwell/Majid 2015; Stoddard et al. 2014). Here, security concerns refers to violent acts associated with the course of the conflict and not the cases of looting or violence resulting from people's reaction to a disaster which is less frequent than suggested by the media (Alexander 2013). Some results arising out of security concerns have been the development of strong security policies, the construction of compounds, or 'bunkerisation' of aid agencies, the development of remote management, and the increasing distance between aid workers and people in need (Donini/Maxwell 2014; Duffield 2012; HPN/OPM 2010; Maxwell/Majid 2015; Smirl 2015). Although this

Fig. 3.2 Challenges for disaster response in HIC. *Source* The author

last trend is generally associated with humanitarian aid actions, including emergency and developmental ones, multiple interviewees (AC1, AC2, AP1, AP2, C1) agreed that in HIC scenarios disaster responders, local and internationals, operate in the same way.

The claim of increasing distance between aid workers and beneficiaries can be contested if this is true only in the case of international actors. International organisations providing assistance commonly transfer security risk to local staff and local NGOs (Stoddard et al. 2006), resulting in local actors becoming closer to people in need and international actors becoming more distant. Another aspect of this 'localisation of the response' via national staff as a response to insecurity is the strengthening of remote management and remote programming (Donini/Maxwell 2014; Stoddard et al. 2006). Remote management entails "the practice of withdrawing international (or other at-risk staff) while transferring increased programming responsibilities to local staff or local partner organizations" (Egeland et al. 2011: xiv). Without being confused with the decentralisation of decision-making, remote management is supposedly a temporary managerial adaptation that occurs from outside of the affected country, but other times from the capital with respect to affected regions and territories (Donini/Maxwell 2014; Egeland et al. 2011). The localisation of the response as a way of outsourcing security risk raises multiple questions about the ethics of relocating this risk to local actors, the accountability of the process, and the possible impacts for humanitarian principles, to name a few (Donini/Maxwell 2014; Egeland et al. 2011; Stoddard et al. 2006).

The security challenge has also increased the inclusion of the private sector in HIC scenarios, particularly regarding access and securitisation. The case of Somalia is an example of the intervention of private groups: after the Black Hawk episode (helicopters from the U.S. were shot down), the only means to ensure access and provide security to the humanitarian sector was outsourcing that responsibility to private corporations (Maxwell/Majid 2015). The interviews conducted revealed another example of using the private sector including hiring private trucks and charter flights for the distribution of goods in South Sudan and the use of private financial service providers to transport cash needed for cash transfer programmes, paying salaries and for services, and buying local goods (AC2, AP1, AP2, C1). However, despite all soft and hard security measures, security is a constant concern for aid workers. Among others, books and chapters like Neuman and Weissman (2016), Roth (2011), Fink et al. (2014), Stoddard et al. (2006, 2014) or a report from IFRC (2011) provide a description of the experiences and what it means for aid actors to work in dangerous settings.

Mobility and access to different territories is challenging for all actors, from local people to international institutions (Hilhorst/Pereboom 2016). In addition to the safety issues already mentioned, roads are often not clearly mapped or in poor condition in countries or regions affected by HIC. Roadblocks, hijackings, check-points, landmines, and ambushes are also a general concern (Menkhaus 2010; Pottier 2006). If public transportation is available, it tends to be unstable, unsafe and irregular, especially between cities. Oil shortages and the high prices for fuel are further obstacles. This leads to the impediment of free movement for seeking help, insecurity during long walks, as well as reduced access to respondents and providers of humanitarian aid (Caccavale 2015; Duffield 2012; Grünewald 2012; Hilhorst 2016).

The expected temporary solution to access issues was based on the principle of 'humanitarian negotiated access' which is underpinned by the humanitarian principles of humanity, impartiality, neutrality, and independence. But nowadays access is fragmented and humanitarian institutions have to resort to their negotiating capacities, to hiring private security and helicopters, or to finding alternative ways of access[6] (Donini/Maxwell 2014; Duffield 2012; Grünewald 2012; Healy/Tiller 2014; Maxwell/Majid 2015). Actors like the United Nations Office for the Coordination of Humanitarian Affairs (OCHA) and the International Committee of the Red Cross (ICRC) play a role in negotiating access, but the increasing numbers of groups fighting each other and humanitarian organisations on the ground makes the coordination and negotiation of access in a unified way highly challenging (Donini 2012; Hilhorst/Pereboom 2016; Schwendimann 2011). In addition, when negotiating, humanitarian actors usually see themselves 'negotiating in practice that which is non-negotiable in principle' as many times they have to accept or deal with conditions that in other situations they would not have to confront (Mancini-Griffoli/Picot 2004: 11). In HIC scenarios, as also pointed out by some respondents (AC2, C1, AP2), these difficulties of access already existed before the disaster in the affected territories and produced a deficiency of goods and services that disable local responses (Grünewald 2012). The disaster can occur in a highly vulnerable situation which makes it more sensitive than it would be in a place with lower levels of conflict.

Technology plays a role in circumventing access and security issues. Airdrops or aerial delivery of aid and the use of drones to obtain information are strategies invoked by these issues (Bastian et al. 2016; Belliveau 2016; Emery 2016; Giugni 2016). The use of satellite imagery is more and more popular when responding to multiple disasters including drought, floods, earthquakes, and tsunamis (Harvard Humanitarian Initiative 2011; National Research Council 2007).

Another significant challenge that occurs at different stages in the disasters response-cycle is the lack of, reduced, or fragmented information on the country or some regions of it. First, it complicates the process of coordination and planning of aid and response (Comfort et al. 2004; Wood et al. 2001a). Secondly, any attempts at assessment and accountability of the response are frustrated (Wood et al. 2001a). These issues affect governments, local institutions, humanitarian aid agencies, and the international community in different ways. This information deficiency, though, is not exclusive to HIC settings and is also present in other disaster response settings. Some HIC countries, like Afghanistan, have a long history of research and aid operation and it is easier to access to some of the necessary data, although in some regions controlled by non-governmental parties this information may not be up-to-date. At the other end of the spectrum, in South Sudan the level of

---

[6]Mentioned and reaffirmed in two interviews (AC1, AP1).

information about disasters and aid operations is still low despite its protracted conflict history.

Local people, who are usually asking where to go, what to do, and wondering what is actually happening (particularly when affected by rapid-onset disasters), are confronted with a lack of, or reduced access to, information to help them make informed decisions. The level of rumours in these settings can be high and hence produce more collateral impact. In a different vein, not addressed in this article due to space constraints, data and information are political. The lack of, use, and ways in which information is produced, framed and managed is not neutral and usually responds to multiple agendas and interests, even in the humanitarian aid and disaster response spheres (Cottle 2014; Herman/Chomsky 2002; Olsen et al. 2003; Robinson 1999; Wanta et al. 2004).

Reaching people affected or in need of aid is also challenging for two reasons besides the previously mentioned ones. In HIC settings the levels of internally displaced people (IDPs) is usually high, meaning that it is not always clear how many people could have been present and affected by a disaster. Except for those in refugee/IDP camps, settlements, or in 'protection of civilians' sites (PoCs), the location of people can be in some cases difficult, especially in rapid-onset disasters. Although the fact that IDPs concentrated in PoCs may be advantageous for disaster response because aid agencies are already there at the time of the disaster and the access to affected territories may be easier, these places also represent a second set of challenges in HIC scenarios. Refugee camps and PoCs tend to be more permanent settlements (Jansen 2013, 2015; Lilly 2014), thus, the boundaries between the response and the post-disaster phase become blurred, making the initial task of meeting people's needs and reducing suffering more complex.

Reaching out to people to provide them with aid is usually described as an on-going process that lasts until they can regain certain levels of self-sufficiency or recovery. However, special cases such as the Angola 2013 drought showed that sometimes aid can be delivered only once and in a limited way and then people were left without help because of the denial of the existence of the disaster (Tran 2013). Both dynamics – people being displaced and settling down in refugee camps, settlements, or PoC sites – defy the notion that disaster response and relief is a temporary action. Once again, the border between the effects of the disaster and the conflict become blurred, making it difficult to know if people are moving and seeking refuge due to the conflict, or the disaster, or a combination of both. Interviewing one aid beneficiary living in a refugee settlement (B1) provided me with a good example of the first scenario. At the beginning of our conversation the person mentioned that the main reason to escape from her country was the conflict and the killings. However, after several minutes talking and building trust, she mentioned that, in reality, the main reason to flee was the drought and the incapacity to grow their own food because of the conflict. The drought compounded with the conflict (that prevents the normal trade of goods in the markets) was the real cause of her flight.

Refugee camps play a role in most HIC scenarios with their own political and social dynamics, which are not restricted to the camp itself. As Jansen asserts, "the relation between refugees and aid actors does not stop at the camp's boundaries" (2015: 1). This implies another challenge for the response, because despite the existence of camps, it is not always clear how to reach people in need. In fact, many refugees or disaster-affected persons stay outside the camps, living in neighbouring areas.

Usually, HIC scenarios are also in economic crisis, including recession, instability, inflation or breaking up of the supply chain of goods and services (Grigorian/Kock 2010; Rother et al. 2016). The lack of or difficulty in accessing services and commodities under chaotic conditions expose external aid workers to the challenge of being self-sufficient, especially in cases like South Sudan or Syria. Responders must be able to bring with them everything that they need to provide relief or "have robust local supply chains, pre-planned and with a positive rather than negative impact on local economy" (Norton et al. 2013: 84) so as to not burden the limited supplies available. For local people, HIC scenarios may also include the imposition of substantial tax payments, as was the case in Somalia with Al-Shabaab (Maxwell/Majid 2015: 6). On top of the economic burden imposed by the conflict, socio-environmental disasters usually have a serious economic effect on the population (Keen 1998; Spiegel et al. 2007). On the one hand, they increase their expenses substantially since they should replace what was lost, and on the other hand, they may stop receiving income as many productive activities are affected and people stop working. Consequently, post-disaster recovery, reconstruction and rehabilitation processes may be delayed until the levels of conflict decrease (GFDRR et al. 2016; McGrady 1999). Protracted conflicts thus produce protracted recovery and reconstruction.

Because of the poor access to commodities and services in some regions (due to the economic crisis, disruption of supply chains and roads, and minimal purchasing power), the reliance on aid will be longer than disaster response in non-conflict zones, and the process of dependency and protracted crisis will be reinforced. Dependency here is used to mean that the response may result in a large web of interdependencies and co-shaping amongst multiple actors which becomes embedded in people's everyday lives (Harvey/Lind 2005; Hilhorst/Jansen 2010). This is not positive or negative, but shows the challenge of responding in a complex social context. Accessing services and goods depends on the capacity of each actor involved to move around in the social-humanitarian arena rather than only on a market-oriented strategy of buying and hiring. In many cases of disaster response, actors find it difficult to make the time to build or understand this larger social context.

Related to the economic crisis and the protracted state of the crisis where HIC occurs, disaster response is also many times confronted with corruption, bureaucratic procedures that are not always clear or are always changing, and lack of transparency. These issues affect both aid and society actors, making the response more expensive, slower or less efficient. In a similar vein, as discussed by Keen (1998), violence plays an economic role in civil wars and also in all HIC settings. Violent crisis are far from irrational: they are a rational response to the interest, frequently economic, of some actors (Keen 1998). Disaster response models cannot be blind to this reality and they must start including this situation of 'rational', economically driven violence in the model development, especially when responding in HIC scenarios.

In relation to a more developmental model of action, these complex situations drove the development of the Sphere Project and its 'Humanitarian Charter and Minimum Standards in Humanitarian Response' handbook, with a first edition at the end of the nineties (Sphere Project 2011). Part of this learning process and thinking about acting in a HIC was the occurrence of 'new wars' such as the military interventions of Africa and Asia, or the ex-Soviet conflict of the early nineties (Davey et al. 2013). These wars were 'new' insofar as they represented an intensification of attacks on civilians thus weakening and destabilisation governments' legitimacy and bringing new challenges to the humanitarian sector (Newman 2004).

Particularly relevant for a comprehensive understanding of disaster response in HIC is the human security approach. Human security brings the focus to people involved in everyday practices, even in areas of conflict, focusing on the role of military interventions or the state as the single protector of citizens (Gasper/Gómez 2014a). It also emphasises that complex situations have multiple stressors and, for instance, drought as a disaster may be the cause of more suffering than military interventions (Gasper/Gómez 2014b). Therefore, in HIC scenarios disaster response may play a vital role in reducing peoples' suffering even though disaster response may be seen as a side issue due to the conflict.

Coping with these challenges draws on the capacities of aid-society actors to negotiate with other players, legitimise their actions and presence, as well as change and adapt their actions and strategies according to the context. This last point is of utmost relevance because to act, people and institutions must have the power and legitimacy to do so. Power is relational and legitimacy is part of power relations (Beetham 2013). Without legitimacy, power relations are coercive, and with legitimate power the compliance and acceptance of others is ensured (McCullough 2015). But this can be a difficult task in HIC scenarios, where the level of legitimacy of multiple actors is at stake. For example, the government is not always legitimate and its actions may be seen as coercive by other groups. However, power exerted coercively is also common in these settings and for some actors a valid way to legitimise their actions. Legitimacy and power relations are complex, highly nuanced processes with multiple dynamics. The following section will discuss them in more detail and how they unfold in HIC settings during disaster response.

3.4 Responding to Disasters in HIC Scenarios

3.4.1 Legitimacy and Power

Situating the notions of *legitimacy, power* and *negotiation* at the core of disaster response in HIC scenarios is not a naïve proposal. It is to state that beyond the moral drivers and technical aspects of these dynamics, disaster response is not only political, but also relational. It depends on a large aid-society arena where, as mentioned by Warner (2013: 83), disasters convey political capital, legitimacy and 'may serve humanitarian but also utilitarian political instrumentality'. Moreover, these concepts have long-standing political and sociological relevance, requiring a better understanding of what they mean here and how they are used to study disaster response.

Legitimacy is a concept that has been addressed by different schools of thought. One group of thinkers conceptualise legitimacy as belief or voluntary agreement on the part of a community that a rule or institution must be obeyed (e.g. Levi/Sacks 2009; OECD 2010; Stel et al. 2012). This perspective for Bauman (1992) and Beetham (2013) does not allow tracing of the relational aspect of legitimacy that involves the actor seeking legitimacy from those actors who legitimise it. The body of research using the first definition above is more associated with the study of states' and governments' legitimacy, especially as service providers. Other literature on legitimacy describes a process in which non-state actors find legitimacy in the citizenry, as holders of legitimacy in fragile states (La-Porte 2015; McCandless 2014), even in the case of armed non-state actors (McCullough 2015). Another approach studies legitimacy in a more focused manner, for example, NGOs' legitimacy based on a four-fold model: the market model, the social change model, the new institutionalism model, and the critical model (Thrandardottir 2015).

An alternative definition of legitimacy, and adopted in this chapter because it provides a better fit for the questions addressed is the one provided by Lamb (2014: 34): "worthiness of support, a sense that something is 'right' or 'good' or that one has the moral obligation to support it". This broader definition of legitimacy is contextual and can apply to all sorts of actors. The term 'conferee' is used in this Lamb's approach for the person who is being assessed for legitimacy and 'referee' is the person who judges the conferee as worthy of legitimacy. It must be stressed that this definition of legitimacy is used here within an actor-oriented perspective and so, as asserted by Pattison, "rather than the focus being on whether a particular action is justified, the concern is with the justifiability of the agent undertaking the act" (2008: 397). This notion is crucial for disaster response results in HIC, as will be discussed later.

The legitimacy of an actor may change depending on who the referee is (McCullough 2015), and the referee and conferee may also contest or negotiate the legitimation process of the other (Hilhorst/Jansen 2013; Lamb 2014). To study the legitimacy of aid-society actors, this multi-directional aspect of legitimation is of the utmost relevance, as each actor may need to seek legitimacy from different audiences, requiring different strategies. For instance, an NGO must seek legitimacy at the same time, and by different means, from a donor, from the government of the country where they are responding and from the beneficiary communities.

In the case of (international) humanitarian interventions, for example, two main legitimating factors justify the worthiness of support of their actions: the humanitarian motivations and the humanitarian outcomes (Bellamy 2004). These perspectives indicate that the disinterested, impartial and ethical call to prevent suffering (motivation view) or the capacity of an intervention to produce humanitarian benefits (outcomes view) are the primary legitimising factors for humanitarian action and disaster response. The outcomes view must be complemented with the *effectiveness* approach, so that not any outcome is valid, but only the successful ones (Pattison 2008). Beyond these factors, there are multiple secondary and singular factors legitimising aid actions (Bellamy 2004). To recognise these other factors, including those associated with society actors, a more complex approach is necessary.

In line with the above, Lamb (2014) proposes a framework to assess legitimacy based on its multidimensional, multilevel and bilateral aspects. The first step is to identify legitimacy for what, according to whom, and by what criteria. Then, there are multiple indicators to be obtained and ways of analysing them. Without going into details of the methodology, his development and the variety of approaches articulated account for the relevance and complexity that the study of legitimacy involves. It is not only contextual but also dynamic and embedded in a large set of power relations.

Power is another concept with multiple theories explaining it (e.g. Dahl 1957; Foucault 1983; Parsons 1964; Weber 1964). The focus here is on power as a social construction, implying the capacity or ability of any subject to achieve outcomes and make decisions, as described by Giddens (1984: 257). Giddens' approach to power relates to the capacity of multiple actors to act. He presents an operationalisation of the concept of power based on who provides that capacity, who exerts power, and how it is produced and reproduced. This toolset proved to be useful in exploring further aid-society action in HIC and processes of legitimation.

The exercise of power, in Giddens' view, relates to two kinds of resources: first, allocative or economic resources, such as control over material things, including means of material production and reproduction, and secondly, authoritative resources, like control or organisation of other people's actions, relationships, and social time-space (Giddens 1984). People's actions are, therefore, based on their power and interest to act. But power is relational as people are embedded in social relationships and, as a result, their power interacts with the allocative and authoritative resources of others (den Hond et al. 2012). Power is, therefore, "generated in and through the reproduction of structures of domination" (Giddens 1984: 258), but this does not mean that power is associated with conflict only by producing oppression, struggle or division (Giddens 1984). It is just a medium to produce change that may, or may not, clash with others' interests. In places like HIC, many actors tend to feel powerless (for example, two interviewees: one beneficiary (B1) and one practitioner (AP2)) and at the same time many others need to gain power to respond to conflict and socio-environmental disasters.

Giddens also argues that "there is never a situation in which there is absence of choice" (as cited in den Hond et al. 2012: 239) and therefore people always make decisions on their actions, even if they are difficult, limited or constricted by the context. It is not uncommon to find in the media or hear by different society actors the idea that during HIC or disasters people are forced to act in specific ways or that the surrounding conditions predetermine their actions. Giddens' notion of power

allows defending the contrary: people always have agency, and they are active in
the construction of their social reality (Giddens 1984). For instance, aid beneficiaries are far from passive and empty recipients; rather, they develop strategies to
legitimise their position and pursue their objectives (Hilhorst 2013b).

The role of institutions here is key, as power is mediated by them. Institutions
have greater time-space extension than individuals and may also have higher levels
of allocative and authoritative resources (Giddens 1984). For example, in the case
of nation-states, Giddens calls them 'power containers' (cited in Best 2002).
Although institutions' capacity to take a decision and make changes are sometimes
bigger than individual actors, they are also more constrained, because they are
embedded in bigger social power systems (Best 2002). Therefore, as argued by
Hilhorst (2013a: 7), it is necessary to understand "how power constellations are
negotiated and how they are subject to change".

Power enables the actors to act, and *legitimacy* is the concurrence that these
actions receive from other players. However, we should not oversimplify these
relationships, as many actions may be legitimate for some people, but not for
others. The capacity of some actors, individuals, or group to position their legitimation over the legitimation of others also requires the use of power. Power and
legitimacy, in these terms, are a two-way dynamic, where both are mutually used by
and for the other. Moreover, as warned by Beetham (2013: 39), legitimacy is not
merely the legitimation of power, "[it] is not the icing on the cake of power, which
is applied after (…) and leaves the cakes essentially unchanged. It is more like the
yeast that permeates the dough, and makes the bread what it is". In a humanitarian
arena, these institutional and aid-society actors' power to respond results in a
complex set of processes that shape not only the response but also the actors
involved in it. And humanitarian aid, from this perspective, is like "a conduit
between places and people, facilitating relief and reconstruction assistance as well
as political legitimacy and, hence, the political and economic stability of a place"
(Kleinfeld 2007: 170 in Hilhorst/Jansen 2010: 1119).

3.4.2 Strategies to Respond in HIC Scenarios: Legitimacy and Negotiation in Practice

Due to the complexity of HIC and the challenges discussed, not all aid or society
actors are able to access the places affected and to respond, and some need to
negotiate and legitimise their actions. However, exploring the process in which aid
and society actors relate, negotiate and legitimise themselves is a challenging task.
In the first place, it is challenging because none of these aid-society groups is
homogeneous and there can be significant differences among their actors. Secondly,
not only do each of the actors engage in multiple relationships at the same time but
also these relationships change over time. Even in the case of the same relationships, the strategies and legitimacy processes may change. Aid-society relationships

are multidimensional, multilevel and bilateral. Thirdly, the literature has two main biases. Firstly, the literature is mostly written from a top-down approach; discussing international humanitarian agencies (mainly INGOs and the UN apparatus) legitimacy and negotiation in relation to local-national governments and with armed groups. Secondly, the literature focuses on the frameworks enabling humanitarian interventions in foreign territories, and hence a focus on international actors.

This last point includes debates about international law and the rule of law (e.g. Beal/Graham 2014; Hehir 2011; Zifcak 2015) and the feasibility of the use of force or protected interventions (e.g. Malanczuk 1993; Recchia 2015; Seybolt 2008). Another body of literature discusses the role of the UN Security Council and the responsibility to protect (e.g. Chesterman 2002; MacFarlane et al. 2004; Newman 2002; Troit 2016; United Nations 2015a). In both literatures, there is also a cross-cutting debate about the differences between legality and legitimacy (e.g. Chesterman 2002; Newman 2002; Zajadlo 2005).

Notwithstanding the two biases just mentioned, there is an emergent and growing literature on (i) the internal legitimacy of humanitarian interventions, as the process in which national-local governments legitimate aid actions to their own citizenry (e.g. Buchanan 1999; Vernon 2008); (ii) humanitarian aid, legitimacy and parallel governments (e.g. McCullough 2015; McHugh/Bessler 2006); and (iii) the active involvement of aid beneficiaries and volunteers in negotiating and legitimating their actions.

Trying to separate aid and society actors' legitimacy strategies is intricate. They share many of the strategies, many others are interrelated, and also from a referee and conferee point of view, the legitimising strategies of one may or may not be judged as legitimate by the other. Despite how intricate this exercise might seem, it is possible to observe some broad sets of strategies in aid or society actors. The following paragraph will describe some examples of them in HIC scenarios for disaster response.

In addition to the humanitarian motivations and the humanitarian outcomes mentioned above, and regarding the legitimacy of aid actors for what, according to whom, and by what criteria, the right to intervene in cases of large-scale humanitarian crisis and disaster is well-recognised by the international community (Bellamy 2004). This international legitimacy has two main pillars: international law and the United Nations Security Council (Chesterman 2002; Hehir 2011). As presented by Bellamy through the examples of Somalia and Haiti, "the Security Council identified human suffering and governance issues as threats to international peace and security and therefore legitimate objects of intervention" (Bellamy 2004: 218). These two pillars are widely used in cases of violent armed conflict but their utilisation and validity is less clear for disaster response in places not affected by conflict. In fact, in 2014, with regard the Ebola outbreak[7] in Liberia, Sierra Leone,

---

[7]In this example, the Ebola epidemic outbreak is also considered a socio-natural disaster under the definition of a disaster presented before.

and Guinea, the UN Security Council held its first meeting ever to deliberate on an intervention in a public health crisis (Cohen 2014; UN News 2014).

The humanitarian principles of independence, neutrality and impartiality have also been set out for aid actors as "a magic key to the humanitarian space with an attitude of ultra-pragmatism" (Magone et al. 2011: 3). They act like a shield behind which any action is valid and legitimate, sustained by ideas about what is good, ethical, and moral (Hilhorst/Jansen 2010). In fact, in the interviews, the humanitarian principles were emphasised by all respondents as the main factor legitimising their actions. The principles, moreover, may legitimise aid actors' presence in HIC by presenting themselves as detached from political struggles (Leader 2000), a situation also mentioned by one of the two aid beneficiaries interviewed (B2): 'we accept them (the humanitarian actors and disaster respondents) because we know they are here to help all of us without caring about the conflict'. However, the other aid beneficiary (B1) problematised this assumption by asking the question: 'how is possible that they don't care about what the others are doing?' I have seen this last question frequently raised during fieldwork in countries affected by HIC, not only by beneficiaries but also by NGOs. Some practitioners expressed the view that, although they follow the principle of neutrality, they will never voice it because that could be seen as lack of caring or that they do not stand against the actions of one or other of the fighting parties.

Not only in HIC scenarios, some aid actors see the principles as a universal legitimator, imposing them on others (Leader 2000). If other people respond without following the principles, they are not seen as part of the humanitarian space (Collinson et al. 2012; DeChaine 2002; Hilhorst/Jansen 2013). But they certainly remain part of the humanitarian arena, as discussed earlier. This is equally valid for aid and society actors: they both seek to be seen as following the principles in order to be valid actors in the arena (Hilhorst/Pereboom 2016). State-contesting armed groups also use the principles as an action framework and legitimator (McHugh/ Bessler 2006). In HIC scenarios this can reach another level, where the principles are also seen as the borderline of what is ethically expected, and accepted, in social action, especially in war time. As Leader states (2000: 3), "the principles assume at least an acceptance that war has limits, that the belligerents are concerned with political legitimacy, and that all states have an interest in preserving respect for the IHL".[8]

The process of professionalising humanitarian action and disaster response (in part to respond to the challenges, in part to increase the efficacy and efficiency) opens a new legitimator for aid actors and enables multiple actors to respond to disaster in HIC settings. For instance, water managers for droughts, or professional rescuers in cases of earthquakes, validate and legitimate their actions as professionals in those fields. Likewise, appropriate behaviour by staff members of aid and societal institutions also lays down a legitimacy base. Accountability and actions assessment is also a relevant legitimising factor, especially for aid actors to its

---

[8]International Humanitarian Law (IHL).

donors and beneficiary governments (Donini/Maxwell 2014; Wood et al. 2001a). Professionalism also legitimises actors to be present in HIC scenarios, especially considering the security risk. Disaster response and humanitarian aid organisations have increasingly hired security managers, developed security protocols, and focused on strengthening security managing (Donini/Maxwell 2014; Roth 2011; Stoddard et al. 2006).

In an arena like HIC, where resources and access are restricted, the professionalism stamp of some actors legitimises them over others (Hilhorst/Jansen 2010). This has also been relevant as legitimation between the same aid actors in competition for funds and personnel (Mosse 2013; Wood et al. 2001a). As a result, it can be difficult for societal (local) responding organisations to validate themselves against the more professional aid (international) actors in the arena (Hilhorst/Jansen 2010). That is why society actors develop different legitimacy strategies to aid actors. One situation that I have seen in HIC zones and confirmed by both practitioners interviewed (AB1 and AB2) is the growing trend for local NGOs to hire professional (sometimes international) grant managers and accountants to seek funds from the so-called 'big donors'.

The concept of 'gratuity', or debates about what can be paid or not, is a more hidden factor of legitimation. Aid actors provide their response to disaster not for profit or with commercial interest and any attempt to do it for profit can be criticised. This applies to initiatives offering help in exchange for work and also to the reduction of taxes in exchange for donating money. HIC scenarios allow this phenomenon to be observed in a particular way. For example, it seems to be legitimate for aid and society actors to find protection under state or internationally mandated armed forces (like UN peacekeepers) but not to pay for private armed protection (HPN/OPM 2010). However, and in the spirit of being legitimated by being professional or efficient, some initial debates have taken place about what can be learned from management techniques used by for-profit organisations that could be helpful in an aid and response context (McLachlin et al. 2009).

In a similar vein, but moving towards the strategies of legitimation used by society actors, it is also possible to find legitimated interventions of armed groups in disaster response in HIC scenarios. For example, states may use army intervention or authorise peacekeeping missions with humanitarian agendas (Malanczuk 1993). However, it is again important in these cases to distinguish between legitimacy and legality (Seybolt 2008). These actions may be perceived as legal and legitimate by some actors, but not by others.

In places in HIC, rebel armed groups may also act in the response, but their legitimacy usually pre-exists the disaster and then extends to the response (Arjona 2008; Magone et al. 2011). This legitimacy is in relation to local actors, so, to be legitimised by external actors, armed groups have begun to act in compliance with international legal norms (McHugh/Bessler 2006). State-contesting armed groups may also seek to build legitimacy by engaging with aid actors to counter the non-legitimisation that they meet from official governmental actors (Grace 2016). However, it must not be seen as the intention of aid actors to confer legitimacy on armed-groups by acting with them. This is a well-known dilemma among aid actors usually dealt with by explicit declarations of non-recognition or legitimation of these groups, even when they sometimes need to work with them (Jackson/Davey 2014; McHugh/Bessler 2006). In high intensity conflict settings, it is thus useful to reflect on these controversial and complex situations, especially in cases of disaster response, as it is easy for many actors to name the *natural* disaster as the cause of local problems rather than the real social factors; in this sense, disasters are seen as external and non-related to the conflict.

Society actors may also find in the response arena that their legitimacy is part of what can be broadly called a cultural-community framework. As presented by McCandless (2014), heritage and blood as well as family and tribal bonds, can be legitimating factors rooted in the sense of community. For response activities, this local legitimacy is sometimes more relevant than the official recognition, as local actors are the ones reaching affected people first and local legitimacy, in turn, strengthens their power and general legitimacy, at least de facto. Similarly, some NGOs and other actors (from aid and society) claim their legitimacy through a religious approach (De Cordier 2009). They justify their actions on an ethical basis, but also they act in coordination with local groups of the same religious community, thus gaining access to the response arena (De Cordier 2009; Krafess 2005; Paulson/ Menjívar 2012). In HIC settings this also means stronger social networks to facilitate security and manoeuvre through the challenges.

All the examples mentioned above show how legitimation is crucial for the overall success of humanitarian operations and disaster response yet such endeavours are inherently challenging. The legitimation strategies do not work as 'recipes' and often require negotiation. Magone et al. (2011) state that everything is open to negotiation in the provision of aid, although it is not always a recognised practice.

Humanitarian aid and response is highly politicised and negotiations, along with dealing with political issues, must weigh ethical (e.g. following the principles) and legal considerations (e.g. following the international humanitarian law) as well. For this reason, negotiations may operate under confidentiality agreements or within closed circles (Grace 2016). The response occurs in an arena, and competition among aid and society actors may also lead to privacy or secrecy throughout the negotiations. In cases of rapid-onset disaster, quick action is needed and it can be helpful to bend the rules and operate outside the normal conduits. One academic (AC1) and the consultant (C1) interviewed said that the mindset can be, 'the emergency requires prioritising the aid no matter how it is done'. An analysis of this trend leads to the conclusion that under HIC conditions and in cases of conflict such as those described here, rules and procedures are less relevant and the negotiations occurring in the field are the real enablers of disaster response.

Humanitarian negotiations do not only occur in confidential or closed circles but also in the everyday practices of aid-society actors (Hilhorst/Jansen 2010). In fact, a UN manual on negotiation says that humanitarian negotiations are "those negotiations undertaken by civilians engaged in managing, coordinating and providing humanitarian assistance and protection for the purposes of: (i) ensuring the provision of protection and humanitarian assistance to vulnerable populations;

(ii) preserving humanitarian space; and (iii) promoting better respect for international humanitarian and human rights law" (McHugh/Bessler 2006: 1).

It is through these negotiations and processes of building legitimacy that aid-society actors manoeuvre through the challenges in the HIC arena to respond to socio-environmental disasters. In doing so, the response is shaped and, conversely, the actors' power relationships are shaped, providing an opportunity to study the everyday practices of disaster response.

3.5 Conclusion

Current disaster response models do not incorporate scenarios where socio-environmental disasters, like earthquakes or floods, occur in places affected by violent social conflict. The political and academic attention given to (i) the relation between social conflicts and socio-environmental disaster response, and (ii) the differences between multiple conflict scenarios and disaster response, is still low. Contributing to filling these gaps and proposing a way to deal with them, this chapter explored the process of responding to disasters in places affected by one specific type of conflict: wide-spread violent social conflict.

The chapter proposed the use of high-intensity conflict scenarios (HIC) as an analytical category that would permit the study of disaster response in this particular type of scenario. By an extensive literature review on scenarios matching the HIC profile and using experiences of disaster responses 'which attracted international aid and responders, the chapter tested the value of HIC as an analytical category and answered its main question.

The findings suggest that the features of the HIC scenarios provide a unique opportunity be better understand the response processes. As distinct from terms like 'complex emergency' or 'fragile states', HIC scenarios represent a period in a protracted crisis where, alongside violent social conflict, a particular arrangement of social and political conditions generates a scenario that features complex governance systems, insecurity, access constraints, people displacement, economic instability or crisis, among others. Moreover, as reviewed, there is no one type of HIC setting but a range of possible settings fitting its definition. HIC as an analytical category allowed the study of the large network of actors involved in the process and the mechanisms that they use to cope and respond to disasters under these challenging conditions. The concepts of aid-society, humanitarian arena, and legitimacy played a key role in this.

Aid-society relationships and the humanitarian arena performed as effective analytical tools to explore and enable the study of the large constellation of actors and strategies present in disaster response. Aid-society made visible, and brought attention to, the need to include not only local actors but also donors, evaluators, and the private sector in the scenario, in addition to the best known actors in humanitarian aid like the UN agencies and international NGOs. The notion of the humanitarian arena strengthens the actor-oriented perspective by centering the analysis not in the physical space where the response occurs but on the interaction of aid-society actors, as well as their negotiations and processes that shaped the responses. The legitimacy focus was demonstrated to be a consistent entry point revealing a multitude of strategies used by aid and society actors when responding to disasters. In HIC scenarios, where the coercive use of power is present, the analysis provided a more complex and broad overview of the different ways in which various actors, from UN, armed groups, aid beneficiaries or rescuers (to name a few) enable themselves to act and cope with the challenges that these settings produce. The role played by humanitarian principles, the process of professionalisation of disaster response, international law, and the cultural community background of each actor were all highlights in the analysis. Additionally, the analysis showed the way to explore further the notion of power and the relevance of institutions. These relationships shape the response which, in turn, shapes the aid-society relationships in a symbiotic dynamic.

Regarding the challenges, the analysis revealed that alongside the overt and well-known security and access complications, there is a wider web of social, political, and economic conditions hindering responses in HIC. It allowed the observation of the challenges faced not only by aid but also by society actors in the responding process. Complex governance arrangements during HIC proved to be an overarching challenge. Many other issues are dependent, result from, or are generated by being associated with this factor. Although these results can be expected given the fact that we are discussing places affected by high levels of conflict, it is no less important, especially because the solutions tend to be more technical and focused mainly on the logistics of providing aid and responding rather than political and social change. This last point also highlights the limitations of observing the response only from an aid actor's perspective and reinforces at the same time the need for an aid-society approach to disaster response. Moreover, it accounts for the relevance of studying the relationships of people involved and especially how they manoeuvre through the humanitarian arena. Being legitimate and having the capacity to negotiate, using different resources and strategies is essential in HIC settings (and most probably, in any social arena).

One of the challenges in HIC conditions is the overlap of disaster response and humanitarian aid programmes in responding to the effect of the disaster. It can be said that most of the challenges, legitimacy strategies, and aid-society actors mentioned here are also present in general emergency and humanitarian aid programmes in HIC scenarios. The new contribution of the chapter to the existing literature lies in the fact that responding to disaster in these scenarios requires an understanding of the compound social and political nature of disaster and conflict. Without a more in-depth comprehension of what that means, all disaster response models might fall short in meeting their objectives. However, we must beware that the compound nature of both -general emergency and humanitarian aid programmes- does not lead us to ignore the differences that exist, the special characteristics of one and the other. Maintaining awareness of the differences in the two spheres of activity allows us to understand better what that composite nature means, how the interaction works, and how it affects work on the ground.

This last challenge permeated this research and its analysis. From an analytical point of view, it was difficult at times to assess whether the information presented in the literature and the interviews was clearly about disaster response or about humanitarian aid in general terms. There is still little discussion about the relationship and the differences between both disaster response and humanitarian aid. This, at the same time, strengthens the value of the analysis presented here.

For policy makers, practitioners, and scholars, the concept of HIC offers a richer understanding of disaster response in situations of high level violent social conflict. Multiple documents provide information about the growing levels of insecurity that aid workers face nowadays in their work (e.g. Duffield 2012; Roth 2011; Stoddard et al. 2006, 2014) but how this translates into disaster response is less clear. This chapter contributes information about specific characteristics of HIC, enabling all these actors to assess whether the places in which they are responding match this scenario. For those cases where the response is occurring in HIC scenarios, the chapter systematises the multiple actors that could be found on the ground, facilitating the networking process and participation. Furthermore, it informs aid-society actors about the challenges they may face, allowing better planning and implementation of disaster response. The analysis of legitimacy and negotiation processes and the systematization mechanisms and strategies in place for disaster response, might help practitioners and policy makers in the development, but also evaluation of disaster response.

The extensive literature reviewed leads to the observation that, despite the information gathered here, there is still limited academic understanding on disaster response in HIC scenarios, especially regarding aid-society relationships and disaster governance. The reviewed theoretical frameworks provided a relevant starting point for further research to start filling the gaps. The recurrence of disasters in HIC and the effects of them on local populations and institutions make this task every day more urgent. Global climate change, increasing levels of socio-economic inequality, profound unsolved gender bias, global environmental resources depletion, and the increased rates of violent social conflict are just some factors pointing to the need for better and more comprehensive disaster response. Comprehensive management of and response to complex disasters and crises comes from a full understanding of them. In this regard, it would be fruitful to pursue further research on the topic, in order to continue contributing to disaster response policies and practice by understanding better the special characteristics of responding in HIC and other types of conflict scenarios.

Appendix

Literature Review Sample:

Without grey literature, the sample reaches close to 400 sources (approximate value as some sources are book chapters or short and compound reports).

Interviews:

Type: Semi structured interviews.
Dates: All interviews were conducted in person by the author between November 2016 and March 2017.
Location: The interviews were conducted in The Netherlands, Sierra Leone, Uganda, and South Sudan.
Average duration: 75 min

Thematic Analysis:

Analytical categories

1. Disaster response: formal phases
2. Disaster response: spontaneous
3. Crisis or conflict definitions
4. HIC (high-intensity conflicts)
5. HIC and rival-similar terms
6. Humanitarian aid: definition
7. Humanitarian aid: history
8. Humanitarian aid: actors and organisations
9. Spatio-temporal conditions of the response
10. Type of organisation/s in disaster response
11. Aim/drivers of the response
12. Main problems/constraints
13. Chain of actions
14. Negotiations strategies
15. Networks in disaster and conflict response
16. Legitimacy: definition and theory
17. Legitimacy: whom
18. Legitimacy: what
19. Legitimacy: mechanism and strategies
20. Legitimacy: interrelated
21. Power relationships: definition and theory
22. Funding or financing process
23. Media role and interference
24. State – government/s: role in the response
25. State – government/s: collaboration with aid-society actors
26. State – government/s: control and legitimacy over actions
27. State – government/s: relationship with responders
28. The UN role or presence
29. NGOs role or presence
30. INGOs role or presence
31. ICRC-IFRC role or presence, including national societies
32. Local organisations role or presence

33. Lay people/volunteers' role or presence
34. Local partners' collaboration
35. Military role and collaboration
36. Role and presence of other governments
37. Security/protection against other people: who?
38. Security/protection against other people: how?
39. Security/protection for the socio-environmental disaster: who?
40. Security/protection for the socio-environmental disaster: how?
41. Medical care of the affected population
42. Migration and displacement
43. Governmental, Legal or Regulatory frameworks: control, enforcement, supervision, incentives, rights and obligations.

References

Aboagye, D. (2012): "The Political Ecology of Environmental Hazards In Accra, Ghana", in: *Journal of Environment and Earth Science* 2(10): 157–172.

Atmar, M.H. (2001): "The politicisation of humanitarian aid and its consequences for Afghans", in: *Humanitarian Exchange - HPN*.

Babbie, E.R. (2013): *The practice of social research* (Belmont, Calif: Wadsworth Cengage Learning).

Bankoff, G. (2001): "Rendering the World Unsafe: 'Vulnerability' as Western Discourse", in: *Disasters* 25(1): 19–35.

Bankoff, G.; Frerks, G.; Hilhorst, T. (Eds.), (2004): *Mapping vulnerability: disasters, development, and people* (London; Sterling, VA: Earthscan Publications).

Barnett, M. (2011) *Empire of humanity: a history of humanitarianism.* Cornell paperbacks, 1st print, Cornell paperbacks (Ithaca, NY: Cornell Univ. Press).

Bastian, N.D.; Griffin, P.M.; Spero, E.; et al. (2016): "Multi-criteria logistics modeling for military humanitarian assistance and disaster relief aerial delivery operations", in: *Optimization Letters*, 10(5): 921–953.

Bauman, Z. (1992): "Book Reviews: The Legitimation of Power (David Beetham)", in: *Sociology*, 26(3): 551–554.

Beal, A.L.; Graham, L. (2014): "Foundations for Change: Rule of Law, Development, and Democratization", in: *Foundations for Change: Rule of Law, Development, and Democratization*, 42(3): 311–345.

Beetham, D. (2013): *The legitimation of power.* 2nd Ed. Political analysis (Houndmills, Basingstoke, Hampshire; New York, NY: Palgrave Macmillan).

Bellamy, A.J. (2004): "Motives, outcomes, intent and the legitimacy of humanitarian intervention", in: *Journal of Military Ethics*, 3(3): 216–232.

Belliveau, J. (2016): "Humanitarian Access and Technology: Opportunities and Applications", in: *Procedia Engineering* (Humanitarian Technology: Science, Systems and Global Impact 2016, HumTech2016), 159: 300–306.

Best, S. (2002): *Introduction to politics and society* (London; Ca,: SAGE).

Buchanan, A. (1999): "The Internal Legitimacy of Humanitarian Intervention", in: *Journal of Political Philosophy*, 7(1): 71–87.

Caccavale, J. (2015): "The Struggle for Access in South Sudan", in: *SAVE Research*; at: http://www.saveresearch.net/the-struggle-for-access-in-south-sudan/ (5 August 2016).

Cannon, T. (1994): "Vulnerability analysis and the explanation of 'natural' disasters", in: *Disasters, development and environment*: 13–30.

Chesterman, S. (2002): "Legality Versus Legitimacy: Humanitarian Intervention, the Security Council, and the Rule of Law", in: *Security Dialogue*, 33(3): 293–307.

Cohen, J. (2014): "U.N. Security Council passes historic resolution to confront Ebola", in: *Science*; at: http://www.sciencemag.org/news/2014/09/un-security-council-passes-historic-resolution-confront-ebola (7 November 2016).

Collins, T.W. (2008): "The political ecology of hazard vulnerability: Marginalization, facilitation and the production of differential risk to urban wildfires in Arizona's White Mountains", in: *Journal of Political Ecology*, 15(1): 21–43.

Collinson, S.; Duffield, M. (2013): *Paradoxes of presence. Risk management and aid culture in challenging environments* (London, UK: Humanitarian Policy Group, Overseas Development Institute).

Collinson, S.; Elhawary, S.; Foley, M. (2012): *Humanitarian space: a review of trends and issues*. HPG Report (London: Overseas Development Institute Humanitarian Policy Group).

Comfort, L.K.; Ko, K.; Zagorecki, A. (2004): "Coordination in Rapidly Evolving Disaster Response Systems The Role of Information", in: *American Behavioral Scientist*, 48(3): 295–313.

Cottle, S. (2014): "Rethinking media and disasters in a global age: What's changed and why it matters", in: *Media, War & Conflict*, 7(1): 3–22.

Dahl, R. (1957): "The concept of power", in: *Behavioral Science*, 2(3): 201–215.

Davey, E.; Borton, J.; Foley, M. (2013): *A history of the humanitarian system: Western origins and foundations*. HPG Working Paper (London: Overseas Development Institute Humanitarian Policy Group); at: http://www.odi.org/sites/odi.org.uk/files/odi-assets/publications-opinion-files/8439.pdf.

De Cordier, B. (2009): "The "Humanitarian Frontline", Development and Relief, and Religion: what context, which threats and which opportunities?", in: *Third World Quarterly*, 30(4): 663–684.

DeChaine, D.R. (2002): "Humanitarian Space and the Social Imaginary: Médecins Sans Frontières/Doctors Without Borders and the Rhetoric of Global Community", in: *Journal of Communication Inquiry*, 26(4): 354–369.

Demmers, J. (2012): *Theories of Violent Conflict, An Introduction* (London: Routledge).

den Hond, F.; Boersma, F.K.; Heres, L.; et al. (2012): "Giddens à la Carte? Appraising empirical applications of Structuration Theory in management and organization studies", in: *Journal of Political Power*, 5(2): 239–264.

Donini, A. (ed.) (2012): *The golden fleece: manipulation and independence in humanitarian action* (Sterling, Virginia: Kumarian Press).

Donini, A.; Maxwell, D. (2014): "From face-to-face to face-to-screen: remote management, effectiveness and accountability of humanitarian action in insecure environments", in: *International Review of the Red Cross*, 95(890): 383–413.

Duffield, M. (1994): "Complex emergencies and the crisis of developmentalism", in: *IDS bulletin*, 25(4): 37–45.

Duffield, M. (2007): *Development, security and unending war: governing the world of peoples*. (Cambridge: Polity).

Duffield, M. (2012): "Challenging environments: Danger, resilience and the aid industry", in: *Security Dialogue*, 43(5): 475–492.

Egeland, J.; Harmer, A.; Stoddard, A. (2011): *To Stay and Deliver. Good Practice for Humanitarians in Complex Security Environments* (New York, USA: United Nations Office for the Coordination of Humanitarian Affairs (OCHA)).

Emery, J.R. (2016): "The Possibilities and Pitfalls of Humanitarian Drones", in: *Ethics & International Affairs*, 30(2): 153–165.

Estévez, R.A.; Anderson, C.B.; Pizarro, J.C.; et al. (2015): "Clarifying values, risk perceptions, and attitudes to resolve or avoid social conflicts in invasive species management: Confronting Invasive Species Conflicts", in: *Conservation Biology*, 29(1): 19–30.

Fink, S.; Sistenich, V.; Powell, C. (2014): "Humanitarian Assistance and Disaster Relief", in: *Understanding global health* (New York: McGraw-Hill Education).

Foucault, M. (1983): "The Subject and Power", in: *Michel Foucault, beyond structuralism and hermeneutics* (Chicago: University of Chicago Press).

Galtung, J. (1990): "Cultural Violence", in: *Journal of Peace Research*, 27(3): 291–305.

Galtung, J. (1996): *Peace by peaceful means: peace and conflict, development and civilization.* (London – Thousand Oaks, CA: Sage Publications).

Ganapati, N.E.; Kelman, I.; Koukis, T. (2010): "Analysing Greek-Turkish disaster-related cooperation: A disaster diplomacy perspective", in: *Cooperation and Conflict*, 45(2): 162–185.

Gasper, D. (1999): "Violence and Suffering, Responsibility and Choice: Issues in Ethics and Development", in: *European Journal of Development Research*, 11(2): 1.

Gasper, D.; Gómez, O. (2014b): "Human Security – Twenty Years On", in: *Human Development Reports - UNDP*; at: http://hdr.undp.org/en/content/human-security-%E2%80%93-twenty-years (4 December 2016).

Giddens, A. (1984): *The constitution of society: Outline of the theory of structuration* (Oxford: Polity Press).

Grünewald, F. (2012): "Aid in a city at war: the case of Mogadishu, Somalia", in: *Disasters*, 36: S105–S125.

Harmer, A.; Macrae, J. (2004): *Beyond the continuum: The changing role of aid policy in protracted crises.* (London: Overseas Development Institute).

Harvard Humanitarian Initiative (2011): *Disaster Relief 2.0: The Future of Information Sharing in Humanitarian Emergencies.* (Washington, D.C. -Berkshire, UK: UN Foundation and Vodafone Foundation Technology Partnership).

Harvey, P.; Lind, J. (2005): *Dependency and humanitarian relief: a critical analysis.* HPG research report (London: Overseas Development Institute).

Healy. S.; Tiller, S. (2014): *Where is everyone? Responding to emergencies in the most difficult places.* (London: Médecins Sans Frontières).

Hehir, A. (2011): "Pandora's box?: humanitarian intervention and international law", in: *International Journal of Law in Context*, 7(1): 87–94.

Herman, E.S.; Chomsky, N. (2002): "A Propaganda Model", in: *Manufacturing consent: The political economy of the mass media* (New York: Pantheon Books).

Hewitt, K. (2013): "Disasters in "development" contexts: Contradictions and options for a preventive approach", in: *Jàmbá: Journal of Disaster Risk Studies*, 5(2): 1–8.

Hilhorst, D. (2013a): "Disaster, conflict and society in crises: everyday politics of crisis response", in: Hilhorst, D. (ed.): *Disaster, conflict and society in crises: everyday politics of crisis response*, Routledge humanitarian studies series (New York: Routledge): 1–15.

Hilhorst, D. (ed.) (2013b): *Disaster, conflict and society in crises: everyday politics of crisis response.* Routledge humanitarian studies series (New York: Routledge).

Hilhorst, D.; Jansen, B. (2010): "Humanitarian space as arena: a perspective on the everyday politics of aid", in: *Development and Change*, 41(6): 1117–1139.

Hilhorst, D.; Jansen, B. (eds.) (2013): "Humanitarian space as arena: A perspective on the everyday politics of aid", in: *Disaster, conflict and society in crises: everyday politics of crisis response*, Routledge humanitarian studies series (New York: Routledge): 187–204.

Hilhorst, D.; Pereboom, E. (2016): "Multi-mandate Organisations in Humanitarian Aid", in: *The new humanitarians in international practice: emerging actors and contested principles*, Routledge humanitarian studies series (London - New York, NY: Routledge): 85–102.

Homer-Dixon, T.F. (1994): "Environmental Scarcities and Violent Conflict: Evidence from Cases", in: *International Security*, 19(1): 5.

HPN; OPM (2010): *Operational security management in violent environments* (London, UK: Humanitarian Practice Network, Overseas Development Institute).

Hyndman, J. (2011): *Dual disasters: humanitarian aid after the 2004 Tsunami* (Sterling, VA: Kumarian Press).

Ide, T. (2015): "Why do conflicts over scarce renewable resources turn violent? A qualitative comparative analysis", in: *Global Environmental Change*, 33: 61–70.

Jansen, B. (2013): "Two decades or ordering refugees. In: *Disaster, conflict and society in crises: everyday politics of crisis response*", Routledge humanitarian studies series (New York: Routledge): 114–131.

Jansen, B. (2015): "'Digging Aid': The Camp as an Option in East and the Horn of Africa", in: *Journal of Refugee Studies*, 29(2): 149–165.

Kaufman, L.; Rousseeuw, P.J. (2005): *Finding groups in data: an introduction to cluster analysis* (New York: Wiley).

Keen, D. (1998): *The economic functions of violence in civil wars*. Adelphi paper (Oxford: Oxford Univ. Press).

Keen, D. (2008): *Complex emergencies* (Cambridge: Polity).

Kelman, I. (2006): "Acting on Disaster Diplomacy", in: *Journal of International Affairs*, 59(2): 215–XV.

Kelman, I. (2012): *Disaster diplomacy* (Abingdon, New York: Routledge).

Krafess, J. (2005): "The influence of the Muslim religion in humanitarian aid", in: *International Review of the Red Cross*, 87(858): 327–342.

Lamb, R.D. (2014): *Rethinking Legitimacy and Illegitimacy: A New Approach to Assessing Support and Opposition across Disciplines* (Lanham: Rowman & Littlefield).

Leader, N. (2000): "The Politics of Principle", in: *The Principles of Humanitarian Action in Practice* (London, Overseas Development Institute): 6.

Levi, M.; Sacks, A. (2009): "Legitimating beliefs: Sources and indicators", in: *Regulation & Governance*, 3(4): 311–333.

MacFarlane, S.N.; Thielking, C.J.; Weiss, T.G. (2004): "The responsibility to protect: is anyone interested in humanitarian intervention?", in: *Third World Quarterly*, 25(5): 977–992.

Magone, C.; Neuman, M.; Weissman. F.; et al. (eds.) (2011): *Humanitarian negotiations revealed: the MSF experience* (London: Hurst & Co).

Malanczuk, P. (1993): *Humanitarian intervention and the legitimacy of the use of force* (Amsterdam, The Netherlands: Het Spinhuis).

Maxwell, D.G.; Majid, N. (2015): *Famine in Somalia: Competing Imperatives, Collective Failures, 2011–12* (London: Hurst & Company).

McCullough, A. (2015): *The legitimacy of states and armed non-state actors: Topic guide.* (Birmingham, UK: GSDRC, University of Birmingham).

McLachlin, R.; Larson, P.D.; Khan, S. (2009): "Not-for-profit supply chains in interrupted environments: The case of a faith-based humanitarian relief organisation", in: *Management Research News*, 32(11): 1050–1064.

Menkhaus, K. (2010): "Stabilisation and humanitarian access in a collapsed state: the Somali case", in: *Disasters*, 34: S320–S341.

Mosse, D. (ed.) (2013): *Adventures in Aidland: the anthropology of professionals in international development* (New York, NY: Berghahn).

National Research Council (2007): *Successful response starts with a map: improving geospatial support for disaster management* (Washington, D.C.: National Academies Press).

Nel, P.; Righarts, M. (2008): "Natural disasters and the risk of violent civil conflict", in: *International Studies Quarterly*, 52(1): 159–185.

Neuman, M.; Weissman, F. (2016): *Saving lives and staying alive: humanitarian security in the age of risk management* (London: Hurst & Company).

Newman, E, (2002): "Review Article: Humanitarian Intervention Legality and Legitimacy", in: *The International Journal of Human Rights*, 6(4): 102–120.

Newman, E. (2004): "The "New Wars" Debate: A Historical Perspective is Needed", in: *Security Dialogue*, 35(2): 173–189.

Oberschall, A. (1978): "Theories of social conflict", in: *Annual review of sociology*, vol. 4: 291–315.

Okoroma, F.N. (2011): "Towards effective management of grey literature for higher education, research and national development", in: *Library Review*, 60(9): 789–802.

Olsen, G.R.; Carstensen, N.; Høyen, K. (2003): "Humanitarian Crises: What Determines the Level of Emergency Assistance? Media Coverage, Donor Interests and the Aid Business", in: *Disasters*, 27(2): 109–126.
Parsons, T. (1964): *The social system* (London: Free Press).
Pattison, J. (2008): "Legitimacy and Humanitarian Intervention: Who Should Intervene?", in: *The International Journal of Human Rights*, 12(3): 395–413.
Paulson, N.; Menjívar, C. (2012): "Religion, the state and disaster relief in the United States and India", in: *International Journal of Sociology and Social Policy*, 32(3/4): 179–196.
Pottier, J. (2006): "Roadblock Ethnography: Negotiating Humanitarian Access in Ituri, Eastern DR Congo, 1999–2004". in: *Africa* 76(02): 151–179.
Recchia, S. (2015): "Soldiers, Civilians, and Multilateral Humanitarian Intervention", in: *Security Studies*, 24(2): 251–283.
Robinson, P. (1999): "The CNN effect: can the news media drive foreign policy?", in: *Review of International Studies*, 25(02): 301–309.
Roth, S. (2011): "Dealing with Danger: Risk and Security in the Everyday Lives of Aid Workers", in: Fechter, A.-M.; Hindman, H. (eds.): *Inside the everyday lives of development workers: the challenges and futures of Aidland* (Sterling, VA: Kumarian Press): 151–168.
Rother, M.B.; Pierre, M.G.; Lombardo, D.; et al. (2016): *The Economic Impact of Conflicts and the Refugee Crisis in the Middle East and North Africa* (Washington, D.C.: International Monetary Fund).
Schwendimann, F. (2011): "The legal framework of humanitarian access in armed conflict", in: *International Review of the Red Cross*, 93(884): 993–1008.
Sekhar, C.S.C. (2010): "Fragile States: The Role of Social, Political, and Economic Factors", in: *Journal of Developing Societies*, 26(3): 263–293.
Seybolt, T.B. (2008): *Humanitarian military intervention: the conditions for success and failure* (Solna, Sweden: Stockholm International Peace Research Institute).
Shreya, M.; Vivekananda, J. (2015): *Compounding Risk: Disasters, fragility and conflict.* (London, UK: International Alert).
Smirl, L. (2015): *Spaces of aid: how cars, compounds and hotels shape humanitarianism* (London: Zed Books).
Sphere Project (2011): *The Sphere Project: humanitarian charter and minimum standards in humanitarian response* (Rugby: The Sphere Project).
Spiegel, P.B.; Le, P.; Ververs. M.-T.; et al. (2007): "Occurrence and overlap of natural disasters, complex emergencies and epidemics during the past decade (1995–2004)", in: *Conflict and Health*, 1(1): 2.
Stoddard, A.; Harmer, A.; Haver, K. (2006): *Providing aid in insecure environments: trends in policy and operations* (London, UK: Humanitarian Policy Group, Overseas Development Institute).
Thrandardottir, E. (2015): "NGO Legitimacy: Four Models", in: *Representation*, 51(1): 107–123.
Todd, D.; Todd, H. (2011): *Natural Disaster Response Lessons from Evaluations of the World Bank and Others*. Evaluation Brief (Washington, D.C.: The World Bank: Independent Evaluation Group).
Troit, V. (2016): "Social Sciences Helping Humanitarism? A Few African Examples. Édition bilingue", in: *Transition humanitaire au Sénégal/ Humanitarian transition in Senegal*, Devenir humanitaire; Fons Croix-Rouge Francaise (Paris: Éditions Karthala): 166–178.
UNISDR (2013): *From Shared Risk to Shared Value: The Business Case for Disaster Risk Reduction. Global Assessment Report on Disaster Risk Reduction* (Geneva, Switzerland: United Nations Office for Disaster Risk Reduction [UNISDR]).
Vasilescu, L.; Khan, A.; Khan, H. (2008): "Disaster management cycle–a theoretical approach", in: *Management & Marketing-Craiova*, (1): 43–50.
Vernon, R. (2008): "Humanitarian intervention and the internal legitimacy problem", in: *Journal of Global Ethics*, 4(1): 37–49.

Wanta, W.; Golan, G.; Lee, C. (2004): "Agenda Setting and International News: Media Influence on Public Perceptions of Foreign Nations", in: *Journalism & Mass Communication Quarterly*, 81(2): 364–377.

Warner, J. (2013): "The politics of 'catastrophization'", in: *Disaster, conflict and society in crises: everyday politics of crisis response*, Routledge humanitarian studies series (New York: Routledge): 76–94.

Weber, M. [Parsons T (ed.)] (1964): *The theory of social and economic organization* (London: Collier-Macmillan).

Weiss, T.G. (2007): *Humanitarian intervention: ideas in action*. War and conflict in the modern world (Cambridge; Malden, Mass: Polity Press).

Wisner, B. (2010): "Climate change and cultural diversity", in: *International social science journal*, 61(199): 131–140.

Wisner, B. (2012): "Violent conflict, natural hazards and disaster", in: *The Routledge handbook of hazards and disaster risk reduction* (London - New York: Routledge): 65–76.

Wisner, B.; Blaikie, P.; Cannon, T.; et al. (2003): *At Risk: Natural hazards, people's vulnerability and disasters* (London; New York: Routledge).

Wood, A.P.; Apthorpe, R.J.; Borton, J. (eds.) (2001a): *Evaluating International Humanitarian Action: reflection from practitioners* (London–New York: Zed Books).

Wood, A.P.; Apthorpe, R.J.; Borton, J. (2001b): "Introduction", in: *Evaluating International Humanitarian Action: reflection from practitioners* (London–New York: Zed Books): 1–18.

Zajadlo, J. (2005): "Legality and Legitimization of Humanitarian Intervention New Challenges in the Age of the War on Terrorism", in: *American Behavioral Scientist*, 48(6): 653–670.

Zifcak, S. (2015): "What Happened to the International Community? R2p and the Conflicts in South Sudan and the Central African Republic", in: *Melbourne Journal of International Law* 16(1): 52–85.

Other Resources

Alexander, D. (2013): *Paradoxes and Perception: Four essays on disasters*. IRDR Occasional paper 2013-01 (London: IRDR, University College London); at: https://www.ucl.ac.uk/rdr/publications/irdr-occasional-papers/irdr-occasional-paper-2013-01.

ALNAP (2015): *The State of the Humanitarian System* (London: ALNAP/ODI); at: http://www.alnap.org/resource/21036.aspx (26 September 2016).

Arjona, A. (2008): *Armed Groups' Governance in Civil War: A Synthesis*. Program On States And Security (New York: Ralph Bunche Institute for International Studies); at: http://conflictfieldresearch.colgate.edu/wp-content/uploads/2015/05/Arjona.FINAL_.9.29.pdf.

Collier, P.; Hoeffler, A. (2001): "Data issues in the study of conflict" (Uppsala:); at: https://pdfs.semanticscholar.org/e777/8066cc0569d63bac2548427b7630e6e2255b.pdf (3 July 2016).

FTS (2016): "List of affected countries", in: *Financial Tracking Service - OCHA*; at: https://fts.unocha.org/countries/overview (23 August 2016).

Gasper, D.; Gómez, O. (2014a): "Evolution of thinking and research on human security and personal security 1994-2013"; at: http://repub.eur.nl/pub/76016/ (3 February 2016).

GFDRR; World Bank; UNDP; et al. (2016): *Disaster Recovery in Conflict Contexts: Thematic Case Study for the Disaster Recovery Framework Guide*; at: http://documents.worldbank.org/curated/en/310171495026430229/Disaster-recovery-in-conflict-contexts-thematic-case-study-for-the-disaster-recovery-framework-guide.

Giugni, P. (2016): "What You Need to Know about Humanitarian Airdrops", in: *InterCross*; at: http://intercrossblog.icrc.org/blog/what-you-need-to-know-about-humanitarian-airdrops (4 October 2016).

Grace, R. (2016): *Humanitarian Negotiation: Key Challenges and Lessons Learned in an Emerging Field*. White Paper Series, ATHA; Harvard Humanitarian Initiative; at: http://atha.se/presentations/negotiation/index.html.

Grigorian, D.; Kock, U. (2010): *Inflation and Conflict in Iraq: The Economics of Shortages Revisited* (Washington, D.C.: International Monetary Fund); at: https://www.imf.org/external/pubs/ft/wp/2010/wp10159.pdf.

Harris, K.; Keen, D.; Mitchell, T. (2013): *When disasters and conflicts collide: Improving links between disaster resilience and conflict prevention* (London: Overseas Development Institute Humanitarian Policy Group); at: https://www.odi.org/publications/7257-disasters-conflicts-collide-improving-links-between-disaster-resilience-and-conflict-prevention (8 June 2016).

HIIK (2016): *Conflict Barometer 2015* (Heidelberg: The Heidelberg Institute for International Conflict Research); at: https://hiik.de/conflict-barometer/bisherige-ausgaben/?lang=en.

Hilhorst, D. (2016): *Aid–society relations in humanitarian crises and recovery*. Inaugural Lecture, (The Hague, The Netherlands: Institute of Social Studies of Erasmus University Rotterdam).

IFRC; UNDP (2015): *The Checklist on Law and Disster Risk Reduction* (Geneva: International Federation of Red Cross and Red Crescent Societies and United Nations Development Programme); at: http://www.ifrc.org/PageFiles/115542/The-checklist-on-law-and-drr.pdf.

IFRC (2011): *Protect. Promote. Recognize: Volunteering in emergencies* (Geneva: IFRC): at: http://www.ifrc.org/PageFiles/41321/Volunteering%20in%20emergency_EN-LR.pdf.

Jackson, A.; Davey, E. (2014): *From the Spanish civil war to Afghanistan: Historical and contemporary reflections on humanitarian engagement with non-state armed groups*. HPG Working Paper (London, UK: ODI [Overseas Development Institute]); at: https://www.odi.org/sites/odi.org.uk/files/odi-assets/publications-opinion-files/8974.pdf.

La-Porte, M.T. (2015) The Legitimacy and Effectiveness of Non-State Actors and the Public Diplomacy Concept. Available from: http://dadun.unav.edu/handle/10171/38773 (accessed 8 November 2016).

Lilly, D. (2014): "Protection of Civilians sites: a new type of displacement settlement?", in: *ODI HPN*; at: http://odihpn.org/magazine/protection-of-civilians-sites-a-new-type-of-displacement-settlement/ (12 July 2016).

Loeb, J. (2013): *Talking to the other side: humanitarian engagement with armed non-state actors in Darfur, Sudan, 2003-2012*. HPG Working Paper (London, UK: ODI [Overseas Development Institute]); at: https://www.odi.org/sites/odi.org.uk/files/odi-assets/publications-opinion-files/8590.pdf.

Mancini-Griffoli, D.; Picot, A. (2004): *Humanitarian Negotiation* (Geneva, Switzerland: Centre for Humanitarian Dialogue); at: http://www.hdcentre.org/wp-content/uploads/2016/07/Humanitarian-Negotiationn-A-handbook-October-2004.pdf.

McCandless, E. (2014): "Non-state Actors and Competing Sources of Legitimacy in Conflict-Affected Settings", in: *Building Peace* (Washington, DC: Alliance for Peacebuilding); at: http://buildingpeaceforum.com/2014/09/non-state-actors-and-competing-sources-of-legitimacy-in-conflict-affected-settings/ (20 October 2016).

McGrady, E.D. (1999): *Peacemaking, Complex Emergencies, and Disaster Response: What Happens, How Do You Respond?* (London: DTIC); at: https://www.cna.org/CNA_files/PDF/D0002860.A1.pdf (3 June 2016).

McHugh, G.; Bessler, M. (2006): *Humanitarian Negotiations with Armed Groups* (New York: United Nations, OCHA); at: https://docs.unocha.org/sites/dms/Documents/Humanitarian-NegotiationswArmedGroupsManual.pdf.

Norton, I.; von Schreeb, J.; Aitken, P.; et al. (2013): *Classification and Minimum Standards For Foreign Medical Teams In Sudden Onset Disasters* (Geneva, Switzerland: World Health Organization [WHO]); at: http://www.who.int/hac/global_health_cluster/fmt_guidelines_september2013.pdf.

OCHA (1999): *Orientation Handbook on Complex Emergencies* (New York: United Nations, Office for the Coordination of Humanitarian Affairs [OCHA]); at:<http://reliefweb.int/report/world/ocha-orientation-handbook-complex-emergencies>.

OCHA (2011): *OCHA and slow-onset emergencies*. Occasional Policy Briefing, OCHA; at: http://www.oecd.org/dac/incaf/44794487.pdf (22 September 2014).

OECD (2010): *The state's legitimacy in fragile situations: unpacking complexity* (Paris: OECD); at: http://dx.doi.org/10.1787/9789264083882-en (8 November 2016).

Parker, R.S. (2006): *Hazards of Nature, Risks to Development: An IEG Evaluation of World Bank Assistance for Natural Disasters* (Washington, D.C.: The World Bank); at: http://elibrary.worldbank.org/doi/book/10.1596/978-0-8213-6650-9.

RPN (1997): *The role of the military in humanitarian work* (Oxford, UK: Refugee Participation Network); at: http://www.fmreview.org/sites/fmr/files/FMRdownloads/en/RPN/23.pdf.

Stel, N.; de Boer, D.; Hilhorst, D.; et al. (2012): *Multi-Stakeholder processes, service delivery and state institutions; Synthesis report*. Peace Security and Development Network; at: http://library.wur.nl/WebQuery/wurpubs/fulltext/341507.

Stoddard, A.; Harmer, A.; Ryou, K. (2014): *Unsafe Passage: Road attacks and their impact on humanitarian operations*. Aid Worker Security Report, Humanitarian Outcomes; at: https://aidworkersecurity.org/sites/default/files/Aid%20Worker%20Security%20Report%202014.pdf.

Tran, M. (2013): "Angola 'in denial' over impact of severe drought", in: *The Guardian*, 22 October; at: https://www.theguardian.com/global-development/2013/oct/22/angola-in-denial-severe-drought (2 November 2016).

UN News (2014): "UN Security Council to hold emergency meeting on Ebola crisis", in: *UN News*; at: https://news.un.org/en/story/2014/09/477452-un-security-council-hold-emergency-meeting-ebola-crisis (7 November 2016).

UNISDR (2009): "Terminology: Basic terms of disaster risk reduction" (Geneva, Switzerland: United Nations International Strategy for Disaster Reduction [UNISDR]); at: http://www.unisdr.org/we/inform/terminology.

United Nations (2015a): "Background Information on the Responsibility to Protect", in: *Outreach Programme on the Rwanda Genocide and the United Nations*; at: http://www.un.org/en/preventgenocide/rwanda/about/bgresponsibility.shtml (19 October 2016).

United Nations (2015b): *Sendai Framework for Disaster Risk Reduction* (Sendai, Japan: UNISDR); at: http://www.unisdr.org/we/coordinate/sendai-framework.

Walle, B. van de; Turoff, M. (2008): "Decision Support for Emergency Situations", in: *Handbook on Decision Support Systems 2* (Berlin - Heidelberg: Springer): 39–63; at: http://link.springer.com/chapter/10.1007/978-3-540-48716-6_3 (10 November 2016).

Young, H.; Maxwell, D. (2009): *Targeting in complex emergencies: Darfur case study*. (Medford, MA, Tufts University: Feinstein International Center Report); at: http://fic.tufts.edu/assets/targeting-in-comp-emer-2009.pdf (20 July 2017).

Chapter 4
The Fragile State of Disaster Response: Understanding Aid-State-Society Relations in Post-conflict Settings

Samantha Melis

Abstract Natural hazards often strike in conflict-affected societies, where the devastation is further compounded by the fragility of these societies and a complex web of myriad actors. To respond to disasters, aid, state, and societal actors enter the humanitarian arena, where they manoeuvre in the socio-political space to renegotiate power relations and gain legitimacy to achieve their goals by utilising authoritative and material resources. Post-conflict settings such as Burundi present a challenge for disaster response as actors are confronted with an uncertain transition period and the need to balance roles and capacity.

Keywords Natural disasters · Fragile states · Post-conflict · Humanitarian aid
Humanitarian arena · Power relations · Legitimacy · Disaster response
Burundi

4.1 Introduction

Disasters[1] caused by natural hazards are a disruptive force with grave social, political and economic impact. In 2015, 574 reported disasters killed almost 32,550 people and affected over 108 million people, with over 70,3 billion US dollars in damage (IFRC 2016). Climatologists predict that extreme weather and climate events will increase in both frequency and intensity in the coming years (Field/ IPCC 2012). However, not all populations are equally affected. Fragile and conflict

---

Ms. Samantha Melis, M.A., M.Sc., Ph.D. Candidate. International Institute of Social Studies (ISS), The Hague, The Netherlands. E-mail: melis@iss.nl. This chapter was made possible by a VICI grant of the Netherlands Organisation for Scientific Research NWO, grant number 453-14-013.

[1]Disasters are "a serious disruption of the functioning of a community or a society involving widespread human, material, economic or environmental losses and impacts, which exceeds the ability of the affected community or society to cope using its own resources" (UNISDR 2007). Although socio-natural disasters, disaster and natural disasters are used interchangeably, they are all seen in their socio-political context.

© Springer Nature Switzerland AG 2019
H. G. Brauch et al. (eds.), *Climate Change, Disasters, Sustainability Transition and Peace in the Anthropocene*, The Anthropocene: Politik—Economics—Society—Science 25, https://doi.org/10.1007/978-3-319-97562-7_4

affected states experience a greater impact from disasters. Between 2004 and 2014, 58 per cent of deaths from disasters occurred in countries listed in the top 30 of the *Fragile States Index* (Peters/Budimir 2016: 5). Therefore, it is crucial to recognise the compounding risk factors of natural hazards and conflict-affected societies in fragile states, and to better understand the way in which different aid, state and society actors respond to the disasters that ensue.

A natural hazard[2] does not always result in a disaster. Natural hazards contribute to the risk for disasters[3] to occur; however, it is the social, political and economic context that determines whether natural hazards become a disaster (Blaikie et al. 1994; Hewitt 2013). Therefore, the term 'socio-natural disasters' is a more accurate concept that recognises the profound social nature of disasters. In the aftermath of a socio-natural disaster, the way different actors respond is strongly rooted in the socio-political context. When a disaster strikes in a post-conflict setting, the response will be heavily affected by this.

Socio-natural disasters in conflict-affected countries add a layer of complexity to disaster response management. In fact, Kellett/Sparks (2012) show that each year from 2005–2009, over 50 per cent of people affected by disasters lived in fragile or conflict-affected states, reaching 80 per cent in some years (Kellett/Sparks 2012: 31). When trying to understand the everyday politics of disasters, examining the response phase after a disaster is particularly valuable. Disaster response is highly political (see Olson 2000), and occurs within an intricate socio-political context that affects the implementation of the response by different aid, state and society responders. The myriad of actors in the 'humanitarian arena' (Hilhorst/Jansen 2010) use the space opened by the disaster to advance their goals, whether by competition or cooperation. Different actors deploy discourse as a strategy in their search for resources and authority, and to assert their power, gain legitimacy and renegotiate the arena's values and structures. Post-conflict countries can prove to be especially challenging environments in this regard. While the conflict is assumed to have been largely resolved, the legacy of violent conflict and underlying conflict dynamics continue to have an impact on both social and political processes.

Post-conflict settings present a particular challenge for disaster response as they often undergo an uncertain transition that is characterised by continuous political and societal changes, while relationships are still rooted in its conflict history. This chapter presents a literature review, illustrated with findings from initial fieldwork

---

[2]EM-DAT (EM-DAT 2016) classifies hazards in different sub-groups; namely, geophysical, meteorological, hydrological, climatological, biological and extra-terrestrial. Examples of hazards are earthquakes (including tsunamis), volcanic activity, extreme temperatures, storms, floods, landslides, droughts, wildfires, epidemic etc.

[3]The speed of onset can be either slow or rapid. While rapid onset disasters are seen as the result of a sudden event, OCHA (2011) defines slow onset disasters, such as droughts, as an emergency that develops from a combination of events over time. Also, some disasters such as floods are often the accumulation of several events. In theory, slow on-set disasters could be mitigated and prevented by early response, however, in practice, most responses to slow on-set disasters resemble those of rapid onset disasters, with large influx of aid, primarily food aid, and short-term solutions focusing on saving lives (OCHA 2011: 4).

in Burundi, on disaster response in post-conflict societies. The transitional nature of the state in these settings and the fragility of governance poses specific challenges in the context of disaster response by a multitude of actors.

This chapter explores the role of aid, state, and societal actors who manoeuvre in the humanitarian arena[4] and identifies several challenges to disaster response in post-conflict settings. These challenges call for a new research agenda to develop effective policies that situate the response to disasters in its conflict context.

4.2 Methodology

This chapter is based on a literature review which combines peer-reviewed articles and books from the humanitarian assistance, disaster response, and (post-)conflict literatures. It further includes reports from humanitarian agencies and knowledge institutes focusing on humanitarian aid. In addition to the literature review, a 3-week pilot study was conducted in Burundi in August 2016. In Burundi, the author conducted 31 interviews with actors from different INGOs (International non-governmental organisations), NNGOs (National non-governmental organisations), national, regional and local humanitarian agencies, relevant national, regional and local state representatives, religious institutions, affected communities and IDP camp representatives. These semi-structured interviews included open-ended questions on the organisation of the response, coordination with other actors and the main challenges encountered. The interviews focused on the response to the 2014 flood in Bujumbura and the 2015 floods in Rumonge and Bujumbura. The chapter combines literature and data collected until December 2016, using primarily WorldCat, Google Scholar, Scopus and sEURch.[5] Data has been analysed with NVivo through a thematic content analysis and coding of the interview notes.[6]

---

[4]Aid actors are those actors who have development and emergency assistance as their core mandate, such as various *United Nations* (UN) agencies, local, national, and international *non-governmental organisations* (NGOs), the *International Committee of the Red Cross* (ICRC), regional inter-governmental organisations' agencies responsible for humanitarian assistance, and donor agencies providing funding and coordination. State actors are formally part of state institutions, whether on a national, regional, district, or local level, including national government agencies in charge of crisis response. Society encompasses a vast array of groups and identities, such as civil society, media, the private sector, volunteers, traditional leaders, beneficiaries, citizens and individuals.

[5]To minimise the selection bias of each individual search engine, the combination of these engines was used.

[6]Nodes included 34 emergent categories of different challenges encountered, such as coordination, beneficiary selection, mistrust, differences in response, relations between actors, communication, accountability, etc.

4.3   Disasters: A Socio-environmental Force

Although disasters in conflict-affected countries are linked to vulnerabilities, the main policy frameworks do not include a direct relation to the (post-)conflict context. The disaster management cycle is an organisational and policy tool that deals with disasters and categorises different phases wherein the activities of projects and interventions take place.[7] With the *2005 Hyogo Framework for Action* and the following *2015 Sendai Framework for Disaster Risk Reduction*, important steps were taken towards a focus on disaster prevention and preparedness. Policy frameworks take the extent of peoples' vulnerability, their exposure to natural hazards, and the nature of the hazard, with mitigation factors such as capacity and resilience, into account when determining the risk of socio-natural disasters (INFORM 2016; Wisner et al. 2012). With the *Disaster Risk Reduction* (DRR) paradigm, emphasising risk reduction includes not only a focus on mitigating hazards, planning preparedness and response and post-disaster recovery, such as the disaster management cycle, but also on resilience and the root causes of vulnerabilities. However, for political reasons, there is no explicit reference made to conflict in the Frameworks, obscuring the interconnections between conflicts and disasters.

There is much inequality in the way people are affected by disasters in low- and middle income countries.[8] This poses a risk for post-conflict settings, with conflict compounding the effects of disasters, leaving people more vulnerable to hazards and weakening institutional response capacities (Wisner 2012). Nel/Righarts (2008) found that disasters, especially rapid-onset disasters in low- and middle-income countries significantly increase the risks of recurring civil conflict. A history of conflict also leaves a torn societal fabric and mutual mistrust, which negatively impacts disaster risk reduction activities that require a community to combat environmental degradation.

Various factors influence the vulnerability of people to natural hazards, and in turn, to socio-natural disasters. Vulnerability has been used as a concept in different contexts, each with a specific interpretation of its characteristics (see Brauch 2005). In this chapter, vulnerability is seen as a socio-political concept in relation to natural hazards. Factors influencing vulnerability are often related to the characteristics of different socio-economic groups and a group's recovery capacity is affected by the access to resources and coping mechanisms (Blaikie et al. 1994). Vulnerability is

---

[7]The disaster management cycle is mostly focused on the disaster itself, as it includes measures taken before, during and after the disaster to "avoid a disaster, reduce its impact or recover from its losses" (Khan et al. 2008: 46). The pre-disaster stage includes activities for mitigation and preparedness, and the post-disaster stage starts with emergency response and moves into rehabilitation and reconstruction (Khan et al. 2008: 47). Although the cycle presupposes a linear timeline, in practice the phases overlap.

[8]From 1996 to 2015, low income countries experienced five times more deaths per 100.000 inhabitants compared to high income countries, while high income countries feature on the top ten list for economic losses (UNISDR/CRED 2016).

strongly determined by the socio-political processes, context and history, and thus continuously in flux. Further, factors such as gender strongly affects the vulnerability of disaster-affected groups (Ariyabandu/Fonseka 2009). As Wisner et al. (2012: 27) show, root causes linked to political, economic and social structures, including gender discrimination, affect the access to resources for certain people, leading to marginalization and ultimately increased vulnerability. Therefore, as Hilhorst et al. (2004) note, although vulnerability is often considered as a characteristic or a property, it is actually an outcome of social relations. Bohle et al. (1994) also identify vulnerability as a social outcome of three dimensions: human ecology (the environmental risk in relation to people's resources), social entitlements (and the way they are secured, or expanded), and the structural characteristics of the macro-structure of political economy in which the first and second dimensions are situated (Bohle et al. 1994: 40). Vulnerability, then, is determined by the position of an individual or group in each of these dimensions. In post-conflict settings, vulnerability, besides dependent on the ecological relation between people and nature, is also an outcome of the conflict history and the evolving socio-political changes, and the more structural transnational relations.

In addition to vulnerability, resilience is another key concept in DRR, and a challenge in post-conflict settings. The capacity to adapt is an important factor in the resilience of people at risk for disasters. The focus on resilience in DRR is not new (Alexander 2013; Manyena 2006). Resilience focuses more on strengths and capabilities, compared to vulnerability, which emphasises victimhood and needs. Resilience is often seen as the ability to 'bounce back' (IFRC 2016). However, as Paton (2006: 7) argues, 'bouncing back' does not reflect the reality of a disaster, where communities' pre-disaster state is fundamentally changed and new realities need to be faced. Therefore, 'resilience' means the capacity to adapt to the new reality and capitalise on new opportunities (Paton 2006: 8). This capacity to adapt of individuals and communities affected by natural hazards differs for different social groups. In post-conflict societies, where inequalities and marginalisation are often widespread, bouncing back would legitimise unequal pre-conditions. Therefore, a more transformative interpretation of resilience is necessary.

The aspiration to 'build back better' has been an attempt to better link relief, rehabilitation and development, and address the root causes of inequalities and marginalisation. Disaster response does not just require quick recovery efforts, but improvements to the previous state. However, different actors have their own interest agendas, questioning whose 'better' is being built (Fan 2013). On a local level, strengthening capacities can be easier than reducing vulnerability (Wisner et al. 2012: 29). Nevertheless, on a more structural level, 'strengthening capacity' implies addressing the social, political and economic root causes of vulnerability, which can be challenging, not only for the states signatory to the Sendai Framework, but also for humanitarian agencies in post-conflict settings as they aim to stay neutral and impartial as part of their humanitarian principles. Therefore, the IFRC (2016) recognises that addressing resilience in conflict-affected settings is against the humanitarian principles of impartiality and neutrality. Still, as addressing resilience is a crucial part of DRR and disaster response, humanitarian agencies face challenging dilemmas when responding to disasters in post-conflict settings, balancing neutrality with politics to save lives.

4.4 The 'Post-conflict State': Fragility and Disaster Response

Conflict affects the vulnerabilities and resilience of communities to disasters and continues to impact the post-conflict period as well. The term 'post-conflict' presupposes an end to conflict and the beginning of a peaceful period. However, this assumption is contradicted by the reality of most post-conflict societies, where tensions and even violence continue. Since the post-conflict discourse determines the way in which actors operate in these settings after a disaster, it is essential to understand the particularities and the risks involved in this discourse.

The term 'post-conflict'[9] was first used in the political discourse by UN Secretary General Boutros Boutros-Ghali in 1992 in *An Agenda for Peace*. Boutros-Ghali (1992) defined post-conflict peace-building as "action to identify and support structures which will tend to strengthen and solidify peace in order to avoid a relapse into conflict" (Boutros-Ghali 1992: para. 21). On request of the Security Council, Boutros-Ghali presented recommendations on how to move forward with conflict prevention, peacemaking, peacekeeping and, in particular, *post-conflict peace-building,* after the Cold War period (Hozić 2014: 22). His recommendations, divided into four post-conflict pillars of security, justice, democracy and development, are still illustrative of the aims of current post-conflict interventions.[10] If a disaster occurs in this transitional period, it will impact, and be affected by, these processes. For example, in security sector reform, the security forces such as the army and police may play a central role in emergency response after a disaster. However, these armed actors were usually involved in the conflict, affecting the

[9]Others have preferred the term post-war, which directly refers to a period after the end of a war, which makes it easier to define than 'conflict'. As the 'post' discourse refers to an outcome of the preceding period, war also does not do justice to the complexity of the 'post' situation: war was not the only or primary defining factor, but already an outcome in itself. Also, a post-war period can be a pre-war period and it does not reflect the reality of having a history of multiple conflicts and wars, or a conflict with less than 1000 battle-related deaths annually. This chapter sees both the post-conflict and post-war terms as not truly reflecting the processes and state after peace agreements or other types of political settlement. As post-conflict is a policy term used by the humanitarian actors, this chapter will continue using it to facilitate understanding of the type of period one is referring to.

[10]Some of the recommended actions are: disarmament, restoration of order, repatriation, capacity building of security personnel, monitoring elections, promoting human rights, reforming or strengthening governmental institutions and promoting formal and informal political participation (Boutros-Ghali 1992, para. 55).

trust between them and disaster-affected communities. This problematises their involvement in the response.

After a conflict, different state and non-state actors shape the formation of the state. As governance systems and other pre-conflict conditions were often themselves part of the drivers of the conflict and 'fragility', reconstruction should include reforms and the redistribution of rights and entitlements, promoting an agenda of change (Brinkerhoff 2005). Much of the external reconstruction efforts centre around the state, which has strong implications for the scope and type of interventions, where the emphasis is on statebuilding to prevent a 'relapse' and build stronger and more effective governance institutions to provide services and protection to its citizens. Therefore, statebuilding is often at the core of peace-building efforts.[11] However, the DRR frameworks are centred around a strong state. Non-state actors need to balance the formal role of the state, its actual capacity, and the statebuilding agenda when responding to a disaster.

The state is often seen in a centralised way along the lines of the Weberian notion of the state, which has a monopoly in the use of legitimate physical force in a defined territory (Weber 1978: 164). However, in many conflict-affected countries, the state's use of physical force is not seen as legitimate, or there are other groups that have a degree of legitimacy in their use of violence. In these cases, the 'state' continues to exist, but institutions or governance bodies are contested. Post-conflict countries are often considered part of the 'fragile states' discourse, where 'failed' or 'fragile' states are seen to pose a threat to international security and regional stability, making them a priority in both humanitarian and development policy (François/Sud 2006; Fukuyama 2004; Krasner/Pascual 2005).

One of the main challenges to respond to disasters in post-conflict settings is the state's capacity to respond and the strategies adopted by the aid and society actors to deal with this. Often, post-conflict states are considered fragile states. In previous definitions of 'fragile states', the state was seen to lack the capacity or be unwilling to provide basic functions for its citizens (DFID 2005; OECD/DAC 2007). Or in the Weberian fashion, it has lost its monopoly to use legitimate force (Weber 1978). However, not all developing states are considered 'fragile' even though most lack these capacities (Putzel 2010: 2). Brinkerhoff (2016) underlines the importance of recognising the multidimensional character of fragile states, that states are not uniformly fragile but can have stronger and weaker aspects, and that both structural conditions and agency influence fragility. State institutions often continue to operate, in one form or the other, in times of crises and 'significant pockets of capacity' remain functional (Brahimi 2007: 16). To determine the extent of fragility, Rocha Manocal (2013) identifies capacity, authority and legitimacy as the three key dimensions of the state and argues that 'fragile states' often have weaknesses in one or more dimensions (Rocha Monocal 2013: 389). Call (2011) and Brinkerhoff

---

[11]However, liberal peace theorists have strongly critiqued statebuilding interventions focused on the construction of a liberal democratic state through strengthening markets and through promoting democracy, civil society, and the rule of law (Barnett et al. 2014; David 2001; Paris 2004; Chandler 2013).

(2016) identify similar 'gaps'. In disaster response, some state institutions might be stronger than others. Furthermore, certain institutions could use the response itself to strengthen their capacity through the influx of aid and their authority and legitimacy by providing assistance to the disaster response.

As these dimensions are relational, in disaster response they are affected by the relationship between state, aid and society. The complexity and degrees of fragility within a post-conflict state pose a challenge to disaster response, wherein weaknesses in different institutions, and their development in the transitional period, affect both their capacity to respond and the strategies the other actors take to deal with them, collaborating or competing with the state in the humanitarian arena, using their own material and authoritative resources.

Another challenge in disaster response in post-conflict states is that the state is not a uniform entity, but rather a composition of a variety of state institutions, formal and informal. Traditional socio-political orders also continuously interact and share authority and legitimacy with state institutions (Boege et al. 2008, 2009; Lund 2006; Meagher et al. 2014). Hybrid governance is a tautology, as the definition of governance is the way in which governments and non-state actors relate to each other, including the blurred roles and responsibilities and interdependency in complex organisational forms (Colebatch 2014).[12] Nevertheless, the concepts of state and political hybridity are analytically useful, as it shows the power dynamics between different actors and the difference between institutions. This is particularly important for post-conflict settings, where institutional multiplicity is the rule rather than the exception. In situations of institutional multiplicity, actors lay claim to authority and all have sources of legitimation (van der Haar/Heijke 2013). As the state institutions usually cannot provide the required services, other authorities often coexist to provide core functions of the state, as a 'mediated state' (Menkhaus 2007). If they are not in competition, they even gain legitimacy from the state acting as local mediator. However, when humanitarian aid arrives after a disaster, they can also compete for resources and legitimacy. This will affect their legitimacy and reflect on the legitimacy of the state. In disaster response, the different political orders operate in a specific manner, with or without collaboration, and pursuing their own and common goals and interests.

When a disaster strikes, the responsibility of the state to respond is a core tenet of the DRR frameworks. In a context of transition, where state building is ongoing in an atmosphere of mistrust, this poses major challenges. One of the dynamics in this situation is that pre-existing aid relationships with the state come into play, affecting the way to which extent humanitarian actors cooperate and support the state in the response.

---

[12]As Reyntjes (2016: 358) notes, hybrid governance is not only applicable to fragile states settings, but is universally applicable.

This relationship between the post-conflict state and humanitarian actors is partly shaped by the peace process. Peace agreements and political settlements,[13] or the process towards it, are seen as the starting points of the post-conflict period, even if post-conflict countries are considered to be vulnerable to the conflict-trap, with increased chances of recurring conflict and violence after a settlement (Collier et al. 2008; Collier/World Bank 2003; Nathan/Toft 2011; Walter 2004).[14] The success or failure of settlements and the 'relapse' into conflict, the proliferation of 'new' conflicts, or the creation of a partial peace, depend on a variety of factors (Fortna 2004; Nilsson 2008; Sørbø 2004). Peace agreements can be seen as a certain type of social contract, where the focus lies on the mutually accepted agreement of political power and legitimacy by society to promote stability and peace (Hellsten 2009: 96). However, mediated peace agreements add much complexity as external power relations and politics come into play. The strategy of peace conditionalities can be an instrument for donors to influence and promote the peace process and the consolidation of peace (see Barnett et al. 2014; Boyce 2002; Frerks/Klem 2006), but the pressure of 'deadline diplomacy' and the threat of sanctions beg the question of ownership of the peace settlement and negatively affect the confidence building between parties in conflict (Nathan 2006).

Society plays a large role in both the conflict and in strengthening state-society relations. In the end, change needs to come from within. The citizens and leaders of post-conflict countries are ultimately responsible for governance reforms, with a supporting role for the international agencies (Brinkerhoff 2005; Chandler 2013). Local actors have the power to resist, disregard or adjust the peace processes and to present alternative forms of peace (Mac Ginty 2010). It is important to realise that the question of who sets peacebuilding or state building goals is related to a myriad of factors, including power relations and resources of the actors involved.

Disaster response in post-conflict societies may also be affected by the history of the conflict and its causes. Often, structural conflict continues after a political settlement, where underlying tensions and attitudes are still present, but violent behaviour has ceased (Galtung 1996). This, in turn, affects the way different actors respond to a disaster and the response itself impacts these dynamics. Vulnerability is often reproduced by the disaster, with the increasing needs of marginalised communities and unsuccessful recovery increasing marginalization (Wisner et al. 2012: 30). This can instigate sentiments of 'relative deprivation' (Gurr 2011) in such a way that tensions are aggravated. Aid and recovery increase people's expectations; when these expectations are not met, or when the situation declines

---

[13]Although peace agreements are types of political settlement, political settlements are also more than that. Here, the terms are used somewhat interchangeably to denote the political arrangement (either mediated or not) that defines the start of the post-conflict period. Peace agreements are usually mediated by external actors, either regional or international, and political settlements can also take the form of victory of one party over the others or a divided peace.

[14]Although the methods and numbers Collier uses for his arguments have been critiqued (Suhrke/Samset 2007), his work does show the vulnerability of post-conflict countries to conflict.

further, grievances are strengthened. Grievances can also be increased by a weak government response (Drury/Olson 1998; Gawronski/Olson 2013).

The ways in which these different factors play out is affected by the understanding and narratives that different actors weave around them. These narratives will impact both response implementation and the relations between actors. To understand the challenges of disaster response in post-conflict settings, it is therefore crucial to focus on the interplay of actors and the way they manoeuvre in the humanitarian arena.

4.5 Humanitarian Arena: Actors, Power and Legitimacy

The humanitarian response can be conceptualised in different ways, each affecting the way in which actors relate to each other and how they use their power and capacity to legitimise disaster response. The humanitarian world is described variously as a system (ALNAP 2015; Walker/Maxwell 2008), an empire (Barnett 2011; Donini 2012), or an arena (Hilhorst/Jansen 2010).

As a system, a network of complementary parts, consisting of UN agencies, INGOs, NGOs, donors and other (local) actors, functions by the guidance of an internal logic of principles, standards, norms, values and interests (ALNAP 2015). Although there is a type of systems logic to humanitarian assistance, in practice this 'systems logic' can be debated. The complementarity of agencies, including coordination mechanisms, falls short. Values, given the variety of local and international actors and donors, are often at odds with each other, particularly in post-conflict settings, where the transitional period signifies profound changes in policies and standards guiding the response on both national and local levels.

The empire view argues that aid is mostly self-interested, instrumentalised and controlled by a powerful few (Donini 2012), which can be seen by the concentration of aid in certain countries (Koch 2007; Koch et al. 2009) and the channelling of funds to a select group of agencies.[15] While aid is sometimes instrumentalised for certain goals, this view is mostly top-down. It ignores the many ways in which aid gets translated and altered throughout the chain of implementation and it underestimates the power and role of local organisations, the private sector, digital humanitarians, civil society, diaspora groups, communities and the state to negotiate the terms of the humanitarian response. Especially in post-conflict disaster

---

[15]Donors are important actors who often delineate humanitarian aid. They are increasingly seen to instrumentalise and politicise humanitarian aid and privilege agendas of stabilisation (ALNAP 2015: 13). Government donors channel most of their funds, two-thirds, to multilateral agencies, primarily UN agencies, with six UN agencies receiving 46 per cent of the total funds. Then INGOs 19 per cent, of which ICRC received almost two-thirds (GHA 2016: 66). Only 1.2 per cent is channelled directly to governments, with non OECD-DAC (The Organisation for Economic Co-operation and Development – Development Assistance Committee) donors channelling 70 per cent of their funds to governments (GHA 2016: 73).

response, the large influx of aid and organisations often poses a great challenge to coordination and control. Also, there is no central governance and organisations are self-regulated, creating a 'network based governance' in practice, without an overarching 'empire' (ODI-HPG 2016: 62).

Both the system and the empire do not sufficiently consider the complexity of changing relations and (post-conflict) contexts, and the reflexivity of the actors involved. The arena concept, on the other hand, brings out that multiple actors, including the communities and neighbours, operate in the 'humanitarian arena' and negotiate the conditions and practices of aid (Hilhorst/Jansen 2010). Principles and politics are given meaning in practice and are neither entirely imposed from the top or from outside. Aid is not just determined by humanitarian agencies, but all actors shape humanitarian action. Power, in this theoretical approach, is given and per- formed by the actors involved, legitimising their response. Without denying the importance of power inequalities, where especially local actors face barriers when it comes to financing and partnerships, it is emphasised that all actors manoeuvre in the arena.

4.5.1 Aid-State-Society Power Relations in Post-conflict Settings

In the humanitarian arena, humanitarian, state and societal actors[16] have various claims to legitimacy, capacity and authority to respond to disasters. In humanitarian assistance for disaster response, international law, tools, and standards are important to gain access and distribute aid effectively.[17] However, actors also need to have the power to do so. Disaster response in post-conflict contexts is defined by the interrelations between the different responders and the way in which they exert their power to manoeuvre in the arena and legitimise their actions.

---

[16]Although a distinction between aid, state and society is made, they are considered mutually constitutive and often problematic to identify as separate entities in practice. However, as DRR roles are generally different for aid agencies, states, and societal actors, this distinction is upheld to facilitate analyses of the processes within and relations between different groups of actors in the humanitarian arena.

[17]While international humanitarian law is applicable to armed conflict and occupation, disaster response does not have an overarching legal framework. Instead, it relies on various multilateral treaties, resolutions, declarations, guidelines and bilateral agreements as instruments, known as "international disaster response laws, rules and principles" (IDRL) (ICRC 2007: 15). In practice, much depends on the individual state's integration of disaster response in their national law, and their willingness and capacity to accommodate interventions after a disaster. In post-conflict countries, these policies cannot be seen separately from the Sustainable Development Goals. The Core Humanitarian Standard and the Sphere standard are recognised by humanitarian actors as standards to uphold and the Good Humanitarian Donorship principles and practices provide guidelines for donors to follow.

Often, power is understood in the Weberian sense of coercion and authority (Weber 1978). However, power is complex, interrelational, and performed in discourses, actions and resistance, and strengthened by both material and authoritative resources (Berger/Luckmann 1966; Bourdieu 1989; Foucault 1984; Frerks 2013; Gaventa/Cornwall 2008; Giddens 1984; Hayward 2000; Jabri 1996; Latour 1984; Lukes 2004; Weber 1978). Aid, state, and society actors are both autonomous and dependent on each other, to varying degrees. While the renegotiation of the power relations occurs continuously in everyday politics, actors change and opportunities can be more apparent after a disaster. Pelling/Dill (2010) argue that after a disaster, when the social contract between actors is contested or breaks down, a "space for negotiation on the values and structures of society" is opened (Pelling/Dill 2010: 27). Aid, state and societal actors renegotiate in this vacuum, and power can be redistributed. In post-conflict settings, the political stakes may be more complex and often higher, as this space for negotiation coincides and affects ongoing reshaping of power relations in the transition after a peace process. The room to manoeuvre depends on the various material and authority resources that actors control.

Material resources can contribute to each of the actors' bargaining power and are related to their institutional capacity to respond. Material resources in relation to the state can be understood as the ability of the state "to provide its citizens with basic life chances" (Rocha Monocal 2013: 389). Humanitarian agencies also rely on their material resources to provide humanitarian assistance. The extent to which the state is autonomous or dependent on another actor for the control over services indicates its capacity and level of power in the arena. All the actors can influence the resources of others. States can influence the material capacities of aid agencies by, for example, enforcing bureaucratic rules and regulations for organisations to obtain visas, pay taxes etc. On the other hand, humanitarian agencies can affect the state's resources by including or excluding the state as an intermediary recipient of aid. And as Hilhorst and Jansen (2010) have shown, local authorities and affected populations can also strengthen their material resources by manoeuvring within the humanitarian arena. They can, for example, control the list of beneficiaries and 'define' the rules of aid allocation. They can block the influx of material resources, through blockades or protests. Power relations are seen in the degree of autonomy and dependency each of these actors have and the way they use their resources to respond to disasters.

Authority is another resource actors use to manoeuvre in the humanitarian arena. Authoritative resources in relation to the state can be seen as the security and "the extent to which the state controls its territory and national law is recognised" (Rocha Monocal 2013: 389). These resources are reflected in the extent to which the state, non-state actors, or traditional authorities have control over others. In post-conflict settings, state authority can be fragile, becoming more dependent on the resources of other actors and thereby losing authoritative resources. Humanitarian agencies also need authoritative resources to negotiate safe access for humanitarian assistance. And societal actors exercise control over different social groups, by controlling who interacts and negotiates with humanitarian agencies and the state after a disaster. They also hold a certain power over agents from humanitarian organisations and the state when they need access to certain communities or sites. Information is also an authoritative resource, as it can be used for strengthening control. All actors can withhold information to gain more control or material resources. Actors who have access to information while others are restricted also have more control over the type of information that is shared.

With authoritative and material resources, the actors can manoeuvre and act in the arena, but their legitimacy determines whether their actions are accepted or resisted. Legitimacy is a term that has been mostly used in relation to the state, where it entails a normative belief of a political community that rules or institutions should be obeyed. The concept is sometimes extended to include performance as well (Levi et al. 2009; OECD 2010; Papagianne 2008; Rocha Monocal 2013; Stel et al. 2012; Weber 1978). However, legitimacy can also be seen as more symbiotic and multidimensional (Beetham 2013; Lamb 2014; Lister 2003). Besides a normative dimension, legitimacy is constructed through beliefs and practices. To Lamb (2014), legitimacy is 'the worthiness of support', a sense that something is right and should morally be supported, and illegitimacy as 'the worthiness of opposition'. It does not only pertain to the state, but also to other organisations, institutions or entities. Lamb (2014) emphasises that not only the entity of perceived legitimacy, or the 'conferee' is important, but especially the one who is evaluating, or 'the referee'. As Lister (2003) notes, it is important to understand 'which legitimacy matters' and the relative 'weights' of different actor referees (Lister 2003: 184). These weights can be related to the material and authoritative resources of the actors. The entity will therefore adjust their strategies to gain legitimacy appropriately, depending on the legitimacy the entity attributes to the referee. A state or humanitarian agency might act differently to one societal group than another. Humanitarian agencies often grant more legitimacy to more vulnerable groups when distributing aid after a disaster.

4.6   Disaster Response in Post-conflict Burundi

This section applies the core concepts and relationships discussed to the case study of Burundi, a country recovering from a civil war that formally ended in 2005 when Pierre Nkurunziza was sworn in as president.[18] Burundi has also been affected by different types of disasters and is considered to be fragile.[19] Fragile transitional governance, the statebuilding agenda and post-conflict politics impact disaster response and the relations between the aid and state actors.

---

[18]The Arusha Peace and Reconciliation Agreement was signed in 2000, while the CNDD-FDD signed a power sharing agreement in 2004 and Palipehutu-FNL signed a cease-fire in 2006.

[19]Alert ranking in Fragile states index 2015 (FFP 2015).

Burundi is exposed to a variety of natural hazards, including droughts, floods, landslides, torrential rains and earthquakes. On 9 February 2014, heavy torrential rains caused flooding, mud- and landslides in five communes (districts) in the capital of Bujumbura, killing 64 people and leaving an estimated 12,500 people homeless. In March and November 2015, and throughout the rainy season of 2015–2016, heavy rains and floods associated with the El Nino affected over 30,000 people, destroying over 5,000 houses and resulting in 52 deaths. In Rumonge, 276 households were settled in two IDP camps (UNOCHA 2016). The response to the floods in Bujumbura differed from the one in Rumonge, but the main challenges were indicative of disaster response in post-conflict settings.

The (centralised) state is responsible for disaster preparedness, risk reduction and response, which has been institutionalised through the Sendai Framework. In general, the roles and responsibilities of the state in disaster response are as follows: (1) declaring the crisis and emergency appeal for assistance, (2) providing assistance to and protection of the affected people, (3) coordinating the response and (4) ensuring a conducive legal environment (Harvey 2009: 6). These roles are part of the rules of the game within the humanitarian arena, but the way they are given shape in practice can be contested by the different actors, particularly in conflict-affected countries, including Burundi. In each of these roles, material and authoritative resources, and the way in which the different actors use their power vis-à-vis each other, affect the legitimacy of the state and other responders, posing challenges to post-conflict disaster response.

4.6.1 Appeal and Assistance

In Burundi, even though an appeal was made after the 2014 Bujumbura floods and November 2015 Rumonge floods, there was no emergency appeal for the floods on 29 March 2015 in Gitaza, Muhuta,[20] where the president called for national solidarity and on the communities to help each other.[21] The main reason for this difference is that the state likely did not want foreign interventions in the highly politicised pre-electoral period.[22]

Whether an emergency is declared and an appeal is made directly affects aid actors' material and authoritative resources to respond in post-conflict settings.

---

[20]Previously part of Bujumbura Rural, but became officially part of Rumonge after the creation of the latter province on 26 March 2015.

[21]See at: http://www.iwacu-burundi.org/gitaza-le-president-nkurunziza-en-appelle-a-la-solidarite-nationale-pluies-torrentielles/. The victims of the March floods were displaced in host families or lived in make-shift shelters, with limited support from the Ministry of Solidarity, local churches, political parties, the Burundi Red Cross and UN agencies. After the installation of IDP camps following the November floods, some victims from March were also included.

[22]Author's interview with International Humanitarian agency representative 3, 30 August 2016; this is also a view expressed by other actors in informal conversations in the same research period.

With the post-conflict statebuilding and governance agenda, humanitarian agencies will find it to difficult justify assistance without the state's request. Resource mobilisation and room for control of the response are thereby limited. While the post-conflict agenda increases the authoritative resources of the state as control over other actors, it does not directly benefit the material resources and capacity to provide assistance. Still, the state-centred agenda will foreground the assistance that the state provides to the affected communities, who are more reliant on existing social capital, people's relationship to the local state institutions and the degree of access to their services. This foregrounding of the assistance can be used for political reasons, particularly in post-conflict settings, wherein actors need to establish their legitimacy in the transitional period.

In Burundi, during the pre-electoral period, the president visited the affected sites in March, while ministers came in November.[23] However, even with a declared emergency, the way in which aid agencies respond is also dependent on the relationship between the donors and the state. After the protests and presidential elections in 2015, various donors suspended their cooperation and bilateral aid to Burundi, around half of whose national budget consists of aid. Although humanitarian assistance has been exempt, the number of organisations responding to the disaster in 2014 is less than in 2015.

How the second role – to provide assistance and protection – is translated into practice exposes the evolving relations between the actors and their legitimacy in the humanitarian arena. The state's capacity and resources to respond are not always present and the state is often partisan in the conflict or responsible for social inequalities, complicating equal protection. The relations between state, aid and society actors crystallise in the way they cooperate to provide assistance, using their different material and authoritative resources to legitimise their actions and themselves.

Societal actors are the first responders after a disaster. After the Bujumbura floods, the affected people were assisted by neighbours and family members who provided clothes and food, the local churches and youth and women groups, followed by the Burundi Red Cross.[24] Afterwards, other communities and private initiatives organised themselves around the country and sent their donations to Bujumbura. However, beneficiaries of aid are often only seen as victims and vulnerable people or as people who want to profit from the system.

In the humanitarian arena, communities actively seek survival and co-shape the realities of aid delivery, even if their manoeuvring power might be limited by the lack of resources or organisation. As Hilhorst/Jansen (2010) note, beneficiaries actively seek out aid and strategise to acquire it. They see "the humanitarian encounter as an interface where aid providers and aid-seekers meet each other" (Hilhorst/Jansen 2010: 1122). Even when recipients portray themselves as passive,

---

[23]Author's interview with local government representative 2, 22 August 2016.
[24]Author's interview with local actor 4, 25 August 2016, Focus Group with community actors 2, 29 August 2016.

this is agency in the sense of what Mats Utas (2005) calls 'victimcy'. People socially navigate the humanitarian arena by representing themselves in such a way to actively claim aid, using their agency contextually to foreground or background aspects of their identities which they consider most appropriate or effective for that specific situation, gaining legitimacy as being 'worthy of support'. This can take the form of foregrounding 'vulnerability' or 'identity', increasing access to goods and services and engaging in a type of 'forum shopping'. In Burundi, people from neighbouring communities presented themselves as victims to different organisations to access aid. A local government official acknowledged this practice and accepted that all Burundians are vulnerable, so when a non-affected person asked for aid during the distributions, they would also receive it.[25]

The state institutions are not always included in the response by aid agencies. During the civil war in Burundi, humanitarian aid often by-passed the state and depended on local authorities, who frequently took advantage of the supplied aid, weakening local governance and consequently reinforcing patrimonial systems (Uvin 2008). After the flood of 2014, local leaders were also given the responsibility to identify recipients, and creative list-making was rampant, through which favouritism privileged political party members who had a sum of money to share.[26] This view was also shared by the local communities and reinforced by numerous accounts of corruption on the level of the volunteers and their close collaboration with local governance structures, where lists were enlarged through the addition of names and some aid distributors requesting money for the materials.[27] Various actors, from INGOs to individuals and community groups, responded directly without coordination and cooperation with the National Platform or the Burundian Croix Rouge. These organisations preferred to distribute aid without cooperating with others, out of fear of corruption.[28] Direct implementation is used as a strategy by the agencies to deal with the lack of trust in the state institutions and the complexity of institutional multiplicity on the local level, where different local leaders have varying degrees of relations with the communities. Therefore, some larger INGOs distributed aid based on their own recipient lists. But the multitude of lists generated a high chance of some names being duplicated, as most actors did not have an overview of what assistance others had provided.[29] However, INGOs self-organised locally, which facilitated the co-creation of one beneficiary list that

---

[25]Interview with local actor 4, 25 August 2016, interview with local government representative 2, 22 August 2016.

[26]Author's interview with Community actors 1, 25 August 2016, and Community actors 2, 29 August 2016.

[27]Author's interviews with Local actor 4, 25 August 2016, NNGO 2, 17 August 2016, community actors 1, 25 August 2016, community actors 2, 29 August 2016, UN representative 3, 30 August 2016.

[28]Author's interview with Local actor 4, 25 August 2016.

[29]Author's interview with Local actor 4, 25 August 2016.

other actors followed.[30] As the government had requested all aid to be transferred through the Platform and the Croix Rouge, actors who did not use the Croix Rouge as a medium defied the state's request. Different organisations and communities expressed the belief that they could not trust the Platform or the Croix Rouge.[31] This affected both the legitimacy of the Croix Rouge and the government, as they were not considered 'worthy of support', strengthening agencies' authority over the response and guarding material resources.

Communication and information are both an essential and challenging part of disaster response. They directly relate to power relations, as information, or the withholding thereof, can be a way to control both people and material resources and to include or exclude groups from formal decision-making. Although information flows from the state to its citizens through various media, the capacity for state institutions to collect, manage and diffuse vital climate information accurately and on time is often weak. In Burundi, the *Geographical Institute of Burundi* (IGEBU) has limited means for the collection and communication of climate data.[32] Furthermore, the roles are not always clear; even when IGEBU has information about an impending hazard, they can only share this information with the state, which has the responsibility to act. If the state does not communicate this information on impending hazard, warning becomes problematic.

Most humanitarian agencies produce strong discourses on beneficiary accountability. However, in practice beneficiary accountability is often challenging, and gaps in quality information, communication flows, and the inclusion of affected people persist (Alexander 2015: 99). Beneficiary accountability has been promoted in humanitarian assistance and included in the Core Humanitarian Standard. It involves 'taking account', through listening and participation, 'giving account', through transparency and information, and 'responsibility', by taking ownership of the successes and failures (Serventy 2015). However, as Hilhorst (2015) argues, patronising forms of accountability, taking the aid agencies as starting points and 'granting' accountability, are dominant, while the 'co-governance of aid' and reciprocal relations are at the core of a more transformative accountability. Without a reciprocal relation, agencies are not giving 'weight' to how the beneficiaries view the legitimacy of their response, depending on a unilateral definition of legitimacy. As Heijmans (2004: 125) notes, most agencies define the situation of victims for them, even though the manner in which local communities perceive and calculate disaster risk, and adapt their strategies, can be different. In this case, communities are not seen as legitimate actors themselves. Especially in conflict settings, the way in which participation is practiced has exposed challenges, such as the reproduction of existing power inequalities, participants who do not have a real voice or

---

[30]Author's interview with NNGO 2, 17 August 2016. Author's interview with Local actor 5, Bujumbura, Burundi, 25 August 2016.

[31]Author's interview with NNGO 2, 17 August 2016, FGD with community 1, 25 August 2016, FGD with community 2, 28 August 2016, UN representative 2, 29 August 2016, UN representative 3, 30 August 2016, Local actor 4, 25 August 2016.

[32]Author's interview with Government representative 4, 18 August 2016.

influence, or the elite's voice being legitimised (Alejandro Leal 2007; Cornwall 2008; Pelling 1998). By limiting accountability, an actor limits the authoritative resources of another.

An important part of beneficiary accountability in Burundi is two-way communication between the aid agencies and the communities. The main feature of this is the complaints mechanism; either through phone communication lines, suggestion boxes or directly with staff in the affected areas. As local leaders often have a strong presence in the community, depending on their power, they can also control other community members' complaints. Further, corruption and mistrust between community members and their leadership is often a restraining factor. After the 2014 flood, affected community members did not feel free to register a complaint, as they regarded the local leaders and humanitarian volunteers as part of the corruption. One person addressed a complaint to an expatriate humanitarian worker and hoped to receive assistance, but when assistance arrived, a volunteer at the distribution level blocked the release of materials to him; so, for complaints, community members would rather "address themselves to god".[33] Also, the criteria for aid reception were not sufficiently communicated, as some potential recipients heard that aid was only for widows and therefore did not make any aid claims. During the emergency response, there was a sense of powerlessness within the community. They did not participate in any of the project phases. In the IDP camps, the committees, who do participate in the distribution of aid, did not have any information on the date or length of continuation of food deliveries by the World Food Programme (WFP), even though they are dependent on them.[34] As humanitarian agencies try to collaborate with local structures such as the local leaders as focal points for information, this selective information exclusion risks reproducing local inequalities by giving more resources to leaders who receive more information and who are therefore seen as more legitimate by some community groups.

4.6.2 Coordination and Legal Environment

The third role – coordination – also faces many challenges. As Harvey (2009) notes, there is often friction between the government's system of line ministries and sectors and the systems set up by humanitarian agencies, who do not always perceive the state as an equal partner and who are unwilling to relinquish responsibilities and power.

Post-conflict countries that are focusing on statebuilding and governance reforms are often not prepared for a large-scale disaster response. In countries where the UN cluster system is not permanently established, the activation and functioning of the clusters is slow. In Burundi, the UN and the government work together following a

[33]Author's interview with Community actors 1, 25 August 2016.
[34]Author's interview with IDP camp 2 and 3, 24 August 2016 and 26 August 2016.

sectorial approach along the lines of the different ministries. All sectorial ministries are part of The Burundian Platform for Disaster Management, which falls under the responsibility of the Ministry of Civil Protection, and is chaired by the Director-General of Civil Protection, under the Ministry of Interior and Public Security. But they do not have autonomy to have their own funds. While disasters often weaken state capacity, they further increase the demands placed on the state's limited resources (International Alert 2015). To gain material resources, the state is dependent on those of the international organisations, creating friction through the interdependency of the authoritative resources of the state and the material resources of the donors and humanitarian agencies.

In post-conflict countries, capacitating state institutions is an important part of the reconstruction endeavour. However, humanitarian agencies do not always fully cooperate and coordinate with the state, particularly if the state is seen as fragile. In Burundi, agencies and communities expressed their mistrust in the state and its institutions.[35] Humanitarians actively manoeuvre in the humanitarian arena. As Hilhorst/Jansen (2012) state that "principles and policies get translated, altered, co-opted or circumvented in everyday practice" (Hilhorst/Jansen 2012: 894). Humanitarian actors themselves take on roles to actively cope with challenging environments. Hilhorst (2016) calls this type of agency 'ignorancy': actors who manoeuvre political spaces by choosing to deploy their technocratic approach to achieve their goals and consciously choose naivety as a strategy.

In Burundi, humanitarian actors with the material resources also adopted a strategy to gain more authoritative resources through coordination. Although the National Platform for Disaster Management, chaired by the head of the Civil Protection Agency, is responsible for inter-agency coordination together with the line ministries and related UN agencies, the UN has attempted to take more of a lead as they found the government was trying to keep too much control over the platform, even though the government lacks the resources to coordinate and implement interventions. Exercising their 'ignorancy' by telling the government that they were probably too busy, the UN gained more authority as they were allowed to coordinate the sector meetings with the ministries attending; the sector meetings is where the decisions are made, while the Platform meetings are used to exchange information.[36]

Burundi's post-conflict state and its relation to aid actors and the communities must be understood through its conflict and peace history. As discussed above, the conflict period and the peace process have an impact on the post-conflict aid-state relationships. In Burundi, this manifests in the relationship between the CNDD-FDD, as the political party in power, and foreign aid actors. Curtis (2013) relates the mistrust of donors in Burundian institutions to the peace process wherein

---

[35]Author's interview with NNGO 2, 17 August 2016, FGD with community 1, 25 August 2016, FGD with community 2, 28 August 2016, UN representative 2, 29 August 2016, UN representative 3, 30 August 2016, Local actor 4, 25 August 2016.

[36]Author's interview with UN agency representative 2, 27 August 2016.

the goals for stabilisation and control were favoured over social justice and a liberal peace, turning a blind eye to governance abuses benefiting the power holders, and allowing some leaders to gain more authority and control. Violence continued throughout the peace process and was often used as a bargaining chip. The main rebel group, the CNDD-FDD, was not involved in the Arusha peace process, but signed a power sharing agreement in 2004 and won the 2005 elections. Curtis (2013) argues that violence and control continued to play a central role in the post-conflict Burundian state. This makes the Burundian government actors less 'worthy of support', or legitimate, in the eyes of the international aid agencies, which legitimises their increasing control over the coordination of the response through a technocratic discourse. Furthermore, as Curtis (2015) notes, international donors were more accustomed to work with the other Burundian parties who were part of the Arusha negotiations.

Internal state authority to coordinate can also be weak. In Burundi, attendance of the focal points of the line ministries in the Platform meetings is usually very low.[37] These focal points are seen to be chosen by favouritism and motivated by the additional premium they receive. After political changes, these staff are exchanged, which does not benefit human resources.[38] On a local level, government staff of technical institutions, such as the *Provincial Department of Agriculture and Livestock* (DPAE), is also often changed, making it more difficult to build capacities and become more professional, with staff having 'two heads', both political and technical.[39] With the multitude of local authorities, each with their own base of authority and legitimacy, coordination roles are challenging.

Finally, the fourth role – the legal environment – can either facilitate or hamper aid assistance, with strict rules and registration defining the allocative and author-itative resources of humanitarian agencies. Although the National Contingency Plan for Disaster Management existed and the National Platform for Disaster Management had been established in 2007, the Platform was not active at the time of the 2014 floods in Bujumbura. Furthermore, the lack of a separate budget for the Platform and the limited means of the government to respond are seen as major constraints by the government.[40] And even though the National Platform has been decentralised on a provincial and district level, its implementation is still in process. Because the material resources are not available for response, the government is more reliant on the aid agencies to provide assistance, affecting the control and authority they can exercise over the response. Still, the state can influence the authority and material capacities of humanitarians by enforcing strict tax or visa

---

[37]Author's interview with Government representative 3, 18 August 2016, and interview with Humanitarian agency representative 1, 17 August 2016.

[38]Author's interview with National Humanitarian agency representative 1, 17 August 2016.

[39]Author's interview with Humanitarian actor 1, 17 August 2016.

[40]Author's interview with government representative 1 on 17 August 2016, and with government representative 3 on 18 August 2016.

rules, or restrict international interventions all together. In Burundi, the state enforced additional taxes for international staff and controlled the visa procedures, delaying immediate staff support when the organisations needed it.[41]

4.7 Conclusion

Disasters are produced by a myriad of social, political, and economic factors and there is much inequality in the way people are affected by them. When a disaster strikes in a post-conflict setting, the response will be shaped by the conflict history. However, the organisations' embeddedness in the conflict context is not included in their policy frameworks. Although each disaster and context are unique, there are specific challenges that are exacerbated by the nature of a post-conflict setting, wherein the transition period creates a tension between the statebuilding agenda and the disaster response.

To respond to disasters, different aid, state and societal actors enter the humanitarian arena, where they manoeuvre in the space opened by the disaster, to renegotiate power relations, using material and authoritative resources, and gain legitimacy, as part of their everyday politics and the political nature of disasters. The way in which disaster response is implemented is affected by the discourses, actions and resistance of the actors in the arena. Post-conflict settings are challenging environments for disaster response, with a complex web of compounding vulnerabilities on both social and political levels. The state, aid and society actors face challenges in dealing with the socio-political fragility of the transition period, wherein the capacity to respond is often diminished. The way in which different actors respond to a disaster, each with their own agenda and resources, is affected by the conflict history and the peace process.

Although this research is limited to a literature review, the Burundi pilot-study allowed an analytical reflection on the theory. The application of the case using the Sendai Framework was an exercise that illustrates the applicability of disaster response theories in post-conflict scenarios. The response to the floods in 2014 and 2015 showed how different actors find strategies to manoeuvre in the arena, using their material and authoritative resources in relation to the resources of others affected by the post-conflict context. The response roles of the state to the disasters have been contentious in a context where the state is considered fragile and in need of capacity building. This perceived capacity gap resulted in the different actors taking on their own strategies to deal with the others, centring around the challenges of capacity, coordination, implementation, mistrust and accountability. To bring in the state as central to the framework has proven to be relevant, as the examples showed that one of the most important challenges is the friction between policy and practice. On the one hand, for policy makers, the state is central to post-conflict

---

[41]Author's interview with UN agency representative 3 on 30 August 2016.

statebuilding efforts and responsible for the implementation of the DRR framework, but on the other, in practice, is often by-passed by donors and humanitarian agencies, who limit the state's material and authoritative resources.

This chapter has focused on the specificities of disaster response in post-conflict settings and although each setting is unique, the general characteristics have identified several core challenges. However, further research is needed to understand the different challenges in other conflict settings and for a new research agenda to fully uncover the complexities and interconnectedness of conflict and disasters. More research is needed to address power relations and the question of legitimacy in relation to its conflict context, for policy-makers to include a better understanding of disaster response in conflict affected settings in the DRR frameworks, and for practitioners to conceptualise and implement disaster response activities that are sensitive to these processes and able to address the challenges they face in co-governed disaster response.

References

Alejandro Leal, Pablo, 2007: "Participation: The Ascendancy of a Buzzword in the Neo-Liberal Era", in: *Development in Practice*, 17(4–5): 539–48.

Alexander, D. E., 2013: "Resilience and Disaster Risk Reduction: An Etymological Journey", in: *Natural Hazards and Earth System Science*, 13(11): 2707–2716.

Alexander, Jessica, 2015: "Informed Decision Making: Including the Voice of Affected Communities in the Process", in: *Humanitarian Accountability Report 2015. On the Road to Istanbul: How Can the World Humanitarian Summit Make Humanitarian Response More Effective?* (Geneva: CHS Alliance): 98–103.

Ariyabandu, Madhavi Malalgoda; Fonseka, Dilrukshi, 2009: "Do Disasters Discriminate? A Human Security Analysis of the Impact of the Tsunami in India, Sri Lanka and of the Kashmir Earthquake in Pakistan", in: Hans Günter Brauch et. al. (Eds.): *Facing Global Environmental Change: Environmental, Human, Energy, Food, Health and Water Security Concepts*, 1215–26 (Berlin – Heidelberg: Springer).

Barnett, Michael, 2011: *Empire of Humanity: A History of Humanitarianism*, Cornell Paperbacks (Ithaca, NY: Cornell Univ. Press).

Barnett, Michael; Fang, Songying; Zürcher, Christoph, 2014: "Compromised Peace-building", in: *International Studies Quarterly*, 58(3): 608–620; at: https://doi.org/10.1111/isqu.12137.

Beetham, David, 2nd ed., 2013: *The Legitimation of Power*. Political Analysis. (Houndmills, Basingstoke, Hampshire – New York, NY: Palgrave Macmillan).

Berger, Peter L.; Luckmann, Thomas, 1966: *The Social Construction of Reality: a Treatise in the Sociology of Knowledge* (Garden City, NY: Doubleday).

Blaikie, Piers M.; Cannon, Terry; Davis, Ian; Wisner, Ben (Eds.), 1994: *At Risk: Natural Hazards, People's Vulnerability, and Disasters* (London – New York: Routledge).

Boege, Volker; Brown, M. Anne; Clements, Kevin P., 2009: "Hybrid Political Orders, Not Fragile States", in: *Peace Review*, 21(1): 13–21.

Bohle, Hans G.; Downing, Thomas E.; Watts, Michael J., 1994: "Climate Change and Social Vulnerability: Toward a Sociology and Geography of Food Insecurity", in: *Global Environmental Change*, 4(1): 37–48.

Bourdieu, Pierre, 1989: "Social Space and Symbolic Power." *Sociological Theory*, 7(1): 14; at: https://doi.org/10.2307/202060.

Boyce, James K., 2002: "Aid Conditionality as a Tool for Peacebuilding: Opportunities and Constraints", in: *Development and Change*, 33(5): 1025–1048.

Brauch, Hans Günter, 2005: *Threats, Challenges, Vulnerabilities and Risks in Environmental and Human Security*. Source 1 (Bonn: UNU-EHS).

Brinkerhoff, Derick W., 2005: "Rebuilding Governance in Failed States and Post-Conflict Societies: Core Concepts and Cross-Cutting Themes", in: *Public Administration and Development*, 25(1): 3–14; at: https://doi.org/10.1002/pad.352.

Brinkerhoff, Derick W., 2016: "State Fragility, International Development Policy, and Global Responses." in: *RTI International*, International Development. Working Paper. October 2016. *Forthcoming*.

Call, C. T., 2011: "Beyond the 'Failed State': Toward Conceptual Alternatives", in: *European Journal of International Relations*, 17(2): 303–326.

Chandler, David, 2013: "International Statebuilding and the Ideology of Resilience: International Statebuilding", in: *Politics*, 33(4): 276–286.

Colebatch, H. K., 2014: "Making Sense of Governance", in: *Policy and Society*, 33(4): 307–316.

Collier, Paul; Hoeffler, Anke; Soderbom, Mans; 2008: "Post-Conflict Risks", in: *Journal of Peace Research*, 45(4): 461–78, at: https://doi.org/10.1177/0022343308091356.

Collier, Paul; World Bank (Eds.), 2003: *Breaking the Conflict Trap: Civil War and Development Policy*. A World Bank Policy Research Report (Washington, DC: World Bank – New York: Oxford University Press).

Cornwall, Andrea, 2008: "Unpacking 'Participation': Models, Meanings and Practices", in: *Community Development Journal*, 43(3): 269–83; at: https://doi.org/10.1093/cdj/bsn010.

Curtis, Devon, 2013: "The International Peacebuilding Paradox: Power Sharing and Post-Conflict Governance in Burundi", in: *African Affairs*, 112(446): 72–91.

Curtis, Devon, 2015: "Development Assistance and the Lasting Legacies of Rebellion in Burundi and Rwanda", in: *Third World Quarterly*, 36(7): 1365–81; at: https://doi.org/10.1080/01436597.2015.1041103.

David, Charles-Philippe, 2001: "Alice in Wonderland Meets Frankenstein: Constructivism, Realism and Peacebuilding in Bosnia", in: *Contemporary Security Policy*, 22(1): 1–30.

Donini, Antonio (Ed.), 2012: *The Golden Fleece: Manipulation and Independence in Humanitarian Action* (Sterling, Virginia: Kumarian Press).

Drury, A. Cooper; Olson, Richard Stuart, 1998: "Disasters and Political Unrest: An Empirical Investigation", in: *Journal of Contingencies and Crisis Management*, 6(3): 153–61.

Duffield, Mark, 2002: "Social Reconstruction and the Radicalization of Development: Aid as a Relation of Global Liberal Governance", in: *Development and Change*, 33(5): 1049–71.

Fan, Lilianne, 2013: "Disaster as Opportunity? Building Back Better in Aceh, Myanmar and Haiti". HPG Working Paper (London: ODI [Overseas Development Institute]).

Field, Christopher B.; IPCC (Eds.), 2012: *Managing the Risks of Extreme Events and Disasters to Advance Climate Change Adaption: Special Report of the Intergovernmental Panel on Climate Change* (New York, NY: Cambridge University Press).

Fortna, Virginia Page, 2004: "Does Peacekeeping Keep Peace? International Intervention and the Duration of Peace after Civil War", in: *International Studies Quarterly*, 48(2): 269–292.

Foucault, Michel, 1984: "The Order of Discourse", in: Michael J. Shapiro (Ed.): *Language and Politics* (New York: New York University Press).

François, Monika; Sud, Inder, 2006: "Promoting Stability and Development in Fragile and Failed States", in: *Development Policy Review*, 24(2): 141–160.

Frerks, Georg, 2013: "Discourses on War, Peace and Peacebuilding", in; Dorothea Hilhorst (Ed.): *Disaster, Conflict and Society in Crises: Everyday Politics of Crisis Response*. Routledge Humanitarian Studies Series 1 (New York: Routledge); 19–37.

Frerks, Georg; Klem, Bart, 2006: "Conditioning Peace among Protagonists: A Study into the Use of Peace Conditionalities in the Sri Lankan Peace Process" (Clingendael: Netherlands Institute of International Relations).

Fukuyama, Francis, 2004: *State-Building: Governance and World Order in the 21st Century*. (Ithaca, N.Y: Cornell University Press).

Galtung, Johan, 1996: "Part II: Conflict Theory", in: *Peace by Peaceful Means: Peace and Conflict, Development and Civilization* (Oslo: International Peace Research Institute; London; Thousand Oaks, CA: Sage Publications): 70–80.

Gaventa, John; Cornwall, Andrea, 2nd ed., 2008: "Power and Knowledge", in: *The Sage Handbook of Action Research: Participative Inquiry and Practice*, in: Reason, Peter; Bradbury, Hilary (Eds.): (London – Thousand Oaks, Ca: SAGE Publications): 172–189.

Gawronski, Vincent T.; Olson, Richard Stuart, 2013: "Disasters as Crisis Triggers for Critical Junctures? The 1976 Guatemala Case", in: *Latin American Politics and Society*, 55(2): 133–149.

Giddens, Anthony, 1984: *The Constitution of Society: Outline of the Theory of Structuration.* (Berkeley, CA: University of California Press).

Gurr, Ted Robert, 2011: *Why Men Rebel*. 40th Anniversary paperback ed. (Boulder, Colo.: Paradigm Publications).

Haar, Gemma van der; Heijke, Merel, 2013: "Conflict, Governance and Institutional Multiplicity: Parallel Governance in Kosovo and Chiapas (Mexico)", in: Hilhorst, Dorothea (Ed.): *Disaster, Conflict and Society in Crises: Everyday Politics of Crisis Response*, edited by Routledge Humanitarian Studies Series 1. New York: Routledge.

Harvey, Paul; Overseas Development Institute, 2009: *Towards Good Humanitarian Government: The Role of the Affected State in Disaster Response* (London: Humanitarian Policy Group).

Hayward, Clarissa Rile, 2000: *De-Facing Power. Contemporary Political Theory* (Cambridge, UK – New York: Cambridge University Press).

Heijmans, Annelies, 2004: "From Vulnerability to Empowerment", in: Bankoff, Greg; Frerks, Georg; Hilhorst, Dorothea (Eds.): *Mapping Vulnerability: Disasters, Development, and People*. (London – Sterling, VA: Earthscan Publications); 115–127.

Hellsten, Sirkku K., 2009: "Ethics, Rhetoric, and Politics of Post-Conflict Reconstruction: How Can the Concept of Social Contract Help Us in Understanding How to Make Peace Work?", in: Addison, Tony; Brück, Tilman (Eds.): *Making Peace Work* (London: Palgrave Macmillan UK): 75–96.

Hewitt, Kenneth, 2013: "Disasters in 'Development' Contexts: Contradictions and Options for a Preventive Approach", in: *Jàmbá: Journal of Disaster Risk Studies*, 5(2), Art.#91, 8 pages.

Hilhorst, Dorothea; Jansen, Bram, 2012: "Constructing Rights and Wrongs in Humanitarian Action: Contributions from a Sociology of Praxis", in: *Sociology*, 46(5): 891–905.

Hilhorst, Dorothea, 2015: "Taking Accountability to the next Level". in: CHS (Core Humanitarian Standard) Alliance (Ed.): *Humanitarian Accountability Report 2015. On the Road to Istanbul: How Can the World Humanitarian Summit Make Humanitarian Response More Effective?* (London: CHS Alliance): 108–110.

Hilhorst, Dorothea; Bankoff, Greg; Frerks, Georg, 2004: "Mapping Vulnerability", in: *Mapping Vulnerability: Disasters, Development, and People* (London – Sterling, VA: Earthscan Publications): 1–9.

Hilhorst, Dorothea; Jansen, Bram, 2010: "Humanitarian Space as Arena: A Perspective on the Everyday Politics of Aid", in: *Development and Change*, 41(6): 1117–1139.

Hozić, Aida, 2014: "The Origins Of 'Post-Conflict'", in: *Post-Conflict Studies: An Interdisciplinary Approach*. Routledge Studies in Peace and Conflict Resolution (London – New York: Routledge, Taylor & Francis Group): 19–38.

ICRC (International Committee of the Red Cross), 2007: "Law and Legal Issues in International Disaster Response: A Desk Study" (Geneva: International Federation of Red Cross and Red Crescent Societies).

IFRC (International Federation of Red Cross and Red Crescent Societies), 2016: *World Disaster Report 2016* (Geneva: International Federation of Red Cross and Red Crescent Societies).

International Alert, 2015: "Compounding Risk: Disasters, Fragility and Conflict", Policy Brief (London: International Alert).

Jabri, Vivienne, 1996: *Discourses on Violence: Conflict Analysis Reconsidered* (Manchester; Manchester University Press).

Khan, Himayatullah; Vasilescu, Laura; Khan, Asmatullah, 2008: "Disaster Management Cycle–a Theoretical Approach" in: *Management & Marketing-Craiova*, no. 1; 43–50.

Koch, Dirk-Jan, 2007: *Blind Spots on the Map of Aid Allocations Concentration and Complementarity of International NGO Aid* (Helsinki, Finland: United Nations University/World Institute for Development Economics Research [UNU/WIDER]).

Koch, Dirk-Jan; Dreher, Axel; Nunnenkamp, Peter; Thiele, Rainer, 2009: "Keeping a Low Profile: What Determines the Allocation of Aid by Non-Governmental Organisations?", in: *World Development*, 37(5): 902–918.

Krasner, Stephen D.; Pascual, Carlos, 2005: "Addressing State Failure", in: *Foreign Affairs*, 84(4): 153–163.

Lamb, Robert D., 2014: *Rethinking Legitimacy and Illegitimacy: A New Approach to Assessing Support and Opposition across Disciplines* (Washington, D.C.: Center for Strategic and International Studies [CSIS]).

Latour, Bruno, 1984: "The Powers of Association", in: *The Sociological Review*, 32(May): 264–80.

Levi, Margaret; Sacks, Audrey; Tyler, Tom, 2009: "Conceptualizing Legitimacy, Measuring Legitimating Beliefs", in: *American Behavioral Scientist*, 53(3): 354–375.

Lister, Sarah, 2003: "NGO Legitimacy Technical Issue or Social Construct?", in: *Critique of Anthropology*, 23(2): 175–192.

Lukes, Steven, 2nd ed., 2004: *Power: A Radical View* (Houndmills, Basingstoke, Hampshire – New York: Palgrave Macmillan).

Lund, Christian (Ed.), 2006: *Twilight Institutions: Public Authority and Local Politics in Africa* (Malden, MA: Blackwell Publications).

Mac Ginty, R., 2010: "Hybrid Peace: The Interaction Between Top-Down and Bottom-Up Peace", in: *Security Dialogue*, 41(4): 391–412; at: https://doi.org/10.1177/0967010610374312.

Manyena, Siambabala Bernard, 2006: "The Concept of Resilience Revisited", in: *Disasters* 30(4): 434–450.

Nathan, Laurie; Toft, Monica Duffy, 2011: "Civil War Settlements and the Prospects for Peace", in: *International Security*, 36(1): 202–210.

Nel, Philip; Righarts, Marjolein, 2008: "Natural Disasters and the Risk of Violent Civil Conflict", in: *International Studies Quarterly*, 52(1): 159–185.

Nilsson, D., 2008: "Partial Peace: Rebel Groups Inside and Outside of Civil War Settlements", in: *Journal of Peace Research*, 45(4): 479–495.

OCHA (United Nations Office for the Coordination of Humanitarian Affairs), 2011: *OCHA and Slow-Onset Emergencies*. 6th Occasional Policy Briefing (New York: OCHA).

OECD (Organisation for Economic Co-operation and Development), 2010: *The State's Legitimacy in Fragile Situations Unpacking Complexity* (Paris: OECD Publishing).

OECD/DAC (Organisation for Economic Co-operation and Development, Development Assistance Committee), 2007: "Principles for Good International Engagement in Fragile States and Situations" (Paris: OECD Publishing, April).

Olson, Richard Stuart, 2000: "Toward a Politics of Disaster: Losses, Values, Agendas, and Blame", in: *International Journal of Mass Emergencies and Disasters (IJMED)*, 18(2): 265–288.

Papagianne, Katia, 2008: "Participation and State Legitimation", in: Call, Charles (Ed.): *Building States to Build Peace* (Boulder: Lynne Rienner Publishers): 49–71.

Paris, Roland, 2004: *At War's End: Building Peace after Civil Conflict* (Cambridge, U.K. – New York, NY: Cambridge University Press).

Paton, Douglas, 2006: "Disaster Resilience: Building Capacity to Co-Exist with Natural Hazards and Their Consequences", in: Paton, Douglas; Johnston, David M. (Eds.): *Disaster Resilience: An Integrated Approach* (Springfield, Ill: Charles C. Thomas): 3–10.

Pelling, Mark. 1998. "Participation, Social Capital and Vulnerability to Urban Flooding in Guyana", in: *Journal of International Development*, 10: 469–486.

Pelling, Mark; Dill, Kathleen, 2010: "Disaster Politics: Tipping Points for Change in the Adaptation of Sociopolitical Regimes", in: *Progress in Human Geography*, 34(1): 21–37; at: https://doi.org/10.1177/0309132509105004.

Peters, Katie; Budimir, Mirianna, 2016: *When Disasters and Conflict Collide: Facts and Figures* (London: ODI [Overseas Development Institute]).

Reyntjens, Filip. 2016. "Legal Pluralism and Hybrid Governance: Bridging Two Research Lines: Legal Pluralism and Hybrid Governance." *Development and Change*, 47(2): 346–366.

Rocha Monocal, Alina, 2013: "Aid and Fragility: The Challenges of Building Peaceful and Effective States", in: Chandler, David P.; Sisk, Timothy D. (Eds.): *Routledge Handbook of International Statebuilding*, Routledge Handbooks (Milton Park, Abingdon, Oxon – New York, NY: Routledge): 387–399.

Serventy, Matthew, 2015: "Collective Accountability: Are We Really in This Together?", in: *Humanitarian Accountability Report 2015. On the Road to Istanbul: How Can the World Humanitarian Summit Make Humanitarian Response More Effective?* (London: CHS [Core Humanitarian Standard] Alliance): 82–91.

Sørbø, Gunnar M., 2004: *Peacebuilding in Post-War Situations: Lessons for Sudan* (Bergen, Norway: Chr. Michelsen Institute, Development Studies and Human Rights).

Suhrke, Astri; Samset, Ingrid, 2007: "What's in a Figure? Estimating Recurrence of Civil War", in: *International Peacekeeping*, 14(2): 195–203; at: https://doi.org/10.1080/13533310601150776.

UNISDR (United Nations International Strategy for Disaster Reduction); CRED (Center for Research on the Epidemiology of Disasters), 2016: "Poverty & Death: Disaster Mortality 1996–2015" (Louvain: CRED – Geneva: UNISDR).

Uvin, Peter, 2008: "Local Governance after War: Some Reflections on Donor Behaviour in Burundi", in: *Praxis*, 23: 109–122.

Walker, Peter; Maxwell, Daniel G., 2008: *Shaping the Humanitarian World*. Global Institutions Series (Milton Park, Abingdon, Oxon – New York: Routledge).

Walter, Barbara F., 2004: "Does Conflict Beget Conflict? Explaining Recurring Civil War", in: *Journal of Peace Research*, 41(3): 371–88; at: https://doi.org/10.1177/0022343304043775.

Weber, Max, 1978: *Economy and Society: An Outline of Interpretive Sociology* (Berkeley, CA: University of California Press).

Wisner, Benjamin, 2012: "Violent Conflict, Natural Hazards and Disaster", in: *The Routledge Handbook of Hazards and Disaster Risk Reduction* (London; New York: Routledge): 65–76.

Wisner, Benjamin; Gaillard, J.C.; Kelman, Ilan; 2012: "Framing Disaster", in: *The Routledge Handbook of Hazards and Disaster Risk Reduction* (London – New York: Routledge): 18–33.

Other Literature

ALNAP (Active Learning Network for Accountability and Performance), 2015: *The State of the Humanitarian System* (London: ALNAP/ODI); at: ALNAP/ODI. http://www.alnap.org/resource/21036.aspx.

Boege, Volker; Brown, Anne; Clements, Kevin; Nolan, Anna, 2008: "On Hybrid Political Orders and Emerging States: State Formation in the Context of 'fragility'"; at: http://edoc.vifapol.de/opus/volltexte/2011/2595/.

Boutros-Ghali, Boutros, 1992: "Report of the UN Secretary-General: 'Agenda for Peace'", United Nations General Assembly; at: http://www.un.org/documents/ga/res/47/a47r120.htm.

Brahimi, Lakhdar, 2007: "State Building in Crisis and Post-Conflict Countries", in: *7th Global Forum on Reinventing Government, Vienna, June*: at: http://unpan1.un.org/intradoc/groups/public/documents/un/unpan026305.pdf.

DFID (Department for International Development), 2005: "Why We Need to Work More Effectively in Fragile States"; at: https://www.jica.go.jp/cdstudy/library/pdf/20071101_11.pdf.

EM-DAT (Emergency Events Database), 2016: "Disasters: General Classification", in: http://www.emdat.be/classification.

FFP (Fund for Peace), 2015: "Fragile States Index 2015" (Washington D.C.: Fund for Peace); at: http://library.fundforpeace.org/library/fragilestatesindex-2015.pdf.

GHA (Global Humanitarian Assistance), 2016: *Global Humanitarian Assistance Report: 2016* (Bristol, UK: Global Humanitarian Assistance, Development Initiatives): at: http://devinit.org/wp-content/uploads/2016/06/Global-Humanitarian-Assistance-Report-2016.pdf.

Hilhorst, Dorothea, 2016: "Aid–society Relations in Humanitarian Crises and Recovery". Inaugural Lecture (The Hague, The Netherlands: Institute of Social Studies of Erasmus University Rotterdam); at: https://repub.eur.nl/pub/97954/Hilhorst-inaugural-lecture-ISS-2016-.pdf.

INFORM (Index for Risk Management), 2016: "Index for Risk Management: Results 2016" (Brussels: Inter-Agency Standing Committee Task Team for Preparedness and Resilience and the European Commission); at: http://www.inform-index.org/Portals/0/InfoRM/2016/INFORM%20Results%20Report%202016%20WEB.pdf.

Kellett, Jan; Sparks, Dan, 2012: "Disaster Risk Reduction" *Spending Where It Should Count* (Somerset, UK: Global Humanitarian Assistance); at: http://devinit.org/wp-content/uploads/2012/03/GHA-Disaster-Risk-Report.pdf.

Meagher, K.; Herdt, T.; Titeca, K., 2014: "Unravelling Public Authority: Paths of Hybrid Governance in Africa"; at: https://biblio.ugent.be/publication/5863789/file/5863810.

Menkhaus, Ken, 2007: "Governance without Government in Somalia: Spoilers, State Building, and the Politics of Coping"; at: http://www.mitpressjournals.org/doi/abs/10.1162/isec.2007.31.3.74.

Nathan, Laurie, 2006: *No Ownership, No Peace: The Darfur Peace Agreement* (London: Crisis States Research Centre); at: http://www.lse.ac.uk/international-development/Assets/Documents/PDFs/csrc-working-papers-phase-two/wp5.2-darfur-peace-agreement.pdf.

ODI-HPG (Overseas Development Institute – Humanitarian Policy Group), 2016: *Time to Let Go: Remaking Humanitarian Action for the Modern Era* (London: ODI); at: www.odi.org/hpg.

Putzel, James, 2010: "Why Development Actors Need a Better Definition of 'State Fragility'"; at: http://eprints.lse.ac.uk/41300/.

Stel, Nora; de Boer, Diederik; Hilhorst, Dorothea; van der Haar, G.; van der Molen, I.; Douma, Nynke W.; Mostert, R. Herman, 2012: "Multi-Stakeholder Processes, Service Delivery and State Institutions; Synthesis Report". Peace Security and Development Network; at: http://library.wur.nl/WebQuery/wurpubs/fulltext/341507.

UNISDR (United Nations International Strategy for Disaster Reduction), 2007: "Terminology – UNISDR"; at: https://www.unisdr.org/we/inform/terminology#letter-r.

UN OCHA (United Nations Office for the Coordination of Humanitarian Affairs), 2016: "Burundi Inter-Agency Monitoring Report" (New York: UN OCHA); at: https://reliefweb.int/report/burundi/burundi-inter-agency-monitoring-report-29-January-2016.

Chapter 5
Climate-Smart Agriculture and a Sustainable Food System for a Sustainable-Engendered Peace

Úrsula Oswald Spring

Abstract In addition to increasing extreme events due to climate change, losses of ecosystem services, soil depletion, water scarcity, and air pollution, in most emerging countries the importation of basic food items, especially corn, soya beans, and wheat, has increased. These countries often purchase genetic modified grains which might affect their biodiversity. The present chapter proposes a *climate-sustainable agriculture with food sovereignty* (CSAFS) that combines the climate-smart agriculture promoted by the Food and Agriculture Organisation (FAO) of the United Nations with the recovery of local food cultures, environmental diversity, and healthy food intake from a gendered perspective. This approach deepens the concept of food sovereignty from Via Campesina, the international movement which coordinates small and medium scale agricultural producers and workers across the globe. This case study of Mexico illustrates the nutritional impact on poor people of industrialised and imported food. In 2018, half of all Mexicans live in conditions of poverty, with informal jobs and insufficient income. The increase of food prices has forced many people to substitute nutritious fresh food with sugar and carbohydrates. This change of diet has increased obesity, diabetes, cardiovascular diseases, cancer and other chronic illnesses. Since 2017 President Trump has initiated a renegotiation of the *North America Free Trade Agreement* (NAFTA) and his administration has charged import taxes on selected Mexican export products. The complexity and urgency of this crisis, aggravated by climate change impacts, obliges the Mexican Government to rethink its agricultural policy, which is now unable to provide healthy food to everybody. The State, the business community and the citizens must design a policy of a sustainable agriculture and a healthy food culture, which may reverse environmental deterioration, increase the capture of greenhouse gases, mitigate climate change impacts, reduce the malnutrition of adults, and improve the chronic undernourishment of small children.

Úrsula Oswald Spring, Research professor, UNAM, CRIM; Secretary General of IPRA, Email. uoswald@gmail.com.

© Springer Nature Switzerland AG 2019
H. G. Brauch et al. (eds.), *Climate Change, Disasters, Sustainability Transition and Peace in the Anthropocene*, The Anthropocene: Politik—Economics—Society—Science 25, https://doi.org/10.1007/978-3-319-97562-7_5

5.1 Introduction

Climate change is seriously affecting tropical regions with a high biodiversity that are global providers of food products. Mexico is one of the most highly food-diverse countries in the world and contributes to three of the five basic global food items (corn, beans and potatoes), with more than other 60 food products such as tomatoes, chocolate, avocado, mamey, squash, beans, amaranth, chili pepper, peanut, pineapple, turkey, papaya, vanilla, etc.

In spite of its rich and biodiverse agricultural production, in Mexico only 14 per cent of the population are adequately nourished (EnsanutMc 2016). Most people, including children, are obese and among indigenous children 13.6 per cent are also undernourished. Despite of Mexico's food diversity, what went wrong with its agricultural policy? The costs to resolve the crisis-level health problems created by an industrialised food culture depending on food imports which have affected health and wellbeing are enormous. Advertisements for foods rich in sugar and carbo-hydrates promoted in the mass media should be controlled by the government through taxes. Additionally, the industrialised food companies, who promote this unhealthy food culture e.g. through soft drinks and fast food, should participate economically to remediate the costs of the health impacts created by their unhealthy food.

Mexico has also become highly dependent on basic food imports. Therefore, the uncertainty on the results of the re-negotiation of the *North American Free Trade Agreement* (NAFTA) may challenge even further Mexico's food security and create food pressures. Mexico's neoliberal economic policy has drastically reduced the basic food production and the governmental control over transnational food companies. Mexico has few options to reassess and change its food sovereignty. Facing this complexity and being aware that almost half of its people live in extreme or moderate poverty (53.4 million people) and 24.6 per cent have not enough access to basic food (Coneval 2017), Mexico must promote a different food security policy which increases its national food sovereignty.

There are also important regional, age and societal differences of food intake, where social classes, ethnic affiliations, and rural-urban settlement challenge the existing food culture. While in urban areas poverty affects 39.7 per cent of the population, among indigenous populations the poverty rate rises to up to 77.6 per cent, of which 34.8 per cent are extremely poor people. The poverty index increases further among indigenous women, up to 85.1 per cent, of which 45 per cent are living in extreme poverty (Coneval 2017). There is a further problem of chronic malnutrition among children below 18 years of age, where healthy food is crucial for physical and brain development. Nine per cent of children and young people suffer from extreme poverty, and 51.1 per cent experience some degree of poverty. Confronted with this basic lack of healthy food and being a country with a long-standing agricultural culture, alternative agricultural policies most overcome the lack of money, reduce the high prices for basic food staples, consolidate nutritional education, and control the manipulation of mass media. This combined approach might provide a healthy food culture inside the country.

5.1.1 Focus of the Chapter

This chapter explores a *climate-sustainable agriculture with food sovereignty* approach of CSAFS at the regional level, which includes the goals of that should improving nutritional food intake of the poorest people (indigenous girls in the mountains, in drylands and in urban slums). This CSAFS goes further than the scientific and technological approach of climate-smart agriculture and global climate policies promoted by FAO (2016a; Chandra et al. 2018), and includes the co-benefits of mitigation, adaptation and food sovereignty (IPCC 2014a) in regions severely exposed to climate change impacts. This CSAFS approach analyses the interactions among environmental and social processes with a special gender perspective, focussing on the restoration of environmentally deteriorated soils (FAO 2015) and ecosystems. This CSAFS approach works to improve food sovereignty that relies on locally and regionally biodiverse resources, sustainable water management (Biswas Tortajada 2011) and ruled urbanisation (Delgado Ramos 2014) that may overcome the growing food insecurity globally (FAO 2002, 2016b) and in Mexico (EnsanutMc 2016). Particularly in the small-scale agriculture approach, women play a crucial role in producing safe and nutritious food.

5.1.2 Structure of the Chapter

The chapter starts with four key research questions (Sect. 5.2). Later it analyses three key concepts (Sect. 5.3): climate-sustainable agriculture with food sovereignty (Sect. 5.3.1), food sovereignty vs. food security (Sect. 5.3.2) and a sustainable-engendered peace (Sect. 5.3.3). An analysis of food insecurity, hunger and malnutrition globally, in Latin America, and in Mexico follows (Sect. 5.4). Mexico is the country with the highest use of soft drinks and with one of the most highly obese populations in the world, whose chronic diseases are related to over-weight. Part five explores CSAFS from a gender perspective (Sect. 5.5). The chapter then examines the deterioration of soils and the resulting carbon foot print (Sect. 5.5.1), followed by an examination of governmental water management and the deterioration of ecosystem services (Sect. 5.5.2). To combat hunger, malnutrition and environmental deterioration, this model suggests a small-scale sustainable agriculture with a gender perspective, as globally half of the food is produced by women in orchards and on small plots of land and in Mexico even higher levels of food are produced at a small scale by women – 64 per cent (Sect. 5.5.3). Part six scrutinises the co-benefits of a CSAFS (Sect. 5.6). As part of a paradigm shift towards a small, sustainable, equal and just society and productive system, an urban and rural circular agriculture is proposed, where women play a crucial role in achieving food sovereignty, especially in the marginal slums and in the rural mountains (Sect. 5.7). The chapter concludes (Sect. 5.8) that a sustainable-engendered peace may be able to overcome the present economic, social, political, and military crises and explores a utopia of a just and equal future with sustainability for humankind and the environment, including the most marginal people located in the most highly biodiverse regions.

5.2   Research Questions

How could a *climate-sustainable agriculture with food sovereignty* (CSAFS) with a gender perspective promote food sovereignty in regions that are highly exposed to climate change impacts? How could a regionally adapted CSAFS exploit the local biodiversity to adapt to increasing climate threats? How could a gender-sensitive food policy reduce hunger and malnutrition, producing healthy food with lower loss, adapting to increasingly adverse climate conditions and promote an engendered-sustainable peace with a just access to natural resources? How could the co-benefits of such a CSAFS approach which combines mitigation, adaptation and resilience improve the food intake of the poorest people (indigenous girls in the mountains, in drylands and in urban slums)?

5.3   Conceptual Considerations

This section briefly reviews the concept of a CSAFS with the interactions among climate change, *land-use change* (LUC), loss of soil fertility, desertification of agricultural land, the deterioration of ecosystems, and its services. The negative feedback loops of this systemic deterioration, aggravated by industrial agriculture, are increasing hunger and malnutrition globally. Via Campesina (2002, 2016) has promoted a change in approach, moving from food security to food sovereignty, which includes native seeds and gender equity. IPCC (2014a) noted that women globally produce food, usually with organic methods and few agrochemicals, therefore they already promote regionally and nationally elements for a CSAFS. When combined with sustainable water management and conflict resolution an engendered-sustainable peace (Oswald 2016) is stimulated.

5.3.1 Climate-Sustainable Agriculture with Food Sovereignty (CSAFS)

Climate-sustainable agriculture was established as a framework by FAO (2016a) to deal with growing food insecurity (FAO 2016b) globally and especially in the global South, which is highly exposed to climate change impacts. The original concept was policy-oriented and included emerging scientific and technological innovations (Chandra et al. 2018). The IPCC (2014a) stated in its fifth assessment report that adaptation, mitigation and resilience can produce co-benefits, however agricultural systems must be transformed from dependence on intensive chemical inputs towards methods that are able to restore the natural conditions of water, soil and air. The IPCC (2014a) further noted that women are crucial for sustainable food production, which is mostly done in orchards and in small plots of land, generally recycling organic waste and using grey water. The *United Nations Convention to Combat Desertification* (UNCCD 2017) claimed that during the past four decades, the world has lost about one third of its arable land or between 13 and 36 billion tons per year of top soil,[1] due to soil mismanagement.

Industrial agriculture based on a productivist approach of the green revolution was able to feed the growing global population for several decades, where large quantities of the produced crops were also used as animal feed and biofuels. Heavy tilling, multiple harvests, inadequate irrigation practices, and abundant use of agrochemicals have increased the yields at the expense of a long-term sustainability and now crop yields are declining in almost all regions, but especially in Africa and in the fragile drylands (UNCCD 2017) of Asia and Latin America.

Growing population, urbanisation, changes in food patterns from grains towards more meat have further impacted fragile tropical soils. Livestock has also eroded lands and, coupled with extreme events related to climate change, there are significant factors responsible for the loss of agricultural land and biodiversity. The two 'Rio Conventions': the *United Nations Framework Convention on Climate Change* (UNFCCC) and the *Convention on Biological Diversity* (CBD) that were signed in Rio de Janeiro in 1992 and the UNCCD followed in 1994 are now analysed together to address the increasing threats to the environment and to humankind. A perspective that linked the problems associated with the UNFCCC, CBD and UNCCD Treaties concluded that poor management of natural resources is also increasing the emissions of *greenhouse gases* (GHG). Together with deforestation and LUC, agriculture is responsible globally for 23 per cent of GHG, thus increasing climate change impacts. Land degradation is reducing further the resilience of affected people and has decreased the sink of CO_2 in soil and natural vegetation, while LUCs are aggravating the chaotic urbanisation, where rural migrants, who have lost their livelihood in agriculture, try to settle and find a new livelihood. Finally, new food customs have globally deteriorated nutrition and

---

[1]This huge difference was assessed by FAO using several simulation models (FAO 2015) and satellite images.

health standards and contributed to obesity and malnutrition among increasing population groups.

Facing this complexity, in order to promote an effective development and to ensure food security in a context of a changing climate, this CSAFS proposal addresses five main objectives: 1. a sustainable increase in agricultural productivity and incomes without further deterioration of soil, water and biodiversity; 2. adaption and resilience to climate change impacts; 3. successful mitigation and sinks of *greenhouse gas emissions* (GHG) through organic agriculture and soil conservation; 4. an integrated gender perspective which ensures the visibility and participation of women in agriculture and food transformation; 5. sustainable agricultural production, safe food transformation, and a nutritious diet for a healthy life.

CSAFS analyses these means and helps stakeholders from local to national and international levels to identify agricultural strategies suitable to be used in their local conditions. CSAFS goes beyond prevailing modern agricultural technologies like genetic modified seeds (Oswald Spring 2011a, b) or precision farming in the form of clusters (Paustian/Theuvsen 2016)[2] and addresses simultaneously multiple objectives: sustainable productivity, food security and sovereignty, enhanced farmer and gender resilience, improvement of livelihood, and limitation of rural-urban migration.

CSAFS is based on a systemic perspective, which includes the management of landscapes, ecosystems and their services, food production, value chains, and livelihoods. From this systems approach (Prigogine/Stengers 1997), CSAFS may establish synergies among food production, sustainability, equality, and equity with a bottom-up involvement of stakeholders and co-benefits for carbon sinks, soil recovery, river basins management, and healthy food intake. CSAFS addresses these trade-offs at a small scale and achieves cost- and co-benefits, including the environment and gender development (UNEP 2016).

CSAFS explores how to improve food sovereignty by linking the production cycle with storage, direct marketing, consumption, food conservation, and nutritional improvement. The whole production-harvest-transformation-intake-nutrition process integrates ten related policies: organic agriculture; reduction of ecological, water, and carbon footprint; integrated water resource management; composting of organic waste and recovery of natural soil fertility and ecosystems with biofertilisers and biopesticides (FAO 2010); efficiency in inputs; family gardens mostly managed by women for improving nutrition; loans to micro-producers including

---

[2]Nestlé has proposed in 2001 an alternative *productivist* model, which is overcoming the green revolution that has stagnated, due to the high prices of hydrocarbons, the contamination of water, soil and air and the effects of agrochemicals on human health. The new paradigm called *life sciences* or *precision farming* is promoted by transnational enterprises, who control GMO-seeds, agrochemicals, storage, supermarket chains and the finances. These enterprises are generating a productivist-commercial monopoly, in which genetic modified organisms, health and food transformation technologies are integrated in clusters for the production and transformation of food. In the view of this author, only *green or organic agriculture* offers an alternative model, where environmental services are combined with food production and where peasants, women and indigenous people are finding alternatives for their survival in rural areas.

women (FAO 2006); local transformation of food through micro enterprises and an economy of solidarity (Richards 2018); consumption and local marketing; reduction of losses throughout the food cycle (FAO 2013a, b); and changes towards a healthy and nutritious diet for all inhabitants, especially for children. CSAFS might increase local employment and reduce costs of inputs by recycling local resources, restoring deteriorated soils, basins and climate, and additionally capturing GHG.

5.3.2 Food Sovereignty Versus Food Security

Food security is the technical and political term to explain "a situation that exists when all people, at all times, have physical, social, and economic access to sufficient, safe, and nutritious food that meets their dietary needs and food preferences for an active and healthy life" (FAO 2002). However, this food security has not allowed humankind to improve their health during the past four decades. On the contrary, *genetically modified organisms* (GMO), polluted environments with toxic agrochemicals, meat with hormones, and industrialised food with trans-feds are increasing obesity and causing degenerative diseases.

For these reasons, Via Campesina (2005) developed an alternative concept of food sovereignty, which includes:

- local production and trade in agricultural products with access to land, water, native seeds, credits, technical support, and financial facilities for all participants;
- access to land, credits, and basic production, especially for women, girls, indigenous peoples and peasants, since women are key food producers worldwide, but are without official and private support;
- inclusion of small land holders (indigenous, women, and peasants) in regional and national rural policy and decision-making processes, which focus on local sustainability, healthy food, and safe livelihoods;
- the basic right to consume safe, sufficient, and culturally accepted food produced with native seeds;
- the rights of regions and nations to establish compensations and subsidies to protect farmers from dumping and genetically modified food;
- the obligation of national and local governments to improve food reserves in the case of drought or crop failure, which has now become more frequent due to climate change;
- cheap and healthy basic food provision in urban poor neighbourhoods and promotion of urban orchards, green roofs, and rain water harvesting;
- governmental guarantees of adequate nutrition for babies, infants, and pregnant women to overcome chronic undernourishment and early-life permanent brain damages in children (Álvarez/Oswald Spring 1993);
- promotion and exchange of locally produced seeds, which are adapted to the present environmental conditions;

- reduction of harvest loss due to climate forecasting, early warning, technical support and adapted seeds;
- clean water and sewage facilities in villages and towns with recycling of treated sewage water in agriculture;
- policies that link environmental services, agriculture, territorial planning, carbon sequestration with a nutritious and safe food for everybody.

As a result of these food sovereignty policies, each citizen should be granted his or her basic rights to life, which includes also the right to stay in the rural area with local productive opportunities (Oswald Spring 2009a, b). However, this understanding of food sovereignty is continuously threatened by multinational food companies who are producing new food products based on biotechnology, agrochemical and veterinarian pharmaceutics, without long-term studies on their potential health impacts.

5.3.3 Engendered-Sustainable Peace and Security

Peace-building is an effort to overcome violent conflicts and wars. The origin of war is related to the patriarchal system of rule (Reardon 1980, 1986; Reardon/ Snauwaerd 2015a, b), which permeates from the household to the economy to global governance. Patriarchy is a hierarchical, violent, and exploitive system of rule that exploits women and all other less powerful people. Patriarchy emerged when agricultural communities produced a food surplus, which resulted in social stratification with a division of labour and the submission of women inside their homes. Male rulers started to justify their power by establishing supposed contacts with supernatural beings to better protect their village or small town from disasters, hunger, diseases, and other calamities. Power and economic wealth was concentrated in these leaders or a small group of leading men. Later, with the surplus they developed arms and trained soldiers, first to protect themselves against invaders and to protect themselves against invaders and to increase their wealth by conquest. They attacked neighbouring villages, cities and kingdoms, took away their goods, and transformed the defeated people into slaves.

Patriarchy has consolidated all over the world during thousands of years and has promoted hierarchical societies, where the labour of women was rendered invisible. Men controlled women within the extended family, and patriarchal unities were consolidated by the rules of patrilinear inheritance and patrilocal establishment of households (Oswald Spring 2016). Patriarchy developed regional differences, whenever it has been globally imposed on women through violence (via conquest, spoliation, rape, and feminicide), discrimination (by laws and rules), subordination (through economic and sexual control), hierarchy (by the notion of paterfamilias, today also extended through global oligarchy), inequality and discrimination (in education, income, leisure, and political access to power and wealth), through exclusion (patrilineal and patrilocal inheritance, today also through exclusive globalisation), and by devaluing female labour in the household, which often remains without visibility and economic value (Burton 2013; Fraser 1994). The conformation of social classes and the consolidation of gender-discriminative social representations reinforces this hierarchical behaviour, where self-constructed and self-assumed gender roles of 'caring for others' developed (Serrano 2009, 2010). These norms were instrumental in the emergence of a gender discipline and the internalisation of socially consolidated gender roles, which vary regionally, but always maintain key elements of patriarchy, such as violence, inequality, exploitation, discrimination, and the lack of visibility of women's work.

The control over women and of the whole society was traditionally imposed by repression, military, war, and political power (Reardon 1986). The dominant male elites' security concerns were the defence of territory and their nations, which were later legally codified in the peace of Münster and Osnabrück in 1648, creating the terms of the Westphalian state. Women were treated as vulnerable, what justified their subordination and their need for protection by militarised men.

However, with the consolidation of globalisation and the establishment of complex security states, the Copenhagen School of Security Studies widened this narrow security understanding which focussed only on political and military dimensions by adding economic, societal, and environmental security (Buzan et al. 1998). Different values were at risk and new sources of threats emerged (economic crisis, livelihood loss and threats to sustainability). UNDP (1994) deepened this security understanding further with the concept of human security and put human beings at the centre of the analysis. In the debates at the United Nations, the values at risk are now also survival, quality of life, and, the well-being of people. The major threats are now understood to come from an exclusive globalisation, undemocratic governments, global environmental change, and climate change. Related risks include GHG emissions and wasteful consumerism.

For the first time, humankind has created these risks, through climate and global environmental change (Brauch et al. 2008, 2011), poverty, and discrimination. But at the same time, human beings are now also the victims of the consequences of the massive increase in the consumption of fossil fuels and thus of the first direct intervention of humankind into the earth system. In addition, states, organised crime, and other violent actors have increased the present causes of insecurity.

Gender security (Oswald Spring 2009a, 2016) belongs also to a deepened understanding of security, as it extends from the individual social construction of masculinity and femininity to patriarchal behaviour in families, communities, nations, regions up to the global level, where mass media reinforce engendered social representations related to the present model of consumerism and gender discrimination. The values at risk are linked to gender relations (Lagarde 1990; Lamas 1996), deeply rooted in the dominant social representations of gender (Jodelet 1991; Serrano 2010). These values are manipulated by an oligarchy (Stiglitz 2010; Yiamouyiannis 2013) and reinforced by religious fundamentalism, hierarchical churches (Gutiérrez 2017), and vertical school systems (Gramsci 1971,

1977). These social structures undermine fundamental values such as equity, equality, solidarity, justice and inclusion (Truong et al. 2014), cultural identity (Arizpe 2015; Serrano 2015), and often even the survival of individuals and of social groups (Oswald Spring 1994). These threats to security were consolidated over thousands of years by patriarchal institutions (Folbre 2006), religious controls (Jasper 2013), financial monopolies (Stiglitz 2010), and a totalitarian and violent exercise of power (Held 2004; Reardon 1996).

In the Charter of United Nations of 1945, international peace and security reflected the understanding of global order after World War II. The conceptualisation of peace started from its negative understanding as the absence of war, towards a more positive peace understanding of peace with justice. Galtung (1967, 1968, 2007) introduced a structural peace concept, where inequality has produced sources of threats and subordination. Elise Boulding (2000) analysed cultures of peace to identify culturally specific strategies to overcome discrimination against gender, other races, cultures, religions and beliefs. Due to the increasing destruction of the environment and of ecosystem services, Kenneth Boulding (1970) related peace to the environment and Ken Conca (1994) linked peace to sustainability with his concept of environmental peace-making.

Different phases of the environmental security concept (Dalby et al. 2009) prepared the arena for the development of an integrated concept of engendered-sustainable peace (Oswald Spring 2016), which is an outcome of several other types of peace and security discussions, thus it widened, deepened and sectorialised our conception of peace (Fig. 5.1; Brauch et al. 2009). With this conceptual background, the CSAFS approach will be discussed both globally and specifically in the case of Mexico. IPCC (2014a) insisted that women produce food generally in orchards or small plots of land. In Mexico, women's agricultural production is about 64 per cent of the total and in different African countries almost all the production and processing of food is in the hands of women.

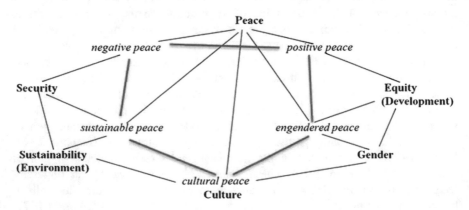

Fig. 5.1 Five pillars of peace. *Source* Oswald Spring et al. (2014: 19)

5.4   Food Insecurity, Hunger and Malnutrition Globally, in Latin America and in Mexico

The United Nations called in 2017 for a special food emergency support as "South Sudan, together with Yemen, Somalia, and Nigeria pose what the UN calls the biggest humanitarian crisis since 1945 as millions flee conflict and drought ..." (Report 2017: 1967). They estimate that about 1.9 million people are internally displaced and 1.6 million have fled to neighbouring countries to survive, due to a severe drought, political instability, and war in Syria, Yemen, South Sudan and others). FAO (2016b: 8) estimated that worldwide 794.6 million people, mostly small children, suffer from hunger. Including hunger with undernourishment, 19.9 per cent of the global population have insufficient nutrition. The highest rate of undernourishment exists in Asia with 22.1 per cent, followed by Africa with 20 per cent, Oceania with 14.2 per cent, Latin America and the Caribbean with 5.5 percent and the industrialised countries with less than 5 per cent.

"In Latin America and the Caribbean, 53 million people lack sufficient food to cover their needs, 7% of children under five years of age are underweight and 16% of these are of low height for their age" (CEPAL 2017: 13). In 2004, the World Food Programme estimated that the costs to overcome chronicle undernourishment of children in Central America and the Dominican Republic would be about 6.7 billion USD per year.

México, which is considered the 16th most important economy globally, has serious problems with food and nutrition. Most of its adults – 72.5 per cent – are overweight or are obese and 25.5 per cent suffer from high blood pressure. Among children between 5 and 11 years old, 33.2 per cent are obese and overweight, while for adolescents from 12 to 19 years the comparable figure is even higher, at 36.3 per cent. Simultaneously, 13.6 per cent of the children below five years are chronically undernourished (EnsanutMc 2016), which implies that 1.5 million small children may be affected by irreversible brain damage due to a chronic absence of sufficient nutrients (Álvarez/Oswald Spring 1986). Systematic loss of purchasing power due to periodic economic crises means that people are unable to buy fresh fruits and vegetables, which are more expensive than sugar and soft drinks, driving these alarming data on nutrition. Additionally, there are also constraints on time to buy and prepare healthy food, and people often lack the education necessary for healthy nutrition given their limited financial resources.

The data on hunger and obesity in Mexico are also the result of a model of consumerism promoted by multinational enterprises, where television and other media have promoted through their advertisements and image dissemination an unhealthy food culture based on an excess of sugar in soft drinks and a lack of fresh vegetable and fruits through industrialised food products. This unhealthy nutrition has produced a dramatic increase of chronic diseases. The Mexican government has abandoned since 1982 its policy of food sovereignty and with NAFTA imported massively subsidised grain from the United States, which have increasingly replaced national corn production. Today, more than 61.4 per cent of corn, wheat, soya beans, barley, sugar, sorghum, oil, pork, beef and chicken meat, and milk are imported, primarily from the US (Trading Economics 2017). This happens not only in Mexico but also in Europe and China which import 21.1 per cent of their food products. In Mexico, Egypt, and Japan, 18.6 per cent of basic food staples are imported, and in Saudi Arabia, the USA, Canada, Brazil, India, Indonesia, Nigeria, Philippines and Morocco 21.7 per cent of food is imported.

This so-called 'virtual water' (Allen 1997) – or traded agricultural goods with high water inputs – implies long distance travel for food items which also contribute to GHG emissions. For Mexico, this requires a transfer of hard currency for food imports, where the stock market for commodities in Chicago determines the prices of basic foods often increased by speculation. In Latin America, Mexico is the major importer of basic grains, followed by Trinidad and Tobago, Venezuela, Peru, and Chile. The largest exporters of agricultural commodities are Argentina and Brazil, and with a lower export capacity Paraguay, Uruguay and Honduras.

Mexico has annually imported an average of 16.973 million tons of grain between 2011 and 2016, basically corn and wheat from the USA. During 2015/2016, this import increased to 20.43 and in 2016/2017 to 20.375 million tons, despite a record harvest in Mexican in 2016 (FAO 2017) due to a rainy Niño year. These large food imports have created a dangerous food dependency, because Mexico imports primarily from one country. In 2017, US President Trump started the renegotiation of the terms of NAFTA. In the NAFTA Treaty of 1994, grain imports were exempted from any import-export tax, which negatively affected local maize production.

Instead of importing basic grains, Mexico encompasses for at least eight million hectares in the tropical region, which are underused due to extensive livestock production. On each hectare about seven tons of corn could be grown, and within a short time, Mexico could again regain self-sufficiency in corn, could also produce a surplus for feeding its livestock, and could even produce enough grain to export. In these tropical regions, there is enough water available during the dry winter season – after the rainy hurricane period – to produce the necessary basic food for the country (Turrent et al. 2013). This change in agricultural policy would also support reduction in poverty in one of the most marginal regions – the South-East – especially among indigenous people and peasants, who have the millennium wisdom to produce corn without an excess of agrochemicals. The present political uncertainty may open a new panorama for innovative food production, which may be able not only to produce adequate basic food, but also to challenge the dominant imposed food culture, which has created obesity and diseases.

5.5 Climate-Sustainable Agriculture and a Nutritious Food Intake from a Gender Perspective

Corporate and 'life science' agriculture with *genetically modified organisms* (GMO) seeds, intensive use of agrochemicals, heavy machines and exhaustive irrigation have caused globally dangerous environmental destruction and an unhealthy food culture. A crucial element for safe and healthy food production is the restoration of a dynamic natural soil with its physical, chemical, mineralogical and biological components. Soils are not only important with regard to their inorganic and organic components, but also their specific texture; the soil composition of sand, clay, and silt is crucial to respond to internal (compaction) and external pressures (wind and water erosion). The biological components of soils are crucial for the assimilation of nitrogen from the air to increase the natural soil fertility[3] and to feed the micro-organisms from organic waste, which retains soil humidity better.

5.5.1 Soil, Carbon Footprint, and Food Production

The moisture stored in tropical and dryland soils is important; the physical properties of these soils, such as texture, structure, porosity, drainage capacity, and permeability allow or impede efficient plant growth. Soils in mountain areas or regions exposed to flash floods are severely threatened by water and wind erosion when tree coverage is removed. Soils may lose their first topsoil, where nutrients are primarily stored. Soils are also crucial providers of ecosystem services. They deliver food, regulate climate and water, break down waste through fungus and other micro-organisms, and offer socio-cultural benefits. Eswaran et al. (2001: 5) estimated that half of the global productivity in agriculture has declined due to soil erosion, loss of natural fertility, and desertification. Drylands are highly exposed to desertification and globally more than one billion people depend on these ecosystems (Table 5.1).

Regionally, North and South America have the highest rate of land erosion due to intensive agricultural use. However, in temperate climates, better land quality has maintained productivity in some areas in spite of the intensive exploitation of soil (e.g. Mississippi delta). FAO (2015) estimates that only 3 per cent of global land

---

[3]Natural fertility falls into two different categories: "macronutrients and micronutrients. Macronutrients are the most important nutrients for plant development and relatively high quantities are required. Macronutrients include: carbon (C), oxygen (O), hydrogen (H), nitrogen (N), phosphorus (P), potassium (K), calcium (Ca), magnesium (Mg), and sulphur (S). Micronutrients, on the other hand, are needed in smaller amounts, but are still crucial for plant development and growth. Micronutrients include iron (Fe), zinc (Zn), manganese (Mn), boron (B), copper (Cu), molybdenum (Mo) and chlorine (Cl). Nearly all plant nutrients are taken up in ionic forms from the soil solution as cations or as anions" (FAO 2015: 35).

Table 5.1 Estimated degradation in drylands (million km^2). *Source* Eswaran et al. (2001: 7)

Continent	Total area	Degraded area[*]	% degraded
Africa	14.326	10.458	73
Asia	18.814	13.417	71
Australia and the Pacific	7.012	3.759	54
Europe	1.456	0.943	65
North America	5.782	4.286	74
South America	4.207	3.058	73
Total	51.597	35.922	70

[*]Comprises land and vegetation

has high quality soil, primarily in areas with a temperate climate. Second and third class soils, which amount to 8 per cent of total global land mass, occur mostly in tropical areas. In 2020, this 11 per cent of available land must feed about 7.6 billion people. One third of the drylands are currently in a rapid process of desertification, but must feed more than one billion people, half of them living in Africa. Soil is also an important reserve for organic carbon storage, thus may sequester GHG emissions.[4]

The climate models that were assessed by the IPCC (2013, 2014a, b) indicate that soils and yield productivity will be affected by more frequent drought conditions, especially in drylands, sometimes paired with occasional flash floods. Both processes will increase soil erosion, while drought will decrease the carbon uptake of plants, due to a lack of soil moisture and more frequent dust storms. The World Health Organisation (WHO 2016) stated that both indoor and outdoor pollution by any chemical, physical or biological agent modifies the physical-chemical composition of the air and may produce more than 100 diseases or 12.6 million deaths each year, of which, 1.7 million deaths per year will be children.

In Mexico, 13 per cent of the territory is used for agriculture and each year about 400,000 hectares of forests are demolished, basically due to land use change (LUC) towards agriculture and livestock. Semarnat and INECC (2012: 56) claimed that 22.73 million hectares (ha) suffer from water erosion; 18.2 million ha from wind erosion; 34.04 million ha from chemical degradation. The loss of natural

---

[4]*Soil organic carbon* (SOC) and *soil organic matter* (SOM) increase the productivity of food, restore degraded soils and increase the resilience of land exposed to climate impacts, thus improve food production. SOC depends on land management and precipitation, thus Lal (2006) estimates yield gains per hectare in the tropics and subtropics for wheat ranging from 20–70 kg and for maize to 30–300 kg/ha. By adding SOM, the natural productivity can be improved and depending on the SOM composition there is a wide variety of impacts on yields. "Soil degradation inherently reduces or eliminates soil functions and their ability to support ecosystem services essential for human well-being. Minimizing or eliminating significant soil degradation is essential to maintain the services provided by all soils and is substantially more cost-effective than rehabilitating soils after degradation has occurred …This increases the area available for the provision of services without necessitating land use conversion" (FAO 2015: 180).

fertility affects 92.7 per cent of the soil used for agriculture; 10.84 million hectares suffer from physical degradation, 68.2 per cent of soils are compacted, and another 25.8 per cent have lost their productive functions. Erosion is especially severe in intensive irrigated regions and in the fragile mountain areas. In Mexico, the northern states of Sinaloa and Chiapas are affected by chemical degradation, Michoacán, Jalisco and Sonora by water erosion, Chihuahua by wind erosion and Veracruz by physical degradation. In all these important commercial agricultural areas, soils are badly managed and have often lost their productive capacity and also their capacity to assimilate carbon dioxide.

5.5.2 Water and Ecosystem Services

Water is a second crucial constraint for food production and ecosystem services. Higher temperatures increase evaporation and evapotranspiration, which affects also the ecological flow required for ecosystem conservation. More intensive extreme events produce flash floods, land- and mudslides, reduce the infiltration into aquifers, and fill dams and lakes with sediments. Plants suffer from water stress. Globally, 36 countries suffer from high water stress with reduction of precipitation, inter-annual and seasonal variability of rainfall, flood occurrence, and severe droughts (Fig. 5.2) that might be exacerbated by climate change. Global population tripled during the last century, but water consumption augmented six-fold, due to new hygienic demands, new productive processes, and the continuing inefficiency in agricultural water use.

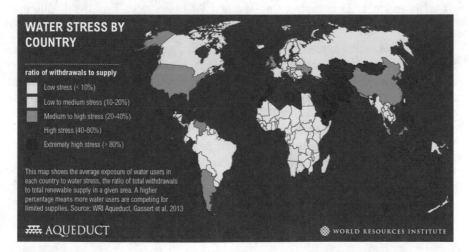

Fig. 5.2 Water stress by country. *Source* Reig et al. (2013). Permission to include this figure was granted by the World Resources Institute, Washington, D.C

Singapore provides an example of adaptability in the face of insufficient water. It suffers the highest level of water stress globally and does not have any proper water supply. It imports 40 per cent of its water from Malaysia, reuses grey water with high technology processes, and desalinates the remaining 10 per cent to meet its total demand. The Singaporeans have developed a water culture to collaborate with water saving technologies.

In Latin America, Chile, Peru, and Mexico account for the highest water stress, together with most islands in the Caribbean. Mexico suffers from six interrelated water constraints:

1. The monsoon occurs from June to September and the rest of the year people and agriculture rely on the extraction from groundwater;
2. The average precipitation is 770 mm/year with significant regional differences: Baja California receives about 199 mm/year and Tabasco 2,588 mm/year;
3. Population growth has reduced the availability in 1950 from 18,035 m^3 down to 3,982 m^3 in 2015 per person and year;
4. An unequal distribution of the water supply with the high economic growth: in the north and centre of the country, where 77 per cent of people live and produce 79 per cent of GDP, only 32 per cent of the water is available. In contrast, in the South and South-east, where 68 per cent of renewable water exists, only 23 per cent of the people live and produce 21 per cent of GDP;
5. There is a sectoral inequality: agriculture uses about 77 per cent of water and produces 4 to 5 per cent of GDP, while industry uses 10 per cent and the domestic sector 13 per cent (Conagua 2014). Palacios/Mejía (2011) estimate that agriculture could reduce water demand by half with efficient irrigation practices;
6. There are also major social constraints: poor people have limited access and do not always get the best quality or safe water. They often suffer from diarrhoea; infant mortality in Guerrero, Chiapas, and Puebla is more than the double the average rate of the country as a whole.

In synthesis, water availability in time and space, permanent supply, quality, per capita, and per sector supply are all factors which deteriorate due to climate change and population growth. With changing climate conditions, Mexico obviously needs a different water management approach and the pressure from multinational enterprises to privatise the water supply and sewage will not solve these six national constraints. Further, 108 aquifers are overexploited; the cost of extraction and the quality of water is deteriorating further with deeper pumping, due to higher temperatures, which dissolve more minerals and pollute the groundwater. Figure 5.3 indicates that sea water intrusion in the North, overexploitation of aquifers with brackish water, and, consequently, salinisation of soils, reduces crop yields.

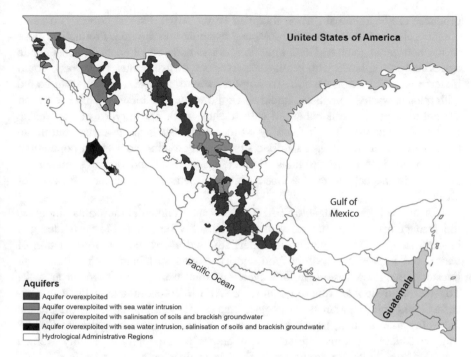

Aquifers

■ Aquifer overexploited
■ Aquifer overexploited with sea water intrusion
■ Aquifer overexploited with salinisation of soils and brackish groundwater
■ Aquifer overexploited with sea water intrusion, salinisation of soils and brackish groundwater
□ Hydrological Administrative Regions

Fig. 5.3 Salinisation of water and soils, and intrusion of sea water into aquifers. *Source* Conagua (2014)

5.5.3   Small-Scale Sustainable Agriculture from a Gender Perspective

Given these socio-environmental conditions, projections of more severe climate change impacts, population growth, and deterioration of ecosystem services, the FAO (2016a) has challenged the dominant paradigm of intensive agriculture. The intensive use of agrochemical inputs and water and heavy machinery, promoted by the so-called 'green revolution', has globally deteriorated most soils. This UN organisation also noted that modern agriculture is producing many crops primarily for biofuel, livestock, dairy and export of exotic fruits and vegetables, and only limited amounts for domestic food markets. Soil and water deterioration have forced many governments and international organisations to promote an alternative paradigm, which may be able to feed growing populations in the poor and climate threatened countries as well as in wealthy countries. The alternative paradigm emphasises traditional domestic agricultural management with small plots of land.

FAO (2014) revealed that women in Africa produce up to 90% of food in orchards and small plots. Almost half of the produced food is lost during harvest, transformation, in supermarkets, and in households (FAO 2013a, b). Organic agriculture in orchards, urban roof gardens, and on balconies offers fresh vegetables and fruits directly for the household. This small-scale sustainable agriculture produces only a minimal carbon footprint (FAO 2010). On the contrary, organic compost can recover deteriorated soils and catch carbon dioxide. Production close to the household reduces GHG emission, waste and food losses. If this organic agriculture is combined with small-scale livestock or fish ponds, their waste could increase the nutrients in the soil. To support the nutrient cycle, animals can be fed with organic waste. Composted liquid and solid waste also increase crop yields and the nutrition of people is enhanced with healthy food. These types of sustainable food production are often handled by women and small-scale peasants, but urban agriculture might also change the food supply of a growing poor urban population, which often lives in urban slums. Further, green roofs and orchards reduce air pollution in megacities, reduce waste collection, and improves the landscape of cities.

In synthesis, this sustainable CSAFS reduces environmental footprints, improves the management of all natural resources, i.e. of water, soil, bio-pesticides and bio-fertilizers, reduces pollution in water, soil and air, reduces the production of waste, and improves a healthy food supply. Mixed sustainable agriculture with livestock integrates waste from one cycle and transforms it into food for animals. Together with stubble from agriculture, CSAFS offers simultaneously fresh agricultural and animal products for people, thus creates an alternative for the present unhealthy food system, based on an excess of sugar and carbohydrates. CSAFS further reduces economic pressures and allows poor people, especially women heads of households, to produce safe food from their orchards or roofs. CSAFS, with governmental support and training, can also reduce tensions and political unrest related to price hikes of food by the international market, thus increasing food sovereignty and promoting peace and security at the local level with healthy livelihoods.

5.6 Co-benefits of Sustainable Agriculture, Ecosystem Services, Carbon Capture, and Conflict Resolution

The CSAFS is a proposal that can preserve and restore flora, fauna, biodiversity, water, soil and air by reducing the extraction of natural resources and re-establishing a dynamic sustainable relation between humans and nature. Today, most deforestation is related to LUC for biofuel production (soya beans, palm oil, sugar cane, corn). In Latin America, the deforestation in Mato Grosso and the north of Argentina is a result of the massive production of soya for biodiesel. In the region of Sao Paulo, deforestation is due to ethanol manufacture from sugar cane (Gao et al. 2011). The situation in Africa is still unclear, where biofuel plantations, especially Jatropha, were only recently introduced and might threaten further the home-grown food supply. Deforestation occurs massively in seasonal dry forests or grassland, thus with geo-referencing methods they cannot be distinguished from natural biota. This means that deforestation in these dry tropical ecosystems is mostly underestimated and can only be assessed when soils are completely eroded.

In Asia, palm oil expansion is responsible for massive deforestation; Indonesia and Malaysia alone produce 85 per cent of the world's supply. The European Parliament estimates that "40% of global deforestation is due to the shift to large-scale oil palm monoculture plantations … [and that] 73% of the world's deforestation results from land clearance carried out for the production of agricultural raw materials" (EU 2017). Based on this report, the European Union assessed that with the biofuel mandate 2020, 8.8 million ha (Mha) might be used for biofuel, "of which 2.1 Mha of land is converted in Southeast Asia under pressure from oil palm plantation expansion, half of which occurs at the expense of tropical forest and peatland" (EU Resolution 2017).

Often these gigantic deforestation processes occur at the cost of the tropical rainforest and the natural habitat of animals and plants. Thus, CSAFS represents an alternative not only for local food production, but also for the ecosystem restoration. However, how should governments deal with industrial animal feed and biofuel? Mixed agriculture, especially in rural regions in Africa and Asia with a high population growth, offers local improvement of food and maintenance of natural soil fertility. Biofuel could be substituted with existing renewable energies, such as sun, wind, marine, and geothermal (Ren21 2017). A regional and local mixture of these existing resources would enable the protection of existing natural areas. Secondary forest plantations in regions with destroyed tropical forests might be used for additional biofuel production.

The protection of natural areas of forests, marine, coastal and grassland would enable the conservation of the ecological heritage of biodiversity, reduce the impact of climate change, and mitigate the massive invasion of pests, plagues, and bush fires related to industrial agriculture and monocultures. In the past, all basic food items (wheat, rice, maize, potatoes) were developed in highly biodiverse regions (Ethiopia, Fertile Crescent, China, India, Meso-America, South America). By maintaining natural areas, greater access to diverse genetic materials is granted to humankind for the future food supply (ISSD 2017). Thus, to encourage a sustainable stewardship of environmental units, people must be allowed to conserve and reproduce a wide diversity of seeds in their local natural conditions, which are adapted to the local climate, soil, and water conditions. All this germplasm might be able to produce new crop varieties that are better adapted to the unknown impacts of climate change (IPCC 2014). This strategy also requires the control of the expansion of GMO seeds, which were developed in temperate climate regions and are not improving yield and combatting plagues and diseases in the tropics.

Thousands of years of agricultural evolution have produced regionally a variety of crop practices. The association of different plants has historically protected the natural fertility and humidity of soils. Traditionally, people and especially women have cultivated medicinal plants, reforested regions with native species, generated biomass for cooking and heating, and reused solid and liquid wastes for conserving soils (FAO 2015). When the natural fertility of the soil is maintained, food sovereignty becomes locally possible. Surplus can be exchanged or traded, which might improve livelihood. By stimulating a greater diversity in production, transformation, and consumption, often the efficiency in the use of natural, technical, human, and financial resources is improved. At the same time, this wisdom has reduced production costs and increased food and incomes.

The globalisation process brought a different productive logic and a vertical and horizontal integration of productive, trade, and consumption activities occurred (Petras 2017). These processes, promoted by international organisations (WB, IMF, WTO, FAO) and multinational enterprises, should improve the economy of scale, reduce food losses, increase income for farmers, and reduce the costs of food for final consumers. However, modern agriculture with large areas of monoculture industrialised primary production for animal feed and biofuel, but failed to produce safe, biodiverse, and cheap food. Agricultural income did not benefit the farmers, but went to the industries of machinery, agrochemicals, seed providers, and irrigation tools. Further, food is traded globally in stock markets and, like the rest of the economy, financial speculation and trade monopolies are increasing basic food prices. Finally, the propaganda of industrialised food and changes in eating habits due to long working hours in remote workplaces have created a fast food culture, which has increased obesity and chronic diseases.

CSAFS should not only address the sustainable production of basic crops, but must also address speculation, trade monopolies (Petras 2017), education for a healthy and nutritious diet, local food culture, and prices that allow people to buy or produce themselves the necessary proteins and carbohydrates for a healthy development and a harmonious life.[5]

Linked to the original research question on how to promote a climate-sustainable agriculture with regional food sovereignty, a second question arises: how a gender-sensitive food policy may reduce obesity and undernourishment as women are mostly responsible for production, transformation, cooking, and intake of healthy food?

5.7 Circular Agriculture with a Gender-Sensitive Food Sovereignty

The paradigm of a circular economy integrates the total cycles related to food management, waste, and nutrition (Zhou 2008). The first step is raising awareness among citizens that the present system of industrialised and imported food is harming personal and family health. It is further destroying natural resources and polluting the environment with GHG emissions and toxic agro-chemicals. Once

---

[5]In developing countries, nutritional and health management for pregnant women and school children may avoid chronic undernourishment and high levels of mother-infant mortality. School breakfasts may offer children a healthy food culture, improve academic and labour involvement, and increase mother and child well-being, while also reducing expenses for diseases and health care.

convinced by the damages caused by the present global food system, a second step is promoting an alternative food culture. It may start with love for nature and concern for the environment. This step includes the reduction and separation of waste, and later, zero waste generation. At this stage, food losses do not exist, remainders of meals are reused, and organic waste is recycled. Zero organic waste reduces the leaching of waste residues into the soil, and eliminates bad odours and health problems related to garbage. Industrial waste can be recycled, and slowly an integrated process of environmental management starts, which includes the four R's: to reduce, reuse, recycle, and re-educate people, not only regarding food waste, but also water, soil, and air deterioration.

In rural areas, the circular economy enables the recovery of eroded land (Alkire et al. 2012). The conservation of soil with stubble and compost better conserves soil humidity, reduces evaporation, and allows plants to resist better drought, higher temperatures, and flooding. Women are fully involved in this production process and husbandry. They also participate in the transformation of the agricultural production and in the preparation of healthy food (Worldwatch Institute 2017). Often, women use grains and domestic leftovers to feed their own animals. The composted manures of chicken, cows, horses, rabbits, fishes, etc. improve the quality of natural fertiliser. Human or animal waste could also be transformed from a bio-digester into a quality organic fertilizer, which improves crop yields and makes plants more resistant against plagues and pests (Jez et al. 2016).

With the promotion of organic agriculture, production costs are reduced, seeds are conserved from the former agricultural cycle, and the local workforce within the family and in the community can produce quality food at lower costs for inputs. Organic agriculture also reduces deforestation and production without agrochemicals allows a recovery of local biodiversity. CSAFS further promotes local employment by more than 30 per cent (Worldwatch Institute 2017). This sustainable organic production takes into account environment and human beings and establishes harmonious relations among ourselves, neighbours, other producers and consumers, the environment, and the government.

With new awareness, people also start to promote eco-designs in houses and parks and renewable energies. They participate in recovering forests and green areas, planting fruit trees in public parks and private gardens. Trained people also cooperate in their neighbourhoods and share experiences of success with other colonies, rural areas, or cities. When this policy is massively promoted, often socially organised clubs and local environmental groups encourage climate adaptation, health improvement, food culture, and sustainable development. By producing food on roofs or balconies, household economy improves, due to less expenditure for purchased food. Additionally, savings for doctors and medicines, thanks to improved health conditions, are added as co-benefits.

A better organised society is better able to demand improved governance and changes towards sustainable policies. Public transportation, car sharing, bicycling, and walking are alternatives which improve the quality of air and personal health. Therefore, citizens change public agendas and promote conditions for better quality of life and livelihoods. Among the substantial constraints for a responsible governance in Mexico are corrupt actors (Morris 2009). They may be publicly denounced and, with social pressure, the system of justice may be reformed. Finally, public pressure and a deeper understanding of the interrelations among governments, decision-makers, and personal and social well-being may force the authorities to change their governmental activities. Governments must move towards an agenda of sustainability, where finally the arena of destruction, personal benefits, and corruption leaves space for greater citizen participation and social control of policies and public work (Maass/Karla 2013).

Finally, a greater harmony within urban and rural communities, less pressure on natural reserves, and better economic conditions reduces conflicts on land and land use changes. A better texture of soils increase the infiltration of rainwater, reduces flash flood impacts, and the supply for water for agriculture. In drylands, traditional systems of water conservation store rainwater, which enables a second cycle of production during the dry season, thus improving the availability of healthy food, surplus, and marketing. All these factors not only create family unities with stable relationships, thanks to safe food and better health, but also reduce conflicts over scarce land and water.

However, conflicts may arise due to climate change-related disasters, extreme droughts, or floods. In these situations, nonviolent conflict negotiations with a hydro-diplomatic approach (Oswald Spring 2011a), may with the involvement of a mediator achieve win-win conditions for all parties involved in the conflict. Both parties may analyse causes and obstacles to achieve an agreement and in a third step strategies raise that could overcome the conflict. The parties explore the feasibility of these ideas and then implement solutions. In this last step, all involved agree, sometimes on a technical fix, sometimes on investments, and in other cases on changes of attitudes (Fisher et al. 2011). The government gets involved when multinational enterprises do not respect the local customs or pollute massively natural resources. Mining and extraction activities are still the key factors for pollution and regional conflicts (Oswald Spring/Serrano 2018), where only the transparent application of international and national laws may control these often-conflictive external enterprises.

In negotiation processes women have a great potential to start the mediation process and later, to propose alternatives in theory and reality to deal with conflicts and to propose win-win solutions for all parties (Reardon/Snauwaert 2015a). From their socialisation process on, they are trained to 'care for others'; thus, they are more sensitive to other demands. In present Mexican society, boys get another socialisation process. As a boy and later as a man, they are trained to show their masculinity, based on a patriarchal understanding of dominance, hierarchy, and violence. Therefore, they are highly vulnerable to involvement in violent acts with other boys and later men (Kaufmann 1999). To change these long-standing socialisation processes and to develop new masculinities and femininities for peaceful conflict resolution, Oswald Spring (2016) proposed an engendered-sustainable peace, where the abilities of women and men might find feasible solutions for all involved parties. The integral management of natural resources, waste, and healthy food processing is a way to overcome several tensions produced by the present neoliberal model, characterised by a lack of income and unstable working conditions. When CSAFS is combined with an engendered peace process, livelihood might increase. People may live with better physical and mental health, which improves the social interactions. Wilkinson/Pickett (2009) insisted that inequality is a key factor of violence and destruction, thus more stable incomes and less inequality thanks to CSAFS may improve social interactions and creativity, especially among youth.

5.8 Conclusions: Towards a Sustainable, Equal, and Just Future with an Engendered Peace and CSAFS

Global environmental change and climate change call for a decarbonised society with a dematerialised production, thus a transition to sustainability (Brauch et al. 2016). This means recycling all materials, increased efficiency in productive and consumption processes, and elimination of waste. However, many citizens look to their government and ask how can their governments grant food security and food sovereignty? Today, most emerging countries depend on grain imports for their food security from the USA, Europe, Brazil, Russia, etc. This is a result of an erroneous food policy in the short-, middle- and the long-term. These food exporting countries have often used agricultural subsidies to export production, sugar cane, and extensive livestock, where in the short-term agribusiness has overexploited aquifers, salinised soils, deforested tropical rain forest, and desertified drylands. Small-range farmers were abandoned without access to credits and female farmers lack training.

Given all these challenges, CSAFS is offering an alternative for governments and their people to recover food sovereignty in the short-run. By restoring soils and watersheds, carbon catch is increased and sustainable management allows access to the 'Green Climate Fund' (2017). CSAFS, together with other economic policies and strong state guidance, is able to reduce and control international and national trade monopolies. Producing food regionally would further reduce transportation costs and storage, thus decreasing consumer prices and increasing profit for farmers through lower costs for inputs. Promoting local food sovereignty also brings dynamism to regions with economic stagnation and might create a virtuous cycle of economic growth and job creation, which may stimulate the service sector (FAO 2017).

CSAFS offers further regionally diverse food products, which generally adapt better to increasingly more adverse climate conditions and may reduce crop losses. Disasters have occurred in the past and will happen more frequently in the future (IPCC 2012). Affected people depend on solidarity during and after a disaster, while social justice and gender equity are still mostly absent in governmental policies. Women and girls are not passive citizens, but dynamic actors able to forge their future. With the empowerment of women and girls, training and credits, they can overcome lacerating female poverty. Women should also be more equally involved in local and regional politics. They can be trained as skilled mediators who are able to promote an engendered-sustainable peace. Further, competent women are key persons to change the present food culture of soft drinks and excess carbohydrates. With a higher income, they could reverse the present level of obesity and, in mountainous and indigenous regions, the undernourishment of small children.

Finally, CSAFS helps to recover the destroyed environment produced by industrial agriculture, multinational mining companies, and neoliberal greed. Getting involved in restoration, human beings understand that they are part of the planet, but they are not its owners. By peacefully negotiating emerging environmental conflicts, economic and human resources can be used to overcome the present model of destruction. By thinking of Mother Earth, ecosystems, and coming generations, Latin American and the Aymara indigenous philosophy explain deeply 'Pacha mama' and 'good living'.

In their indigenous cosmovision, the accumulation of goods is not at the centre of life. Indigenous people focus on sustainable human relations, which includes a stable food culture that is regionally diverse and the living in harmony with nature and one another. The governments of Bolivia and Ecuador have incorporated this cosmovision into their Constitutions, even though they still depend on extractivism to pay their international foreign exchanges. Solón (2018) explained the five elements that consolidate this 'good living':

1. the whole and the 'pacha', which is an integrated movement of the cosmos;
2. a multipolarity between human beings and nature, where the whole community is involved and where social polarisation destabilise the internal harmony;
3. a dynamic equilibrium that promotes a holistic coexistence with ancestral wisdom and scientific knowledge;
4. a complementarity between rules and behaviours inside communities that allow harmonious conflict resolution processes, where each member knows how to live within the community and how to avoid conflicts;
5. a deep decolonisation process, which goes further than a formal independence and includes the overcoming of political, economic, social, cultural and mental blockages, which were promoted by Western socialisation and hundreds of years of colonial and neo-colonial dependency.

Nonetheless, these two countries suffer from a process of co-optation by their national leaders, who have promoted statism and extractivism to maintain their personal privileges, but also to comply with the servicing of foreign debts, and concessions to multinational enterprises. This co-option affects the 'Pacha mama' and blocks the harmonious relationship between nature and humankind. For other emerging countries such as Mexico, the same constraints exist. The best way to achieve a liveable future on in spite of the thunderclouds related to exclusive globalisation, economic crises, global environmental change, and climate change is the paradigm of 'good living'. Safe, organic, and diverse food at home, in the community, in the city, and the country reduces external pressures and grants sovereignty for the nation and for a sustainable future. Yet, countries are totally immersed in the present globalisation process. There is no requirement that this globalisation process must be violent, exclusive, destructive, and discriminative. As patriarchy is the original driver of the present unjust world (Frazer 1994), CSAFS and good living contain still unexplored potential, where from the family and the community level onward, new peaceful relationships can be developed, which may be able to maintain the existing biodiversity, restore water, air, and soil pollution, and produce healthy food and well-being for everybody. Climate-smart agriculture and sustainable food sovereignty are ways to re-establish a more harmonious relationship with Pacha mama.

References

Alkire, Sabine; Meinzen-Dick, Ruth; Peterman, Amber; Quisumbing, Agnes R.; Seymour, Greg; Vaz, Ana, 2012: *The Women' Empowerment in Agriculture Index* (Washington, D.C.: IFPRI).

Allen, J. Anthony, 1997: "Virtual Water: A long-term Solution for Water Short Middle Eastern Economies", Paper for the British Association Festival of Science, University of Leeds, 9 September.

Álvarez Enrique; Oswald Spring, Úrsula, 1993: *Desnutrición Crónica o Aguda Materno Infantil y Retardos en el Desarrollo* [Acute and chronic maternal-child undernourishment and development delays], Aporte de Investigación, No. 59 (Cuernavaca: CRIM-UNAM).

Arizpe, Lourdes, 2015: *Vivir para crear historia. Antología* de estudios sobre desarrollo, migración, género e Indígenas [Live to create history. Anthology of studies on development, migration, gender and indigenous people] (México, D.F.: CRIM-UNAM-M.A. Porrúa).

Biswas Tortajada, Cecilia, 2011: "Water Quality Management: An Introductory Framework", *International Journal of Water Resources Development*, 27(1): 5–11.

Boulding, Elise, 2000: *Cultures of Peace. The Hidden Side of History* (New York: Syracuse University Press).

Boulding, Kenneth, 1970: "The Economics of the Coming Spaceship Earth", in: Jarrett, Henry (Ed.), *Environmental Quality in a Growing Economy. Essays form the Sixth RFF Forum on Environmental Quality held in Washington* (Baltimore: John Hopkins Press): 3–14.

Brahma, B.; Pathak, K.; Lal, R.; Kurmi, B.; Das, M.; Nath, P. C.; Nath, A. J.; Das, A. K., 2018: "Ecosystem carbon sequestration through restoration of degraded lands in Northeast India", *Land Degrad. Develop.*, 29: 15–25.

Brauch, Hans Günter; Oswald Spring, Úrsula; Mesjasz, Czeslaw; Grin, John; Dunay, Pál; Chadna Behera, Navitna; Chourou, Béchir; Kameri-Mbote, Patricia; Liotta, Peter H. (Eds.), 2008: *Globalization and Environmental Challenges: Reconceptualizing Security in the 21st Century* (Berlin-Heidelberg: Springer).

Brauch, Hans Günter; Oswald Spring, Úrsula; Grin, John; Mesjasz, Czeslaw; Kameri-Mbote, Patricia; Behera, Navitna Chadha; Chourou, Béchir; Krummenacher, Heinz (Eds.), 2009: *Facing Global Environmental Change* (Berlin: Springer).

Brauch, Hans Günter; Oswald Spring, Úrsula; Mesjasz, Czeslaw; Grin, John; Kameri-Mbote, Patricia; Chourou, Bechir; Birkmann, Jörn (Eds.), 2011: *Coping with Global Environmental Change, Disasters and Security Threats, Challenges, Vulnerabilities and Risks* (Berlin-Heidelberg: Springer).

Brauch, Hans Günter; Oswald Spring, Úrsula; Grin, John; Scheffren, Jürgen (Eds.), 2016: *Handbook on Sustainability Transition and Sustainable Peace* (Cham: Springer).

Burton, Bruce A., 2013: *The Three D's: Democracy, Divinity and Drama - An Essay on Gender and Destiny* (Amsterdam: SynergEBooks).

Buzan, Barry; Wæver, Ole; de Wilde, Jaap, 1998: *On Security. A Framework of Analysis* (Boulder: Lynne Rienner).

CEPAL, 2017: *El costo del hambre. Impacto social y económico de la desnutrición infantil en Centroamérica y República Dominicana* (Santiago: CEPAL).

Chandra, Alvin; McNamara, Karen; Dargusch Paul, 2018: "Climate-smart agriculture: perspectives and framings", *Climate Policy*, 18(4): 526–541.

Conagua [National Commission of Water], 2014: *Programa National Hidráulico 2014–2018* [National Water Programme 2014–2018] (México D.F.: Semarnat-Conagua).

Conca, Ken, 1994: "In the name of Sustainability: Peace Studies and Environmental Discourse", in: Kakonen, Jyrki (Ed.), *Green Security or Militarized Environment* (Dartmouth: Aldershot): 7–24.

Coneval, 2017: *CONEVAL informa la evolución de la pobreza 2010–2016* [Coneval assesses the evolution of poverty in Mexico 2010–2016]; at: https://www.coneval.org.mx/SalaPrensa/Comunicadosprensa/Documents/Comunicado-09-Medicion-pobreza-2016.pdf.

Dalby, Simon; Brauch, Hans Günter; Oswald Spring, Úrsula, 2009: "Towards a Fourth Phase of Environmental Security", in: Brauch, Hans Günter et al. (Eds.), *Facing Global Environmental Change: Environmental, Human, Energy, Food, Health and Water Security Concepts* (Berlin-Heidelberg: Springer): 787–796.

Delgado Ramos; Gian Carlo, 2014: "Ecología política del metabolismo urbano y los retos para la conformación de ciudades de bajo carbono: una lectura desde América Latina [Political ecology of urban metabolism and the challenges for the conformation of low carbon cities: a reading from Latin America]", *Crítica y Emancipación*; at: http://biblioteca.clacso.edu.ar/ojs/index.php/critica/issue/view/8: 149–173.

EnsanutMc, 2016: *Encuesta Nacional de Salud y Nutrición de Medio Camino 2016* (Cuernavaca: INSP).

Eswaran, H.; Lal, R.; Reich, P.F., 2001: "Land degradation: an overview", in: Bridges, E.M.; Hannam, I.D.; Oldeman, L.R.; Pening de Vries, F.W.T.; Scherr, S.J.; Sompatpanit, S. (Eds.), *Responses to Land Degradation. Proc. 2nd. International Conference on Land Degradation and Desertification* (Khon Kaen-Oxford: Oxford Press); at: https://www.nrcs.usda.gov/wps/portal/nrcs/detail/soils/use/?cid=nrcs142p2_054028.

EU [European Parliament], 2017: *Report on Social and Environmental Impacts of Oil Palm Cultivation* (Brussels: European Union).

EU Resolution [European Parliament Resolution], 2017: "European Parliament resolution of 4 April 2017 on palm oil and deforestation of rainforests (2016/2222(INI))"; at: http://www.europarl.europa.eu/sides/getDoc.do?pubRef=-//EP//TEXT+TA+20170404+ITEMS+DOC+XML+V0//EN&language=EN.

FAO, 2002: *The State of Food Insecurity in the World 2001* (Rome: FAO).

FAO, 2006: *Gender and Agrobiodiversity* (Rome: FAO).

FAO [Food and Agriculture Organization of the UN], 2010: *Sustainable Crop Production Intensification through an Ecosystem Approach and an Enabling Environment: Capturing Efficiency through Ecosystem Services and Management* (Rome: FAO).

FAO [Food and Agricultural Organization of the UN], 2013a: *Food Wastage Footprint. Impacts on Natural Resources* (Rome: FAO).

FAO [Food and Agriculture Organization of the United Nations] 2013b: *Climate Smart Agriculture. Sourcebook* (Rome: FAO); at: http://www.fao.org/3/a-i3325e.pdf.

FAO, 2015: *Status of World's Soil Resources* (Rome: FAO).

FAO, 2016a: *Climate Change and Food Security: Risks and Responses* (Rome: FAO).

FAO, 2016b: *The State of Food Insecurity in the World 2015* (Rome: FAO).

FAO, 2017: "Country brief Mexico, 3 of March 2017"; at: http://www.fao.org/giews/countrybrief/country.jsp?code=MEX.

Fisher, Roger; Ury William; Patton, Bruce, 2011: *Getting to Yes: Negotiating Agreement Without Giving In* (Boston: Harvard University Press).

Folbre, Nancy, 2006: "Rethinking the Child Care Sector", *Journal of the Community Development Society*, Vol. 37, No. 2 (Summer): 38–52.
Fraser, Nancy, 1994: "After the Family Wage: Gender Equity and the Welfare State", *Political Theory*, 22(4): 591–618.
Galtung, Johan, 1967: "Peace research: science or politics in disguise", *International Spectator* 21 (19): 1573–1603.
Galtung, Johan, 1968: "Peace", in: *International Encyclopedia of the Social Sciences* (London-New York: Macmillan): 487–496.
Galtung, Johan, 2007: "Peace Studies: A Ten Points Primer", in: Oswald Spring, Úrsula (Ed.), *International Security, Peace, Development, and Environment*, Encyclopaedia on Life Support Systems, Vol. 39, Chapter 3 (Oxford: UNESCO-EOLSS).
Gao, Yan; Skutsch, Margaret; Masera, Omar; Pacheco, Pablo, 2011: "A global analysis of deforestation due to biofuel development", *Working Paper 68* (Bogor, Indonesia: CIFOR).
Gramsci, Antonio, 1971: *Selections from the Prison Notebooks* (New York: Lawrence and Wishart).
Gramsci, Antonio, 1977: *Cuadernos de la Cárcel* [Selections from the Prison Notebooks] (Mexico, D.F.: Juan Pablos Eds.).
Green Climate Fund, 2017: "Readiness and Preparatory Support Programme: Progress Report"; at: http://www.greenclimate.fund/documents/20182/751020/GCF_B.17_Inf.06_-_Readiness_and_Preparatory_Support_Programme__Progress_Report.pdf/54219665-621e-4cb1-a0ab-7d612f25f114.
Gutiérrez, Luis T., 2017: "Habemus Mamam! Religious Patriarchy is an Obstacle to Integral Human Development", *Mother Pelican*; at: http://www.pelicanweb.org/CCC.TOB.1703.html.
Held, David, 2004: *Global Covenant: The Social Democratic Alternative to the Washington Consensus* (Cambridge: Polity Press).
IISD, 2017: "Summary of the Sixth Meeting of the Working Group to Enhance the Functioning of the Multilateral System of the International Treaty on Plant Genetic Resources for Food and Agriculture: 14–17 March 2017"; at: http://enb.iisd.org/biodiv/itpgrfa/owg-efmls-6/.
INEGI, 2015: *Encuesta Intercensal 2015* [Intercensal survey] (Aguascalientes: INEGI).
IPCC [Intergovernmental Panel on Climate Change], 2012: *Managing the Risks of Extreme Events and Disasters to Advance Climate Change Adaptation* (Cambridge: Cambridge University Press).
IPCC [Intergovernmental Panel on Climate Change], 2013. *Fifth Assessment Report: Climate Change 2013. The Physical Science Basis*; at: http://www.ipcc.ch/report/ar5/wg1/.
IPCC [Intergovernmental Panel on Climate Change], 2014a: *Climate Change 2014. Working Group 2: Impacts, Adaptation and Vulnerability* (Cambridge: Cambridge University Press).
IPCC, 2014b: *Climate Change 2014. Mitigation of Climate Change. Working Group III Contribution to the Fifth Assessment Report of the Intergovernmental Panel on Climate Change* (Cambridge: Cambridge University Press).
ISSS, 2017: *Armed Conflict Survey* (New York: Routledge).
Jaspers, Karl, 2013: "A hundred years of Karl Jaspers' Psychopathology", *Arq Neuropsychiatry*, 71(7) (July): 490–492.
Jez, Joseph; Soon Goo, Lee; Ashley, M. Sherp, 2016: "The next green movement: Plant biology for the environment and sustainability", *Science*, 353(6305) (September): 1241–1244.
Jodelet, Denise, 1991: *Madness and Social Representation* (London: Harvester/Wheatsheaf).
Kaufman, Michael, 1999: "The Seven P's of Men's Violence"; at: http://www.michaelkaufman.com/wp-content/uploads/2009/01/kaufman-7-ps-of-mens-violence.pdf.
Lagarde y de los Ríos, Marcela, 1990: *Los cautiverios de las mujeres. Madresposas, monjas, putas, presas y locas* (México, D.F.: PUEG/UNAM).
Lal, R., 2006: "Enhancing crop yields in the developing countries through restoration of the soil organic carbon pool in agricultural lands", *Land Degradation & Development*, Vol. 17: 197–209.
Lamas, Marta (Ed.), 1996: *El género: la construcción cultural de la diferencia Sexual* [Gender: the cultural construction of sexual difference] (Mexico, D.F.: PUEG/UNAM, Miguel A. Porrúa).

Maass Wolfenson; Karla D., 2013: *Coping with the food and agriculture challenge: smallholders' agenda* (Rome: FAO).

Morris, Stephen, 2009: *Political Corruption in Mexico: The Impact of Democratization* (Boulder: Lynne Rienner Publishers).

Oswald Spring, Úrsula, 2009a: "A HUGE Gender Security Approach: Towards Human, Gender, and Environmental Security", in: Brauch, Hans Günter et al. (Eds.), *Facing Global Environmental Change: Environmental, Human, Energy, Food, Health and Water Security Concepts* (Berlin-Heidelberg: Springer): 1165–1190.

Oswald Spring, Úrsula, 2009b: "Food as a New Human and Livelihood Security Challenge", in: Brauch, Hans Günter et al. (Eds.), *Facing Global Environmental Change: Environmental, Human, Energy, Food, Health and Water Security Concepts* (Berlin-Heidelberg: Springer): 473–502.

Oswald Spring, Úrsula (Ed.), 2011a: *Water Research in Mexico. Scarcity, Degradation, Stress, Conflicts, Management, and Policy* (Berlin-Heidelberg: Springer).

Oswald Spring, Úrsula, 2011b: "Genetically Modified Organisms: A Threat for Food Security and Risk for Food Sovereignty and Survival", in: Brauch, Hans Günter et al. (Eds.), *Coping with Global Environmental Change, Disasters and Security Threats, Challenges, Vulnerabilities and Risks* (Berlin-Heidelberg: Springer): 1019–1042.

Oswald Spring, Úrsula, 2016: "Development with Sustainable-Engendered Peace: A Challenge during the Anthropocene", in: Brauch, Hans Günter et al. (Eds.), *Handbook on Sustainability Transition and Sustainable Peace* (Cham: Springer): 161–185.

Oswald Spring, Úrsula; Brauch, Hans Günter; Tidball, Keith G., 2014: *Expanding Peace Ecology: Peace, Security, Sustainability, Equity and Gender* (Cham: Springer).

Oswald Spring, Úrsula; Serrano Oswald, S. Eréndira (Eds.), 2018: *Risks, violence, security and peace in Latin America, 40 years of the Latin American Council of Peace Research (CLAIP)* (Cham: Springer International Publishing) (in press).

Palacios Vélez, Enrique; Mejía Saez, Enrique, 2011: "Water Use for Agriculture in Mexico", in: Oswald Spring, Úrsula (Ed.), *Water Research in Mexico. Scarcity, Degradation, Stress, Conflicts, Management, and Policy* (Berlin-Heidelberg: Springer): 129–144.

Paustian, Margrit; Theuvsen, Ludwig, 2016: "Adoption of precision agriculture technologies by German crop farmers", *Precision Agriculture*: 1–16; at: https://www.semanticscholar.org/paper/Adoption-of-precision-agriculture-technologies-by-Paustian-Theuvsen/5794f0005ab9830224a6222c44ee90d85d3339c5.

Petras, James, 2017: *Trade Wars and Food Wars: Obama and the Agribusiness Monopolies* (Nyskayina, N.Y.: Global Research).

Prigogine, Ilya; Stengers, Isabelle, 1997: The End of Certainty (Glencoe: The Free Press).

Reardon, Betty A., 1980: "Moving to the Future", *Network* 8(1): 14–21.

Reardon Betty A., 1996: *Sexism and the War System* (New York: Syracuse University Press).

Reardon, Betty; Snauwaert, Dale, 2015a: *Betty A. Reardon: A Pioneer in Education for Peace and Human Rights*, Cham, Springer.

Reardon, Betty; Snauwaert Dale, 2015b: *Betty A. Reardon: Key Texts in Gender and Peace* (Cham: Springer).

Reig, Paul; Maddocks, Andrew; Gassert, Francis [World Resource Institute], 2013: "World's 36 Most Water-Stressed Countries"; at: http://www.wri.org/blog/2013/12/world%E2%80%99s-36-most-water-stressed-countries.

Ren21, 2017: *Renewable Energies* (Paris: Ren21).

Report, 2017: "Famine in South Sudan", in: *The Lancet*, Vol. 389, May 20; at: https://www.thelancet.com/journals/lancet/article/PIIS0140-6736(17)31351-X/fulltext.

Richards, Howard, 2018: "Economy of solidarity: a key for justice, peace and sustainability", in: Oswald Spring, Úrsula; Serrano Oswald, S. Eréndira (Eds.), *Risks, Violence, Security and Peace in Latin America* (Cham: Springer, 2018).

Semarnat-INE [Ministry of Environment], 2014: *México: Quinta Comunicación Nacional sobre el Cambio Climático* (Mexico, D.F.: Semarnat-INE).

Serrano Oswald, S. Eréndira, 2009: "The Impossibility of Securitizing Gender vis à vis 'Engendering' Security", in: Brauch, Hand Günter; Oswald Spring, Úrsula; Grin John, et al. (Eds.), *Facing Global Environmental Change: Environmental, Human, Energy, Food, Health and Water Security Concepts* (Berlin-Heidelberg: Springer): 1143–1156.

Serrano Oswald, S. Eréndira, 2010: *La Construcción Social y Cultural de la Maternidad en San Martín Tilcajete, Oax.* [*The Social and Cultural Construction of Maternity in San Martín Tilcajete, Oax.*], Ph.D. Thesis (Mexico, D.F.: UNAM-Instituto de Antropología).

Serrano Oswald, S. Eréndira, 2015: "Debates teóricos en torno a la seguridad de género: reflexiones para la paz sustentable en el Siglo XXI [Theoretical debates on gender security: reflections for sustainable peace in the 21st century], in: Serrano Oswald, S. Eréndira; Oswald Spring, Úrsula; de la Rúa Eugenio, Diana (Eds.), *América Latina en el camino hacia una paz sustentable: herramientas y aportes* (Guatemala: FLACSO, CLAIP): 31–48.

Stiglitz, Joseph E., 2010: *Freefall. America, Free Markets, and the Sinking of the World Economy* (New York-London: W.W. Norton).

Trading Economics, 2017: Food Imports 1980–2017; at: https://tradingeconomics.com/mexico/imports.

Truong, Thanh-Dam; Gasper, Des; Handmaker, Jeff; Bergh, Sylvia (Eds.), 2014: *Migration, Gender and Social Justice. Perspectives on Human Insecurity* (Berlin-Heidelberg: Springer).

Turrent Fernández, Antonio; Wise, Timothy A.; Garvey, Elise, 2013: "Achieving Mexico's Maize Potential", *Intern. Conference* (New Haven: Yale University) September 14–15.

UNCCD 2017: *Global Land Outlook* (Bonn: UNCCD).

UNDP [United Nations Program for Development], 1994: *Human Development Report 1994* (New York: UNDP).

UNEP, 2016: *Global Gender and Environment Outlook 2016. The Critical Issues* (Nairobi: UNEP).

Via Campesina, 2002: "Food sovereignty", Document distributed during the World Food Summit +5 (Rome: Via Campesina).

Vía Campesina, 2005: *Agreement on Gender in Via Campesina* (Sao Paulo: Via Campesina).

Via Campesina, 2016: *La Via Campesina, Building an International Movement for Food and Seed Sovereignty*; at: https://foodfirst.org/la-via-campesina-building-an-international-movement-for-food-and-seed-sovereignty/.

WHO, 2016: "An estimated 12.6 million deaths each year are attributable to unhealthy environments"; at: http://www.who.int/news-room/detail/15-03-2016-an-estimated-12-6-million-deaths-each-year-are-attributable-to-unhealthy-environments.

Wilkinson, Richard; Pickett, Kate, 2009: *The Spirit Level: Why More Equal Societies Almost Always Do Better* (London: Allen Lane).

Worldwatch Institute, 2017: "Organic Farms Provide Jobs, High Yields", 23 June; at: http://www.worldwatch.org/node/3975.

Yiamouyiannis, Zeus, 2013: *Transforming Economy: From Corrupted Capitalism to Connected Communities* (Kindle-Amazon).

Zhou, Shen-Feng, 2008: "Research on Development Model of Circular Agriculture"; at: *Research of Agricultural Modernization*; at: http://en.cnki.com.cn/Article_en/CJFDTOTAL-NXDH200801013.htm.

Chapter 6
Ethnology of Select Indigenous Cultural Resources for Climate Change Adaptation: Responses of the Abagusii of Kenya

Mokua Ombati

Abstract The consequences of climate change, and the need to adapt and spur livelihood challenges. During periods of (un)expected climate change, traditional African communities applied indigenous cultural resources to secure the agrarian sector which almost exclusively supported their livelihoods. This study combines insights from the theories of cultural functionalism and interaction rituals to provide a descriptive interpretation of select indigenous cultural resources the Abagusii community of southwestern Kenya employed to respond and adapt to manifestations of climate change. The study proffers ways of repositioning this hitherto undervalued knowledge in partnership with contemporary climatological science to provide the 'magic potion' which will enable adaptation to the ever-enduring challenge of climate change in contemporary Africa.

Keywords African indigenous knowledge · Climate change · Adaptation
Abagusii

6.1 Cultural Resources for Climate Change

If cultures are adaptive systems through which human communities develop behavioural patterns to suit particular ecological settings, then the ways in which people respond to climate change bear culturally specific connotations since

> Culture frames the way people perceive, understand, experience, and respond to key elements of the worlds which they live in [...]. Individual and collective adaptations are shaped by common ideas about what is believable, desirable, feasible, and acceptable (Roncoli et al. 2009: 87).

Mr. Mokua Ombati, Ph.D. Candidate, Anthropology and Human Ecology Department, Moi University, Eldoret, Kenya; Email: keombe@gmail.com

© Springer Nature Switzerland AG 2019
H. G. Brauch et al. (eds.), *Climate Change, Disasters, Sustainability Transition and Peace in the Anthropocene*, The Anthropocene: Politik—Economics—Society—Science 25, https://doi.org/10.1007/978-3-319-97562-7_6

In that respect, African communities have, over time, developed unique and complex systems of culture and knowledge with regard to their daily relationships and interactions with the natural ecology. Anthropology's potential contribution to climate research is a study of a people's collective knowledge in culture, the detailed descriptions, comprehension and analytical evaluation of the mediating layers of cultural meanings and social practices, and responses. Accordingly, this study examines select indigenous cultural resources of the Abagusii community of southwestern Kenya. The Abagusii have a long history of marking, adapting and responding to climate change, and the community's history of climate change response and adaptation dates back to the time before the introduction of Christianity and other foreign cultural elements by the Europeans. The study investigates the cultural resources of the Abagusii in terms of their nature and form, socio-cultural settings and details of how and where these cultural practices were performed, the rules and codes of performance, the arenas, facilities, materials and equipment for performance, the participants, their significance and symbolic import to individuals, families, villages and to the entire Abagusii community.

6.1.1 African Indigenous Knowledge

African indigenous knowledge as used in this study is adopted from the definitions of Berkes (2012) and Steiner (2008) to refer to an African community's totality of knowledge, skills, information, attitudes, conceptions, beliefs, rituals, norms, values, capabilities, ideas, practices and ways of solving problems which have been accumulated and handed down through generations. It includes a community's holistic understandings and traditional approaches in education, art, technology, ecology, agriculture, health and medicine, ideology, politics, organisation, institutions, spirituality and worldview. Orlove et al. (2010) contend that such knowledge is place-based and rooted in local cultures, and is generally associated with a community's strong interactions to their natural environments. Such knowledge tends to be the result of cumulative experience and observation, tested in the context of everyday life.

An abundance of common terminologies used for this knowledge includes but is not limited to indigenous technical knowledge, indigenous technology, traditional knowledge, traditional ecological knowledge, local knowledge, farmers' knowledge, folk knowledge, ethnoscience and indigenous science (Nakashima et al. 2012; Orlove et al. 2010; Guthiga/Newsham 2011). Although each of the terms may have somewhat different connotations and reference groups, they all share sufficient meaning to be utilised interchangeably to refer to the Abagusii community's indigenous knowledge throughout this study.

6.1.2   African Indigenous Knowledge in History and Literature

While most African communities including the Abagusii have advanced indigenous methodologies, technologies and knowledge for responding to climate change and weather variability, such as instances of drought and prolonged rainfall failure, such knowledge cannot precisely be accounted for. This is partially because African cultural and historical traditions, wisdom and knowledge have, over the years, been accumulated, preserved and transmitted inter-generationally through verbal com-munication and religio-socio-cultural expressions. This is usually in the form of "beliefs and practices, myths and folktales, songs and dances, liturgies, rituals, proverbs, pithy sayings and names, sacred spaces and objects," and through other cultural activities such as ceremonies, sacrifices, dirges, story-telling, riddles, and idiomatic expressions (Nche 2014: 1). Unfortunately, no verifiable and reliable written records were generated through time and space during the oral history era.

The advent of colonialism and Christianity in Africa marked a significant turning point in the history of African civilisation and culture. The European foreigners unfortunately, out of either Eurocentric disingenuity or colonial malice and bigotry or both, isolated Africans and items of their civilisation and culture, on which African indigenous knowledge is based, and viewed them with scepticism, mistrust, and suspicion as they doubted that any "rational person" would accept such prac-tices and if so under what circumstances. They wondered how people would simply follow "customs that had, through some minor miracle, been caught in a time warp and catapulted into the 'modern'" (Sanders 1997: 5). On that note, Africans, forms of their knowledge, and their civilisation were systematically framed, wrongly categorised, and variously labelled as not science but fetishism, animism, magic, pagans, primitive and savage (Abu-Zahra 1988; Babane/Chauke 2015). For instance, Europeans accused Africans of being savages engulfed in "sympathetic magic that underlies rain rites", and, condemned the "fallacy of rainmaking and the muddled thinking (or lack of thinking) of those who engaged in it" (Sanders 1997: 5). In an occurrence at the coast of Kenya in 1948, the warden of Gedi National Park, British archaeologist James Kirkman, is reported to have "denied the integrity of the Swahili culture" (Linehan/Sarmento 2011: 312), consistently stating that without Islamic and Asiatic influences, the "coast would have remained a land of mud or grass huts like the rest of tropical Africa" (p. 313). This kind of perspective and reasoning, perpetuating a European myth that civilisation came from outside Africa, tends to ignore the history and culture of the African people, and concen-trates almost entirely on the Eurocentric hypothesis. This clear perversion of African history, knowledge, and civilisation strongly diverges from the reality and values of the people living and working within the constructs of African knowledge and civilisation.

Because of those prevailing circumstances, Africans felt too intimidated to divulge authentic information to European researchers, on African indigenous civilisation and cultural practices, as they considered these to be their deepest and most closely guarded secrets. Consequently, some of the initial studies on African civilisation and cultural practices and more specifically, those carried out during the early stages of colonialism, were "often frustrated by unwilling or unruly informants" (Sanders 1997: 2). Evans-Pritchard (1938) reckons encountering similar difficulties when he attempted to carry out research on rainmaking practices of the Bori clan in Anglo-Egyptian Sudan. It was therefore difficult to investigate African culture, knowledge, technologies and civilisation with a fair clarity of mind and responses.

For those reasons, and many others, earlier reports on African civilisation and culture were, therefore, riddled with Eurocentric and colonial biases. This cavalier Eurocentrism formed a major distortion and "epistemic violence to the extent that it involved immeasurable disruption and erasure of local cultural systems" (Linehan/ Sarmento 2011: 307). The (mis)representations, falsifications and distortions have consistently been used to legitimate the destruction of African history, culture, and civilisation.

All these challenges compromised the veracity of the kind of African indigenous knowledge produced during the colonial period and even after. It is, therefore, certain and entirely understandable under those circumstances that many earlier accounts remain, to say the least, incomplete and even distorted. Discounting the distortions, suspicions, and (re)claiming the science of African indigenous knowledge can only be possible after these glaring concerns and gaps in knowledge and literature have been bridged. That is partially what this study hopes to achieve.

6.1.3   Climate Change

Climate change as used in this study refers to changes in historical weather patterns resulting in extreme weather conditions. The changes alter the quantity, intensity, frequency and distribution of rainfall. This, together with increased intensity and frequency of higher temperatures pose major threats to ecosystems and geophysical cycles. The effects of climate change negatively impact communities at multiple levels, significantly threatening their security, ecological, social, religious, economic, political and human conditions. In particular, and of interest to this study, extreme effects are felt in food production, water availability, intensification of wildfires, mud-streams, changes in epidemic vectors and extinction of pollinators. These, together with droughts, famines, heavy storms, and floods disrupt livelihoods and communal wellbeing (Change 2014; Eakin/Walser 2007). For instance, rainfall variability determines agricultural productivity even as it influences family well-being. If a season brings very little rain, people must travel far distances to fetch water from streams, ponds, and wells, a considerable burden that falls almost exclusively on women and children, who are also responsible for a large amount of agricultural labour.

Of particular interest to this study, while the manifestations of climate change are manifold, climate change in the form of droughts negatively affect water supplies.

Changes in temperature patterns, solar radiation, winds and prolonged periods without adequate rainfall cause droughts, which, often result in water shortages. Drought serves as a trigger for famine, which then causes hunger, disease and malnutrition, leaving people physically weak and reducing productivity. Additionally, the combination of higher temperatures and lack of water in the soil decrease crop productivity. Water scarcity also leads to the depletion of crops and deterioration of soil qualities and properties. The resulting impacts include the loss of lives, livelihoods and the displacement of populations from one degraded ecosystem zone to another (Kalungu et al. 2013).

6.1.4   Climate Change Adaptation

Climate change adaptation as used herein refers to adjustments in ecological, socio-cultural or economic systems in response to actual or expected climatic stimuli and their effects. It further refers to changes in processes, practices and structures to moderate potential damages or to benefit from opportunities associated with climate change (UNFCCC 2014). Thus, adaptations are largely responses to climate change focused on the ability to better suit livelihoods to one's own environment. As the case of the Abagusii shows in this study, communities respond to climate change with adaptation efforts in various realms of life by creating innovative and localised solutions aimed at fostering their resilience.

6.2   Theoretical Framework

This study combines insightful lenses from the theories of cultural functionalism and interaction rituals to analyse how a select number of cultural resources of the Abagusii community capture their climate change adaptation responses. The study employs a deconstructive analysis of these cultural resources to arrive at the symbolic meanings, significations, and inferences espoused. And in the process, the study identifies the ways in which these cultural resources enable suitable climate change adaptation responses among the Abagusii community.

6.2.1   Functionalism Theory

Malinowski's (1926, 1944, 1954) functionalist theory of culture demonstrates that "every type of civilisation, every custom, material object, and belief fulfils some vital function, has some task to accomplish, represents an indispensable part within a working whole" (1926: 132). For Malinowski (1944), culture is "the integral whole […] by which man is able to cope with the concrete specific problems"

(p. 36), "in his environment in the course of the satisfaction of his needs" (p. 150). Malinowski argued that the function of any culture is meeting the organic needs of people (whether the needs are spiritual or economic or social), and to supply a set of general laws in the fulfilment of those needs. Culture is always instrumental to the satisfaction of these needs through its existing social institutions.

Explanations of social phenomena must be constructed within their current contextual manifestations. Hence, observation and recording of indigenous customs indicate their functioning as socially sanctioned means of satisfying human needs (Lesser 1985). Customs may serve both to explain local happenings and, in the process, formulate knowledge, and communicate its shared context, relationships and realities. Organisation of technical skills around symbolisms embodied primarily in traditional knowledge, beliefs, norm systems, and institutions make the continuity of civilisations possible (Firth 1957). The meanings which culture elucidates infer interrelated significances to members of social systems. Together with cultural practices serving the pivotal function of acting as repositories of values and norms, their practical and divine significance is also elemental; primary, basic, fundamental and essential. The basic condition for an orderly social existence depends on the transmission and maintenance of those culturally desirable elemental properties.

6.2.2 Interaction Rituals

The mechanism of climate change adaptation responses among the Abagusii community can also be explained through their deeply ingrained ritual performances. Rituals denote stereotyped sequence of activities involving gestures, words, and objects, performed in a sequestered place, according to set sequences. They are rhythmic, coordinated group activity, and performance that guide behaviour in a common direction. Rituals are characterised by sacral symbolism, formalism, traditionalism, invariance, rule-governance, and specific performances (Rappaport 1999; Bell 1997). Rituals are usually prescribed by the traditions of a given community, and Turner (1973: 1100) underscores "the importance of ritual" in sub-Saharan Africa, noting that the continent is "rich indeed in ritual genres". He conceptualizes rituals as,

> ... designed to influence preternatural entities or forces on behalf of the actor's goals and interests. Rituals may be seasonal, hallowing a culturally defined moment of change in the climatic cycle or inauguration of an activity such as planting, harvesting [...] or [...] held in response to an individual or collective crisis [...] to placate or exorcise preternatural beings or forces... (p. 1100).

Equally, Sanders (2002: 290) indicates that rituals are culturally-appropriate ways of acting upon the world with the intention of making things happen. In the context of rituals consistent with climate change adaptation such as rainmaking, rituals are not simply for "symbolic representation, but of animating the cosmic and divine powers of the universe and effecting change." And this practical engagement with the world enables such rites to "draw their cosmic powers from, a number of separate yet interrelated cultural domains [...]".

Collins (2004: 7) defines interaction rituals as "a mechanism of mutually focused emotion and attention" or simply "focused interactions," in the words of Summers-Effler (2006: 135). Interaction rituals take various forms, including singing, dancing, ululating, marching, laughing, clapping hands, incantations and even conversations. Indeed, the central mechanism of interaction ritual theory is an emotional, sacred atmosphere among people, making them feel that they transcend their everyday life. The contents, medium, and location of the indigenous cultural resources constitute a frame—a schema of interpretation for locating, perceiving, identifying, punctuating, and labelling climate change adaptation responses that occur in the Abagusii life-world.

6.3 Research Context

Understanding prehistoric practices require access to the worldview of the people at that time. This, therefore, requires combining historical evidence and ethnographic data to delve into the Abagusii's cultural systems and processes, and to understand the contexts in which they occurred and are linked to climate change adaptation responses. The "product of ethnographic work is a descriptive reconstruction of the hosts' own construction of their worlds" (Whitehead 2004: 16–17).

The study covered the Abagusii, a Bantu ethnic group, who exclusively occupy the dual counties of Nyamira and Kisii in southwestern Kenya. Kisii is the name that the British colonial administration used for, and is still the common name used to refer to, the Abagusii. The language they speak is Ekegusii and the land they occupy is referred to as Gusii. Abagusii occupy a small area, estimated at 2,196 km^2, west of the Great Rift Valley, and slightly over 50 km east of Lake Victoria, sandwiched between the Luo, Kuria, Maasai and Kipsigis ethnic groups. The Abagusii are settled in the fertile equatorial highlands surrounded by several ridges and valleys, and separated by many year-round rivers and streams (Fig. 6.1).

Gusiiland experiences cool climatic conditions with plenty of highly reliable and predictable rainfall throughout the year that favours Abagusii's main social economic activities. The average annual rainfall is between 1800 and 2000 mm. Guided, for the most part, by the colonial administrative structures and organisation, the geography of the region, as well as settlement patterns, the Abagusii are informally classified and organised into six sub-groups that are geospatially dispersed into seven sub-regions, and further sub-divided into several exogamous and endogamous clan categories. The Manga escarpment, Sameta hills, Rivers Gucha (Kuja), Omogonga (Mogonga) and Risonto (Sondu) are some of the most prominent physical features traversing Gusiiland (Ochieng' 2004; Akama/Maxon 2006) (Figs. 6.2 and 6.3).

Fig. 6.1 Map of Kenya showing all the counties. *Source* Kenya Open Data Project [Copyrighted free use]

This study targeted informants who were knowledgeable on the required information and could provide valid, reliable and first-hand primary data. The data was therefore obtained from persons who either participated in the indigenous practices under study, or witnessed them being practiced at the time the community's indigenous cultural systems were still intact. These were people aged seventy years and above at the time of data collection, but whose memory and communication capacities were good enough to recall and narrate the required information. Consequently, seven informants disaggregated by gender and one each drawn from the seven informally classified broad geographical sub-regions of the Abagusii community were purposively sampled through the snowball sampling technique for the study.

The sample size of seven informants was deemed appropriate based on sequential sampling as outlined by Krathwohl (1993: 139), "sequential sampling

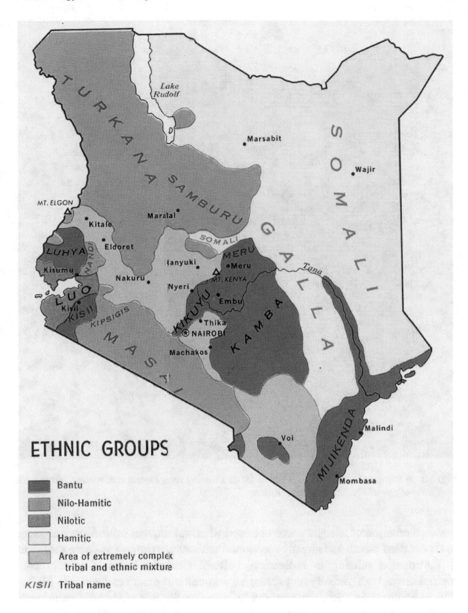

Fig. 6.2 Ethnic map of Kenya. *Source* University of Texas at Austin, Perry-Castañeda Library Map Collection, political map of Kenya (1988); at: http://www.lib.utexas.edu/maps/africa/kenya_ethnic_1974.jpg

allows us to start with a small sample and then continue sampling until some criterion of adequacy is met". The criterion of adequacy, for this study, was reached after successive interviews continued yielding the same information. In other

Fig. 6.3 A topographical sectional view of Gusii showing river Gucha meandering through the hills and valleys of Gusiiland. *Source* Author

words, a criterion of adequacy was obtained when subsequent informants continued to repeat what others had already mentioned without providing any more new data.

Information relating to indigenous cultural elements has traditionally been handed down both verbally and through socio-cultural practices from one generation to another. As such, the required information for this study was more accurately accessed through conversational interviews. Conversational interviews are used where the storytelling genre of the interview is adjusted to a more informal discussion, taking the form of a conversation. The interview dialogue becomes a more-equal two-way process as the interviewer and the interviewee interact in a conversation. However, archival records, written documents, reports and publications, and field observations enriched and provided supplemental information for the study.

The recorded data was transcribed, translated from Ekegusii into English, coded and analysed. Analysis involved data reduction, thematic interpretation and description through a process of discussion, reference and argument in reference to the information provided, perspectives implied or about which could be inferred, and the dynamics suggested about the indigenous resources for climate change adaptation responses.

All informants participated in the interviews after receiving an explanation of the study objectives, procedures, risks, benefits and the voluntary nature of their participation. Study protocols including research authorisation and entry permits were obtained before the research was conducted.

6.4 Ethnology of Select Indigenous Resources for Climate Change Adaptation Responses

6.4.1 Traditional Economy and Climate Change of the Abagusii

The pre-colonial food economy of the Abagusii was built on cultivation of crops and keeping animals. The prevailing favourable climate supported the growth of such crops as millet, sorghum, beans and vegetables of all kinds. The main crop, *obori* (Eleusine) as Bogonko (1977: 40) says, "was first planted with a mild mixture of *amaemba* (sorghum)". Small pink and yellow coloured maize was planted in very small quantities around the Eleusine farm fields. White maize and sweet potatoes were however, not originally subsistence crops of the Abagusii as is currently the case. They were later-day introductions into the farming speciality of Abagusii by the colonial settlers. Imported from Brazil by the Portuguese, white maize production has since surpassed all other crops in the socio-economy and livelihoods of the Abagusii. Other important contemporary crops include cassava, pigeon peas, green grams, onions, bananas, English potatoes and tomatoes.

Later, the Abagusii diversified their cash economy to include coffee farming. Tea and pyrethrum were introduced in 1954 and 1960, respectively. The current cash economy of the Abagusii is dominated by the growth and harvesting of tea at the subsistence farm level. Due to declining availability of land for cultivation of food crops, proceeds from tea sales are increasingly used to buy food items, sustain household livelihoods and food securities. In addition, Abagusii love animals and birds. As Ontita (2007) contends, although they grew *obori* (Eleusine), at the time of their arrival in their present settlement and until the arrival of colonialists around 1905, the Abagusii were de facto pastoralists. They domesticated cattle, sheep and goats that provided milk, meat and blood for food, and, hides and skins for clothing as well as for sleeping mats. They also kept chicken, geese and a variety of other birds. With the livestock and birds, the Abagusii could enter into marriage contracts, as well as engage their neighbours in barter trade (Omosa 2007).

Informant accounts indicate the critical significance of the community's diversification of crops and animals, noting that it was a major climate change adaptation and response strategy, which cushioned them against instances of crop failure and death of animals due to drought and prolonged rainfall failure. During such adverse periods, Abagusii depended on drought resistant crop varieties and animal breeds to sustain their livelihoods. Thus, diversity of crops, birds and animals allowed the community to accommodate climate variability.

Gusiiland experiences two peak rainy seasons. The long rains begin late February and last until June, and the short rains start from August to late November, making the region to have two major ecological zones. That the rains arrive on time —or indeed, that they arrive at all—and fall regularly is, quite literally, a matter of life or death for the community. In addition to its economic importance, rain has a cultural and religious significance in Gusiiland as in other parts of sub-Saharan Africa. Correspondingly, scanty rains are believed to signal that God and the spirits of the ancestors are displeased with the people, while in contrast abundant rains indicate divine and cosmological favour. The relief that accompanies rainfall, over an agricultural season, rests in part on the expectation of good harvests and also reassurance that their world is in good order. In contrast, the distress that comes when rains are delayed stems not only from a fear of hunger and starvation, but also from concerns that the social and religious realms are not quite right. That therefore, explains why, being an agricultural and livestock keeping community, Abagusii adopted and adapted the chronology of their agricultural activities and practices to follow the prevailing climatic weather patterns.

Informants gave vivid detailed accounts of the community's alignment of the agricultural calendar of activities to follow the known climatic weather patterns of the region. In January (*monungu n'barema*), fields were cleared and land preparation began. These activities continued into the dry spell of February (*eng'atiato*) and the month of March (*egetamo*) when twigs were removed or trimmed. In April (*rigwata*), finger millet was sown using the broadcast method. The current practice of planting in March or earlier is an adoption of the highland agricultural calendar.

May was the month of weeding and this was carried out through collective labour groups. The month of June (*ebwagi*) was, and continues to be as the Gusii name implies, a period of scarcity. The month of July (*engoromoni*) was characterised by *ogosuma*, meaning seeking food aid from close family members and even distant relatives. In August (*riete*), the men started making new granaries and old ones were repaired and cleaned up in preparation for the new harvest. Harvesting began in August and continued into the month of September (*ebureti ya kebaki*) when sorghum was trimmed to produce a second flowering. The months of October (*egesunte gia chache*) through December (*esagati*) were a period of rest, a time for festivity that culminated in thanks-giving to *Engoro* (God), at the shrines of worship.

In sum, as informants' accounts indicate and as corroborated by Omosa (2007), the Abagusii had important decisions to make in each climatic season. Due to the advanced knowledge about the changing climatic and weather patterns, rains were, therefore, particularly important and of great value to the community. The knowledge allowed them to make decisions on when and what to plant between the longer cycle and fast-maturing crops, and it also permitted them to decide on the timing of land preparation, cultivation (ploughing), planting, weeding, and harvesting. They, for example, planted fast maturing and drought resistant food crops like cassava, sorghum, eleusine and sweet potatoes during the shorter rain season, and long cycle crops like maize over the long rain season. In that way, they were able to adapt to the variability of climate and secure their livelihoods.

6.4.2 The Abagusii's Traditional Organisation of Work and Labour and Climate Change

The Abagusii used their familiarity and collective knowledge of historical climate patterns to organise the labour of their agrarian activities, in a clear case of adapting their farming technologies and methods to climate variability. Informants reported how traditional Abagusii organised their workload of activities and labour arising from the agricultural crop calendar into collective labour and work groups differentiated variously as *egesangio, ekebosano, risaga* and *ekeombe*. By this arrangement, the community ensured that no family, household or homestead was caught and held in a climatic weather web, which would prevent them from adapting their agricultural crop cycle to the prevailing weather patterns. For instance, any delay in land preparation, or planting, or weeding, or harvesting, would not only affect the current crop in production and yields, but would also interfere with the agricultural crop cycle of the next season, accordingly put families at risk of starvation. This makes clear the fact that the Abagusii anticipated and instituted adaptation responses to the uncertainties of climate variability as a group, and not necessarily on an individual-by-individual basis. This however, does not deny individual climate change adaptation response categories. The differentiation of one labour group from the other was mainly based on the objectives, motivation and composition of each particular group.

The *egesangio* labour group was made up of people from the same neighbourhood who voluntarily cooperated and worked on each other's fields on a reciprocal basis without remuneration. These groups were mainly made up of women, and membership in *egesangio* consisted of contemporaries (*mogisangio*). Whenever these boundaries were exceeded to incorporate others, this was referred to as *ekebosano*.

The *risaga* work group was a cooperative group that performed both routine and non-routine work for a member of the community in exchange for privileges such as local beer and foodstuffs. The composition of this type of group was ad hoc. To attract labour, the homestead with a specific task to be performed, such as massive weeding or ploughing, prepared beer and homestead heads in the neighbourhood sent their household labour force to work. Unlike *egesangio, risaga* was initiated and organised mainly by men.

For the *ekeombe* work group, both men and women worked on each other's land, but not necessarily in a reciprocal manner. The group was organised in such a way that any work done was paid for. This remuneration was kept in a common pool, until, after a certain length of time (usually one agricultural calendar season), it was then shared out equally. The group could also be hired out to work for non-members for remuneration and payment. The *ekeombe* type of work groups is a product of recent transformations, and formed the basis for the cooperative movement and other collective grassroots organisations in contemporary Gusiiland.

In both organised and ad hoc labour groups, input was measured by the number of hours put in and these were equal and compulsory for each person. Whenever a member was indisposed, they were required to send a replacement. Hence, as soon as one decided to participate, one bound oneself to group rules and regulations pertaining to performing the labour tasks.

6.4.3   The Traditional Religious Perspective of the Abagusii and Climate Change

Religion is a complex and variable social phenomenon, which Morris (2006: 1) views as an "institution consisting of culturally patterned interaction with culturally postulated superhuman beings", or the sacred, spiritual beings, divinity, supernaturals, numinals, or occult powers. Religion is characterised by a number of dimensions and attributes including ritual practices, a body of doctrines, beliefs, and traditions, patterns of social relations, a hierarchy of ritual specialists, a dichotomy between the sacred and profane, and an ethos that gives scope for emotional or mystical experience.

It must be emphasised at the outset that religion permeates so fully into all aspects of the life of the Abagusii that it is not always possible to isolate religion from other aspects of Abagusii life. Religious beliefs, as Horton puts it, are "theoretical systems intended for the explanation, prediction and control of space–time events" (Horton 1971: 94). A study of the Abagusii's religious systems, is, therefore, ultimately a study of their complex belief systems, ideologies, norms, values, customs and practices in their innumerable interactions with the environment.

The Abagusii believed in a supreme entity, *Engoro*, the original progenitor and source of prosperity and life, who guided and assisted them, saving them from disasters and calamities. *Engoro*, the creator of the universe, governed the destiny of humanity and of all natural forces, sending rain or drought, plenty or famine, health or disease, peace or war. In this sense, *Engoro* was full of goodness, pureness, impartiality, love and generosity, while *Nyachieni* (Satan) wove evil schemes (Monyenye 1977). This required that the community show reverence to *Engoro* and submit to His will through regular prayers, sacrifices, and rituals. They thus approached *Engoro* in praise, thanks-giving, and repentance, asking for forgiveness. Correspondingly, they offered prayers and sacrifices to mark key milestones in the agricultural crop calendar, before and after land preparation, before and after planting and after harvesting. The religious beliefs of the Abagusii were thus carefully intertwined with their livelihoods.

Engoro was not visible, but occasionally manifested Himself through the wonders of heavenly bodies like storms, earthquakes and lightning from the sky, the sun, the moon and the stars. The sources interviewed explained how human agents acting on behalf of the Supreme Being, including *ababania* (prophets), *abanyibi* (rainmakers), *abanyamesira* (medicine men) and *abaragori* (diviners) mediated in less subtle ways with *Engoro*, countering Satan's evil whims on earth. The community's religious philosophy was, thus, engineered around understanding *Engoro*, and therefore, people always turned to *Engoro* to decipher issues beyond their understanding. For instance, informants recollected how, whenever there was persistent drought, women of impeccable character and men of integrity, officiated by *abanyibi* (rainmakers), sung, danced, and performed supplication *Ribina* ritual (discussed elsewhere in the article) at the top of revered hills and valleys. And *Engoro* would almost always answer their prayers with plenteous rainfall.

6.4.4   The Sacred Ecology of the Abagusii and Climate Change

Many African communities assign sanctity to certain portions of their natural landscape and regard them as worthy of devotion, loyalty, dignity and worship. The Abagusii are no exception.

6.4.4.1   The Abagusii's Shrines of Worship

In the context of sub-Saharan Africa, as Dawson (2009) notes, shrines are sacred spaces, symbolically representative of a group's connection with the spirits of the land and with the ancestors at the cosmological and supernatural levels. Acting as containers, shrines are vessels for the "spirits of ancestors and deities who must be regularly placated and petitioned for blessings, requests for intercession, and divine sanction" (p. VII). Spiritual intercession is habitually sought for "most importantly, the planting or harvesting of a season's agricultural produce" (p. VIII). As vessels, shrines inextricably legitimate a group's existence and linkages with the 'soil', a mystical attachment to the land they occupy, symbolically representing their existence as cultivators and as societies that revere their ancestors.

The Abagusii, had specific sites they considered sacred, which served as shrines for worshipping *Engoro*. These were mostly found on top of hills and at the bottom of valleys, or next to trees considered sacred such as *Omotembe* (Erythrina Abyssinica), *Omosasa* (Brachystegia spiciformis) and *Omogumo* (Fig Tree). God was believed to stay in, beside or on these holy sites and objects of nature. The trees were never utilised as wood fuel and nobody aimlessly went to these sites. The community considered these, holy communal shrines where they went for most ceremonies, vows, reconciliations and covenants. Equally, prayers to give thanksgiving to *Engoro* or to atone for their unbecoming behaviour were made next to or under this sacred ecology. The community also used the shrines as spaces to approach *Engoro* in times of disasters such as famine and drought. Presenting the community to *Engoro* before the holy shrines, the community's spiritual leaders sought divine understanding of the phenomena affecting the community or a section of it at the time.

The shrines most revered by the community comprise Manga and Sameta hills of Nyamira and Kisii counties respectively. The Charachani waterfall in Nyamira County is also held in high reverence by the Abagusii community. Informants gave clear narrative accounts of how the community retreated to these particular shrines and many others spread across the ecological breadth and width of Gusiiland during the famines experienced in the years of 1965, 1972/1974, 1979/1980 and 1983/ 1984. The reported famines occasioned acute food shortages and starvation due to prolonged drought. Livelihoods risked extinction as the devastating consequences of the famine directly challenged the people's very existence. Famine, locally referred to as *egeku* (deadly disaster) was perceived as an unavoidable occurrence, often attributed to some natural (or supernatural) catastrophe that went beyond the Abagusii's control. Famines were simply viewed as acts of God and therefore the people had to ask God's benevolent intervention in disasters.

Informants also reported other famines in the colonial period such as the *langi* famine of 1896. This famine, whose name is borrowed from the Abagusii neighbours, the Luo, resulted in numerous deaths from starvation, disease and hunger. In 1914, Abagusii were afflicted by yet another famine, locally referred to as *nyabiage* or *nyamauga*. During this famine, which informants reported to have been caused by drought, granaries were swept clean. This was closely followed by yet another famine in 1918, known as *kunga* or *enchara ya kengere* or *nyabisagwa*, also caused by a delay in rainfall. The famine is reported to have claimed many lives through starvation, disease, and hunger. The 1931 famine, locally known as *nyangweso* resulted in many Abagusii emigrating to other areas. Alongside this, some families gave away their children in exchange for food to neighbouring communities.

Archival reports, scholarly records and written documents corroborate informants' accounts of these famine episodes. According to Omosa (2007), the 1931 famine was the result of excessive rainfall in the months of March to June in the previous year, exacerbated by the lower than average rainfall of 1929 and the less than abundant harvest of 1928. Informants were particularly adamant that their God, *Engoro* responded with plenty subsequent to their visits to the holy shrines for prayer, worship, sacrifice, ritual, atonement and repentance after these climatic disasters. They, for example, insisted that the *langi* famine of 1896 ended with bountiful harvest of 1897, because God answered their prayers in their holy shrines of worship.

6.4.4.2   The Sacred Manga Shrine

A steep, rocky, ragged, long cliff-line defines the unique ecology of the Manga Hill of Nyamira County. The ridge, an ecology considered sacred, has rich historical mythologies and legend information about the Abagusii's past. Myths and legends on ageless, mysterious caves at the ridge, known locally as *Ngoro ya Mwaga*, encapsulate rich, special memories as *Engoro* (God) and spirits of the ancestors were believed to reside here. The community visited often to pray, ask for purification, be cleansed, and offer sacrifices. Here Abagusii "… sat to determine communal issues, solve conflicts and disputes, offer prayers, sacrifices, ask for blessings and cleansing, and conduct important ceremonies," informant Orina Maroko, whose homestead is a few yards from the shrine, recalls. He adds that the "Abagusii prayed and sacrificed at this ridge during moments of distress (*egeku*) requesting for *Engoro's* intervention". The community "prayed for rain during long droughts and God would immediately answer with plenteous rainfall. It particularly never failed to rain, after the sacrificial prayers", Maroko recounts. Confirming how culturally significant the site is to the Gusii people, Maroko adds:

> In the old days, elderly women came here to mark every harvest. They would sing Gusii folk songs, ululating in praise of the good harvest that would chase away hunger. Then, men would follow with *amarua* (traditional beer) to drink, thanking God for the good harvest.

Because the site is still considered holy to date, all visitors must follow a specific procedure to access it. Visitors are required to tie a knot of a bundle of local green grass before the entrance to appease the spirits so that they do not haunt them, and then collect some pieces of firewood, which they throw into the caves so that ancestors allow access. A waterfall which has been turned into a spring by the locals, *Omosasa* (Brachystegia spiciformis), a tree believed to be the abode for the spirits of the ancestors, and a now invisible lake believed to have been formed after a flood came and swallowed beneath earth a house belonging to a local, Mr. Okari, are some of the sacral mythological mysteries fortifying the shrine.

With mystified trepidation, Maroko recounts that when the region becomes very hot and dry, a mysterious smoke billows out of the caves and then heavy rains are experienced in the area immediately after. Maroko further reports,

> This happens at night. You can see fire burning bushes, hear sounds of cows mooing and women ululating but during the day, you cannot see any damage caused by the mysterious fire. Recently, the ridge was seen burning in the evening and after a few hours later, heavy rains pounded.

While some cultural aspects pertaining to the ridge have changed with time, Orina Maroko says the site's reverence as a place for seeking God's blessings has not waned. Many people still visit *Emanga*, as locals fondly refer to the Manga shrine, to seek God's blessings. Even those who subscribe to the Christian and other modern-day faiths visit the shrine for deep spiritual meditation and prayer. Orina Maroko confirms seeing, "people of all kinds come here almost daily to offer their prayers." This reverence is further manifested in the presence of a modern recollection centre erected at its summit by the Catholic Diocese of Kisii.

From the foregoing, it is clear the Abagusii used spatial ecologies they considered sacred as sites of seeking divine intervention when faced with moments of distress such as prolonged drought, epidemics, famine and floods, among others. In the sacred sites, they sought divine understanding of the phenomena affecting the community at the time. They used the same ecologies to appreciate divine providence and plenty.

6.4.5   The Indigenous Traditions of the Abagusii and Climate Change

Tradition is the enduring aspect of the culture of a people. Gyekye (1997) argues that modernity does not reject what existed in the past, and that every society has inherited part of its cultural values from the past generation. Consequently, he defines tradition as "any cultural product that was created and pursued by past generations and that, having been accepted and preserved, in whole or in part, by successive generations has been maintained to the present" (Gyekye 1997: 221). At the core of the traditional existence of the Abagusii are particular philosophies, ideologies, values, norms, belief systems and customs that form their being as an organised ethnic group. These traditional beliefs, norms, and practices were and to a large extend are still regarded very highly by the Abagusii. Consequently, breaking or disobeying one of them invites direct condemnation by the ancestors and ruining an entire generation of the Abagusii people (*Egesaku kia'Mwamogusii*).

Additionally, the cultural heritage of the Abagusii was rarely taught directly to the members of the community. Instead, it was presented indirectly through cultural expressions and performances. Rituals, folk songs, dances, and other indigenous expressions formed powerful cultural mediums of communication. Songs, dance and rituals are cases in point of the symbolic cultural performances associated with climate change. Climatic changes that occasioned epidemics, famine, floods and/or drought, were locally referred to as *egeku* (deadly disaster), and were attributed to *Engoro* or some supernatural powers which were beyond the people's control and understanding. The attribution of *egeku* as acts of *Engoro* required that the people beseech His benevolent intervention through dance, song and ritual. The Abagusii used hilltops and valley bottoms as shrines to invoke *Engoro's* divine intervention in weather modification during times of climate change. Weather modification rituals, songs, dance and ceremonies were cultural actions, practices, and rites intentionally offered to *Engoro*, for the intended purpose of manipulating or altering the weather, usually, for the reason of increasing the local rainfall supply. They also had the goal of preventing damaging weather, such as hail, hurricanes, drought, or famine from occurring or of provoking damaging weather against enemies.

6.4.5.1 The Ritual of the Ribina Rain Dance

Rainmaking rites and beliefs have been prominent cultural expressions among the Abagusii community since time immemorial. The Abagusii performed ritual songs and dance as a symbol of supplication to a Higher Being. As King'oina (1988: 6) observes, "any mistakes or errors did call for a cleansing ceremony ... to appease the spirits of ancestors". The climax of such ritual ceremonies was the shedding of blood of an animal and sharing the food collectively. The blood signified life and was the ultimate atonement for any sins whatsoever to the giver of life, *Engoro*. Rituals were a covenant between the people and *Engoro* who could reward them for maintaining a good relationship with Him.

Abagusii had formulaic incantations that they used as a kind of prayer. A case in point is the supplication ritual song dance, *Ribina*, used to commemorate climate change. In the ritual song dance, the link between *Engoro* and His spectacular creations is manifest. The ritualistic prayer, and song dance, *Ribina*, was performed at the top of sacred hills or in the bottom of hallowed valleys. The *Ribina* ritual dance marked the end of an agricultural season, a year, and the beginning of a new cycle. It was, for example, observed between the harvest and the next planting period. It also marked instances of severe shortage of rainfall, famine, or drought. In addition, it was performed to commemorate a good harvest. Men and women who participated and officiated in *Ribina* were those considered holy and of high moral character and integrity. Only mature, married and women past childbearing age were allowed to participate in the *Ribina* ritual.

The *Ribina* ritual song dance took place following prior arrangement. A performer delivered *Erungu* (club) to the ridge where the performance would take place and was received by another performer. The reception of Erungu signified acceptance of the request for a performance of the ritual, and preparation for participation got underway. Instruments accompanying the *Ribina* ritual included an *Ekonu* (drum) and an *Esirimbi* (whistle). Participants in the *Ribina* ritual dressed in *Chingobo* (clothes made of animal hides and skins) and decorated themselves with *Amandere* (wild seeds), *Chinchabo*, and *Chinchigiri* (folded metal or iron). The ritual dance took place in the open air and in circles. Two groups participated in the ritual dance. The inner ring stood facing the outer ring. The participants placed their left hands on the chest while the right was held by a partner. After singing, they jumped very high, nodding their heads. Every time jumping stopped, they took two steps backwards and then started jumping again.

Reciting a typical *Ribina* ritual song performed at the hallowed Manga hills of Nyamira County, informants indicates that the ritual dance was usually accompanied by the offering of sacrifices and tithes under the direction of community elders, spiritual leaders, and rainmaking specialists (*Abanyibi*). A typical *Ribina* song dance at the Manga shrine went thus (Table 6.1):

Table 6.1 A typical *Ribina* ritual song, sung at the Manga Shrine. *Source* Song based on field interviews and translated to English by the author

Ekegusii		English (Translation)[*]
1	*Amabera nigo arwera Manga*	Mercy has come from Manga
2	*Ee Amaya arure Manga*	Yes, good has come from Manga
3	*Eeee! Amaya arure Manga*	Yes! Good has come from Manga
4	*Omogunde tureti chia Nyakongo*	Thick clouds from the sides of Nyakongo
5	*Noo omogunde osoka*	That is where the clouds have appeared
6	*Enkanga yarerire Keera ime*	A goose has cried from Keera

6.4.5.2 Symbolic Importance of the *Ribina* Ritual

For efficacy, the convention of the *Ribina* rainmaking ceremony required the personal presence of the people who do the praying, and the spatial and physical location for the ritual. This is because different climatic regions might well require different intervention prayers. Thus, the ritual, the charismatic (suprahuman) leadership of the ritual, the people, the performers, the implements, and the spatial location become connected in a socio-cultural and religio-ecological complex. The connection relates to a feeling of attachment to the sacral quality of the whole complex of the ritual. The use of Manga escapement for the prayer ritual is itself a symbolic reclamation and anointment of the physical landscape into a symbolic sacral landscape.

In the first line of the stanza in the song, *Amabera nigo arwera Manga* (Mercy has come from Manga), the reverence the community has for the tallest Manga hill which is also considered the most outstanding and sacred geographical feature in Gusiiland is acknowledged. Therefore, God's mercies, in the form of rain, for His people, can only be experienced through the sacred Manga hill. '*Keera,*' in line six, is also a spectacular waterfall considered sacred within Nyamira County. The Abagusii believed that their God, Engoro, lived high above them but manifested Himself through various creations, waterfalls being among them. By supplicating God atop the hill, the performers were, in a way, drawing closer to God's mighty presence and also coming to a place where He could possibly be found. The participants placing their left hands on their chests as they performed implied the humblest submission to the creator and maker of humanity and nature.

The *Ribina* ritual like other indigenous rituals was a versatile form of communication; a powerful call for environmental harmony, unity and, sometimes, a vehicle to pre-empt chaos and conflict—both personal and interpersonal—among members of the community, in favour of an explicit order of existence free from evil. In the *Ribina* ritual the Abagusii communicated through symbols and 'word pictures' together with plain language. The images, drawn from traditional settings, were capable of communicating to *Engoro*, the creator of heavens and the earth, their everyday ecological experiences.

The *Ribina* ritual is about the omnipotence of God the Almighty. The ritual also recognises the Creator's supremacy through His extraordinary creation of nature. In the natural ecology of revered valleys and hills, the Abagusii saw God in His various creations. The people were clearly fascinated by God's creations and omnipotence. *Ribina* ritual evoked feelings of veneration for the highest power, particularly for His unlimited power symbolised in His awesome creations and providence. The *Ribina* ritual was performed as a symbol of the cultural, ecological, economic, political, and social order of the community. The *Ribina* ritual incorporated singing and dancing and served as a means of summoning God's benevolence during times of plenty, times of scarcity, and at times of drought and famine (*egeku*).

6.4.6 Cultural Ecology and the Utility of Rituals

Cultural ecology is the adjustment and adaptation of human societies or populations to their environments through cultural means. Ecology aims to understand the relationships between organisms and their wider environment. And this is achieved through making sense of peoples' modes of thought and by explaining particular cultural features and patterns which characterize different ecological zones. Emphasis is often placed on the "arrangements of technique, economy, and social organisation through which culture mediates the experience of the natural world" (Winthrop 1991: 47). As described in reference to the Abagusii community, diverse cultures across the world resort to the cultural resource of ritual to respond to instances of environmental instability within their localities, though the ritual details may differ.

Anthropologist Roy Rappaport (1968) reports an exemplar case of using ritual to respond to environmental instabilities, and to balance ecology for the benefit of humanity, similar to the *Ribina* ritual dance of the Abagusii, in a study of the *Kaiko* pig ritual of the Tsembaga Maring of New Guinea. Studying their animal husbandry practices with pigs, he found that pigs consume the same food as humans in the same environment. Therefore, the Tsembaga must produce a surplus to maintain their pig populations. Pigs are slaughtered for bride price and at the end of war. So, the pigs must be kept at exactly the right numbers. This is accomplished through a cycle of war, pig slaughter for ritual purposes, and re-growth of the pig populations.

As such, "indigenous beliefs in the sacrifice of pigs for the ancestors were a cognised model that produced operational changes in physical factors, such as the size and spatial spread of human and animal populations" (Netting 1996: 269). Thus, religion and culture (the pig ritual) are cybernetic factors that act as a gauge to assist in maintaining equilibrium within the ecosystem.

6.5 Relevance of African Indigenous Knowledge

Is African indigenous knowledge on climate change viable in the face of contemporary climatological science? The greater majority of people in Kenya and indeed the whole of the African continent, as noted by Kalungu et al. (2013), live in rural areas where they are particularly exposed and sensitive to climate change impacts due to their heavy dependency on natural resources for their livelihoods, such as rain-fed agriculture, and the location of their settlements in marginal environments. Their vulnerability is further compounded by many other stressors including extreme levels of poverty, a high pre-existing disease burden, limited technological capacity, inefficient governance, gender inequality, fragmented services provision, low levels of education and training, water and food insecurity, and frequent occurrence of conflicts, wars and natural disasters. They, for example, have little access to contemporary climate change information from the modern-day meteorological agencies to guide decisions on their livelihoods. This is partially explained by the minimal penetration of modern meteorological weather forecasting on which predictions on possible climate change are based in Africa, more so in rural areas. As Orlove et al. (2010) indicate, climate organisations usually operate at the national, regional and international levels, where indigenous knowledge is less easily accessed and more difficult to incorporate.

All these factors and many others, then, dictate that most people in Africa rely on local knowledge to sustain their livelihoods. The majority of them depend on their local understanding of times and climatic seasons which have worked for them for many years. They rely heavily on natural indicators in predicting heavy rains, long dry seasons or higher temperatures and upon these predictions, make decisions on agrarian and farming operations, particularly decisions concerning planting, weeding, ploughing, and harvesting. In the livestock sector, local knowledge allows the pastoralist economy to adapt to climate change vis-à-vis multiple elements comprising their ecology including soils, vegetation, and livestock, and in the process, make appropriate decisions on stock densities, landscape-grazing suitability, and landscape-grazing potential during the wet or dry grazing seasons. This complex local knowledge utilisation illustrates the suitability of indigenous knowledge in mitigating human livelihood adaptation decisions interfacing climate change, ecological variability, and systems of land use (Nakashima et al. 2012). It is, therefore, evident that indigenous knowledge is significant.

Indeed, indigenous knowledge is critical for local communities because it focuses on elements of significance for local livelihoods, security and well-being, and as a result is essential for climate change adaptation responses within their localities. In addition, indigenous knowledge is comparatively less expensive and more effective than other climate change adaptation responses such as providing aid for climate change impact-ravaged communities. IFAD (2016) identifies the advantages of local traditional knowledge as offering a very "cost-effective and reliable system," (p. 13) and the "most effective way to increase the resilience of landscapes and communities to climate change challenges" (p. 11). Equally, forms of indigenous knowledge are readily acceptable to the expected beneficiaries as they can easily identify and relate to it (Guthiga/Newsham 2011). Indigenous knowledge and wisdom which has been accumulated, preserved and transmitted in a traditional and inter-generational context over a long period is one resource that can be harnessed to complement contemporary scientific climatological knowledge. It is therefore, essential that indigenous climatological knowledge be identified, studied, documented and integrated into modern climatological science.

6.6 Positioning African Indigenous Knowledge in the Framework of Contemporary Science

The process of incorporating, mainstreaming, and integrating African indigenous knowledge into contemporary scientific knowledge may take many forms. However, the first step African governments must take is that of protecting indigenous knowledge from knowledge predators and plagiarism. This will take the form of patenting indigenous knowledge as used by local communities, thus allowing them exclusive property and industrial rights over the knowledge and officially acknowledging their originality and contribution to the advancement of knowledge, science, technology, and innovation.

This will often take the form of constitutional guarantees. As the Kenya constitution (2010) says in article 11(2) [b & c], the state must "recognise the role of science and indigenous technologies in the development of the nation [b]; and promote the intellectual property rights of the people" [c]. In addition, the state must take necessary measures to "protect and enhance intellectual property in, and indigenous knowledge of, biodiversity and the genetic resources of the communities" (Article 69(1) [c]). Thus, the first line of strengthening the relationship between modern scientific knowledge and indigenous knowledge in Africa is for governments to recognise and appreciate the inherent benefits of African indigenous knowledge by securing and anchoring their rights and values in sacred national documents, each country's constitutions. African governments can, then, synchronise the indigenous knowledge into local, regional, and national development policies, programmes, and projects.

African governments must also seize the opportunity African indigenous knowledge presents and make a strong case for its use in climate change and disaster risk reduction. They must recognise the need for strengthening indigenous knowledge in their development policies, programmes, projects and strategies and in adapting local livelihoods to the adverse impacts of climate change and climate variability.

In such circumstances, integrating indigenous knowledge into climate change policies, programmes and projects through scientific research and data would lay the foundation for resilient climate change adaptation. African governments must put in place mechanisms of mapping, locating and documenting indigenous knowledge, focusing on its efficacy and complementarity with modern scientific knowledge in climate anticipation, preparedness and forecasting. Such mechanisms must be founded on identifying, validating, and documenting indigenous climate change adaptation responses and weather prediction methodologies that communities have depended on for so long. All these measures and approaches must adopt a bottom-up approach, where communities with years of advanced local knowledge of adapting and responding to certain conditions, such as drought or flood, provide lessons and strategies to other communities. In this way, the potential of folk knowledge and actions will be recognised, shaped, and strengthened.

African governments must advocate for, showcase, and promote indigenous knowledge which have been assessed and proven to be applicable, through dissemination workshops, demonstrations, exhibitions, and conferences at the local, national, regional, and even international levels. Where need be, adjustments in specific African indigenous knowledge systems must be made to suit different contexts.

African governments must promote education, training and advocacy of indigenous knowledge in formal education by incorporating it into the educational curricular as part of the comprehensive climate change response programme from primary school to the university. This should also be complemented with adult education and advocacy programmes targeting and tailored to suit people and communities in their particular ecological zones.

Apparent encouraging trends have been noted at country, continental, and global levels. Universities have started to regard indigenous knowledge as an area worthy of investigation and study, and some like the University of KwaZulu-Natal in South Africa offer full-fledged degree programmes on African Indigenous Knowledge Systems (AIKS). In Kenya, some institutions of higher learning including The Great Lakes University of Kisumu have so far integrated traditional rainmaking in their teaching and research curriculum. Also, in a hybrid weather intelligence system in Kenya, modern meteorologists, local universities, scientists, and researchers partner with traditional rainmakers to conduct weather forecasts and study the adverse effects of climate change. At the international level, a partnership between the United Nations University Institute for Advanced Studies' Traditional Knowledge Initiative (UNUIASTKI) and the Intergovernmental Panel on Climate Change (IPCC) work together in organizing series of workshops that enable the expertise of indigenous people and knowledge on climate change become an integral part of IPCC and be made widely available to the global community.

6.7 Conclusion

This study has ethnologically historicised and brought into perspective a rich cultural heritage of resources used as climate change adaptation responses by the Abagusii community of southwestern Kenya as they struggle to secure their livelihoods. As a result, the study establishes that indigenous cultural resources represent a community's repository of knowledge and wisdom embodied in cultural beliefs, values, norms, songs, dance, rituals, sacrifices and other socio-cultural and economic-political activities and practices. As is the case with other indigenous knowledge in Africa, traditional knowledge on climate change is closely attached to people's culture and their belief systems concerning the land and its flora and fauna. These cultural resources exemplify the past, present, and future aspirations of a community, which it uses to fashion and refashion itself, and transmit and reproduce group practices. They are thus, effective ways of transferring appropriate cultural knowledge, information, and skills to the next generation. In that way, apart from formulating local knowledge, the cultural resources represent a form of media for communicating culturally significant, symbolic, and comprehensible information.

In addition, culture embraces the whole gamut of knowledge, skills and relationships by which people live in any organised society. A community's experiences, life styles, ecological, and socioeconomic conditions shape the worldviews and proclivities of its people. It is therefore, important to be cognisant of a community's cultural formulation in knowledge, experiences, and aspirations when formulating and innovating solutions to contemporary challenges. In that respect, climate change policies, programmes, and projects that fail to take into consideration the different ways in which communities indigenously experienced their past, as well as how they adapted to the corresponding challenges, are inherently bound to fail.

In conclusion, although indigenous cultures are faced with the possibility of extinction due to rapid cultural changes, there is the need for contemporary Africa to (re)examine and accept what indigenous culture offers as a form of information, knowledge, and data, and incorporate and adopt these for deliberate and effective livelihood solutions. As Dei (1994: 9) observes, "there are numerous aspects of Africa's indigenous traditions and collective historical past that can be recovered, reclaimed, and reconstituted by African peoples today as they struggle to reproduce their lives and livelihoods".

References

Abu-Zahra, N., 1988: "The Rain Rituals as Rites of Spiritual Passage", in: *International Journal of Middle East Studies*, 20(4): 507–529.

Akama, J. S.; Maxon, R., 2006: *Ethnography of the Gusii of Western Kenya: A Vanishing Cultural Heritage* (New York: The Edwin Mellen Press).

Babane, M. T.; Chauke, M. T., 2015: "The Preservation of Xitsonga Culture through Rainmaking Ritual: An Interpretative Approach", in: *Studies of Tribes and Tribals*, 13(2): 108–114.

Bell, C., 1997: *Ritual: Perspectives and Dimensions* (New York: Oxford University Press): 138–169.

Berkes, F., 2012: *Sacred ecology: Traditional ecological knowledge and resource management* (3rd ed.) (New York: Routledge).

Bogonko, S., 1977: *Christian Missionary Education and its Impact on Abagusii of Western Kenya 1909–1963* (Ph.D. thesis, University of Nairobi).

Collins, R., 2004: *Interaction ritual chains* (Princeton, NJ: Princeton University Press).

Dawson, C. A. (Ed.), 2009: *Shrines in Africa: History, Politics and Society* (Africa, missing voices series) (Canada: University of Calgary Press).

Dei, G. J. S., 1994: "Afrocentricity: A cornerstone of pedagogy", in: *Anthropology & Education Quarterly*, 25(1): 3–28.

Dornan, S. S., 1928: "Rainmaking in South Africa", in: *Bantu Studies*, 3: 185–195.

Evans-Pritchard, E. E., 1938: "A note on the rain-makers among the Moro", in: *Man*, 38: 53–56.

Firth, R., 1957: *Man and Culture. An Evaluation of the Work of Bronislaw Malinowski* (London: Routledge and Kegan Paul).

Guthiga, P.; Newsham, A., 2011: "Meteorologists Meeting Rainmakers: Indigenous Knowledge and Climate Policy Processes in Kenya", in: *IDS Bulletin* 42(3): 104–109.

Gyekye, K., 1997: *Tradition and Modernity: Philosophical Reflections on the African Experience* (New York: Oxford University Press).

Horton, R., 1971: "African Conversion", in: *Africa*, 41: 85–108.

IFAD (International Fund for Agricultural Development), (2016). The Traditional Knowledge Advantage: Indigenous peoples' knowledge in climate change adaptation and mitigation strategies. *IFAD Environment and Climate Division*.

Kalungu, J. W,; Filho, W. L,; Harris, D., 2013: "Smallholder Farmers' Perception of the Impacts of Climate Change and Variability on Rain-fed Agricultural Practices in Semi-arid and Sub-humid Regions of Kenya", in: *Journal of Environment and Earth Science*, 3(7): 129–140.

Krathwohl, D. R., 1993: *Methods of Educational and Social Science Research: An Integrated Approach* (2nd ed.) (New York: Longman Publishers).

Lesser, A., 1985: "Functionalism in Social Anthropology", in: Alexander Lesser (Ed.): *History, Evolution, and the Concept of Culture, Selected Papers* (Cambridge: Cambridge University Press).

Linehan, D.; Sarmento, J., 2011: "Spacing Forgetting: The Birth of the Museum at Forth Jesus, Mombasa, and the Legacies of the Colonization of Memory in Kenya", in: Meusburger, P.; Heffernan, M.; Wunder, E. (Eds.): *Cultural Memories: The Geographical Point of View. Knowledge and Space*, 4: 305–325.

Malinowski, B., 1954: *Magic, Science and Religion, and Other Essays* (Garden City, N.Y.: Doubleday).

Malinowski, B., 1944: *A Scientific Theory of Culture and Other Essays* (Chapel Hill: University of North Carolina; York: Roy Publishers).

Malinowski, B., 1926: *Myth in Primitive Psychology* (New York: W.W. Norton).

Monyenye, S., 1977: *The Indigenous Education of the Abagusii* (MA thesis, University of Nairobi).

Morris, B., 2006: *Religion and Anthropology: A Critical Introduction* (New York: Cambridge University Press).

Nakashima, D. J.; Galloway McLean, K.; Thulstrup, H. D.; Ramos Castillo, A.; Rubis, J. T., 2012: *Weathering Uncertainty: Traditional Knowledge for Climate Change Assessment and Adaptation* (Paris, UNESCO; Darwin: UNU).

Nche, G. C., 2014: "The Impact of Climate Change on African Traditional Religious Practices", in: *Journal of Earth Science & Climatic Change*, 5(7): 1–5.

Netting, R. M. C., 1996: "Cultural Ecology", in: Levinson, David; Ember, Melvin (Eds.): Four Volumes: *Encyclopedia of Cultural Anthropology* (New York: Henry Holt): 267–271.

Ochieng', W. R., 2004: *A Pre-colonial History of the Gusii of Western Kenya: From C.A.D. 1500–1914* (Nairobi: East African Bureau).

Omosa, M., 2007: "Incorporation into the Market Economy and Food Security among the Gusii: Paradise Lost or Paradise Gained?", in: *African Journal of Sociology* 6(1): 28–60.

Ontita, E. G., 2007: *Creativity in Everyday Practice: Resources and livelihoods in Nyamira, Kenya* (Unpublished PhD Thesis, Wageningen University, Netherlands).

Orlove, B.; Roncoli, C.; Kabugo, M.; Majugu, A., 2010: "Indigenous climate knowledge in southern Uganda: the multiple components of a dynamic regional system", in: *Climatic Change*, 100(2): 243–265.

Rappaport, R. A., 1999: *Ritual and Religion in the Making of Humanity* (Cambridge: Cambridge University Press).

Rappaport, R. A., 1968: *Pigs for the Ancestors: Ritual in the Ecology of a New Guinea People* (New Haven, Connecticut: Yale University Press).

Roncoli, C.; Crane, T.; Orlove, B., 2009: "Fielding climate change in cultural anthropology", in: Crate, Susan A.; Nuttal, Mark (Eds.): *Anthropology and climate change. From encounters to action* (Walnut Creek: Left Coast Press): 87–115.

Sanders, T. D., 2002: "Reflections on Two Sticks: Gender, Sexuality and Rainmaking", in: *Cahiers d'Études africaines,* 166, XLII(2): 285–313.

Sanders, T. D., 1997: *Rainmaking, Gender and Power in Ihanzu, Tanzania, 1885–1995* (Unpublished Ph.D. thesis submitted to the Department of Anthropology, London School of Economics and Political Science, University of London).

Summers-Effler, E., 2006: "Ritual Theory", in: Stets, J. E.; Turner, J. H. (Eds.): *The Handbook of the Sociology of Emotions* (New York: Springer).

Turner, W. V., 1973: "Symbols in African ritual", in: *Science* 179(4078): 1100–1105.

Whitehead, T. L., 2004: *What Is Ethnography? Methodological, Ontological, and Epistemological Attributes* (EICCARS Working Paper Series; University of Maryland, CuSAG).

Winthrop, R. H., 1991: *Dictionary of Concepts in Cultural Anthropology* (New York: Greenwood Press).

Other Literature

Change, I., 2014: *Climate change*; at: https://camelclimatechange.org/index.html (17 September 2016).

CoK (Constitution of Kenya), 2010: Published by the National Council for Law Reporting with the Authority of the Attorney General; at: http://extwprlegs1.fao.org/docs/pdf/ken127322.pdf (2 April 2016).

Eakin, H.; Walser, M. L., 2007: "Human Vulnerability to Global Environmental Change", in: Cleveland, C. J., (Ed.): *Encyclopedia of Earth*; at: https://editors.eol.org/eoearth/wiki/Human_vulnerability_to_global_environmental_change (18 September 2016).

Steiner, A., 2008: *Indigenous knowledge in disaster management in Africa.* United Nations Environment Programme (UNEP) (2 April 2017).

United Nations Framework Convention on Climate Change (UNFCCC), 2014: *"FOCUS: Adaptation".* (18 September 2016).

Chapter 7
Violent Gender Social Representations and the Family as a Social Institution in Transition in Mexico

Serena Eréndira Serrano Oswald

Abstract The Anthropocene has led to significant discussions on the emergence of a new era in human and Earth history, with all its implications. In it, human beings are at the same time the main threat to the Planet and the potential solution, leading to discussions in the social sciences and the humanities addressing the societal consequences of complex interrelations between global environmental change, lack of sustainable development, poor governance, inequality, social challenges, economic crises and risk society. In the midst of these changes and debates, social relations, social dynamics and social institutions have also changed significantly and at a very rapid pace, reflecting changes over the past decades. The family, considered the basic institution of society, is also a historically-bound institution, based on violent dynamics of gender domination, exclusion and subordination in patriarchal societies, that has changed across time and space.

In the past five decades, the period of most intense anthropogenic activities, it is one of the main social institutions that has experienced very visible changes which are redefining social knowledge, social relations and identities in multiple ways. It is not that the family has changed because of the Anthropocene, but rather that broad societal changes reflected in the Anthropocene have also impacted on the family. *Social Representations Theory* (SRT) is an epistemological, theoretical and methodological perspective that has been evolving since the 1960s and that deals with common-sense knowledge, a way of making the unfamiliar familiar, understood as the link between knowledge and practice, and practice and knowledge in everyday life. It also looks into the way in which scientific and expert knowledge is accommodated in lay people's lives on a quotidian basis.

This chapter, based on the linkages between gender and social representations studies, develops a theoretical-conceptual framework to investigate the transitions, challenges and continuities of the family as institution in the current *époque*, in the specific case of Mexico, especially following technological advances and legislative changes that have polarised the public.

Associate Professor at the Regional Multidisciplinary Research Centre (CRIM), National Autonomous University of Mexico (UNAM). Email: sesohi@hotmail.com.

© Springer Nature Switzerland AG 2019
H. G. Brauch et al. (eds.), *Climate Change, Disasters, Sustainability Transition and Peace in the Anthropocene*, The Anthropocene: Politik—Economics—Society—Science 25, https://doi.org/10.1007/978-3-319-97562-7_7

It unfolds in three main sections. First, in the Introduction, there are four inter-related subsections dealing with the general historical and conceptual framework, namely (i) the current historical *époque*: the dawn of the Anthropocene (7.1.1); (ii) the theoretical and methodological model: Social Representations Theory (7.1.2); (iii) the object of study: the family as social institution (7.1.3); and (iv) the context of the study: social and gender violent dynamics in Mexico (7.1.4). The second main section addresses the family, social representations and gender; it presents the theoretical-methodological model. The third and last section is thematic, looking at assisted reproductive technologies and 'homoparenting' (gay and lesbian parenting) in Mexico. There is a brief closing reflection at the end of the chapter.

7.1 Introduction: The Historical and Conceptual Framework

7.1.1 The Current Historical Époque: The Dawn of the Anthropocene

The Anthropocene, a concept popularised by Nobel laureate Paul Crutzen, has generated a field of studies where the influence of human activity on the earth system and nature, since the Industrial Revolution and especially in the past fifty years, is at the same time the main threat and potential solution. It has led to significant discussions on the emergence of a new era in human and earth history, with all its implications, polarisations and contradictions. Although societal debates surrounding the Anthropocene are recent and they have been centered on environmental aspects, the Anthropocene has become a rapidly growing field of social scientific enquiry, incorporating divergent theoretical, conceptual and methodological tools. Currently, it attracts researchers looking into the societal consequences of complex interrelations between global environmental change, lack of sustainable development, poor governance, inequality, social challenges, economic crises and risk society. In the midst of these changes and debates, social relations, social dynamics and social institutions have also changed significantly and at a very rapid pace, reflecting changes over the past decades. Hence, looking at the changes in what has been commonly referred to in the literature as the 'basic social institution', the family becomes pertinent. Although the changes and adjustments in the family as institution do not directly stem from the Anthropocene, they have nevertheless to do with the functioning of society as a whole in this particular historical period, with its epistemological, theoretical, conceptual and practical innovations and contradictions. Although it may be early to look at the direct social and societal impacts of the Anthropocene on the family, this is a field of studies that is just unfolding, and looking at it from the lens of the social changes, challenges and characteristics of this present era captures the pertinence of undertaking a more systematic, although initial, reflection surrounding the shifts in the family as institution in the Anthropocene epoch. Given the very wide scope of such an objective, the reflection must be limited to generating a theoretical and conceptual framework, based on the pertinence of Social Representations Theory, in order to look at two of the most relevant changes in the family as institution in the context of Mexico in the current era of the Anthropocene.

7.1.2 The Theoretical and Methodological Model: Social Representations Theory

Social Representations is an epistemological, theoretical and methodological perspective that has been evolving since the 1960s that deals with common-sense knowledge, a way of making the unfamiliar familiar, understood as the link between knowledge and practice, and practice and knowledge in everyday life. It also looks into the way in which scientific and expert knowledge is accommodated in lay people's lives on a day-to-day basis. A field of social studies, it has evolved over five decades and is based on the seminal work of Serge Moscovici, who was director of the European Laboratory of Social Psychology, and was distinguished with numerous prizes, including sixteen honorary doctorates from around the world.

Although initially rooted in Social Psychology, Social Representations Theory is one of the main critical perspectives of social thought of French origin, which has established itself in Europe, America, Australia and Asia (Wagner et al. 1999; Wagner/Haycs 2005; Marková 2017). It has led to important developments in sociology, public health, social identity, education, gender studies, culture, political science, anthropology, history, media research, the study of emotions, systemic theory, and, increasingly, environmental studies. Given that the focus is common-sense knowledge, the practices of everyday life, and their co-creation and evolution, Social Representations Theory provides a relevant theoretical-methodological model for addressing the changes and continuities of the family as social institution. This chapter focuses on our current historical context and the case of Mexico, but as research model and reflection, although it might be very relevant and useful as a lens for other contexts.

7.1.3 The Object of Study: The Family as Social Institution

The family is one of the oldest and most relevant social institutions. It has always been historically evolving, with periods of rapid changes and readjustments, such as the transition from nomadic to agricultural societies and more recently to industrial and information societies.

Feminism as a broad spectrum of theorisation argues that the family is a socially ordered political institution linked to sex, production, and reproduction. It is part of the political realm and should be subject to principles of justice (Satz 2017).

According to the Human Rights Charter, "the family is the natural and fundamental group unit of society and is entitled to protection by society and the State" (UN 1948, Article 16).

Historically, the family was naturalised as it was linked to the sexual division of labour in which female and not male bodies gestate. This was supported by social, religious and political institutions. Organic and naturalistic explanations of the family have led today, following the linguistic turn, to multiple contextual and biopsychosocial models (Falicov 2006). Nevertheless, the raising of children, care and parenting can be undertaken by anybody. Over the past decades, family structure, composition and meaning have changed rapidly. Currently, single-parent households headed by women and men evidence this. In Mexico in 2010, 84% of single-headed households were female-headed and 16% were male-headed. In 2017, single-headed female households represented 27.4%, a third of all households (INEGI 2017). Social and historical constructions of gender make these female-headed households more vulnerable to higher rates of poverty and precariousness. Although there are natural differences between women and men, debates consider the fact that culture is biologically imprinted as well as biology being culturally determined and that differences do not justify social injustice, namely, social structures of oppression, invisibilisation, alienation, and hierarchisation of people based on gender (Scott 1986; Lagarde 2001, 2004; Lamas 1996, 2002).

The family is one of the social institutions that has seen very visible changes in function, structure and process, and these changes are redefining social knowledge, social relations and identities in many ways. In the context of Mexico, changes over the past fifty years include increased ages at marriage, number of children, use of contraception and spacing of children, couples living together without marrying, increasing rates of divorce, the changing role of women and motherhood, increased interest in parenting by men, higher life expectancy and prolonged life family cycles, urbanisation of family life, education, precarisation of younger generations and the impossibility of their leaving their household of origin, teenage pregnancies, infertility, care for elders and their participation in the economic life of the family, use of technology and information, migration, changing sexual rules, changes in authority and limits, the balance between the individual and the collective, reconstituted families, family services and recreation, organised crime and family life, the types of family ties and relations, the legal acceptance of same-sex families, and the use of human reproductive technologies, amongst others.

Amongst these changes, two of them are of special relevance when considering the changing face of the family as social institution, given their disruptive potential in terms of traditional and naturalistic social representations of the family, of womanhood and maleness and their complementarity. The reproductive function and organisation of the family has shifted, as many families are not interested in reproduction and have or raise no children. However, for those families that accord pre-eminence to reproduction, the setting has rapidly shifted. First, although reproduction remains linked to female bodies, as infertility rates have increased, reproduction often takes place in the laboratory and often gestation takes place outside the body of one of the family members following a transaction for reproductive services. Secondly, the complementary naturalised heterosexual binomial male-female of the 'natural family' has started to crumble, as same-sex marriage regulation and 'homoparenting' (gay and lesbian parenting) have been accepted by the state. Today, there are also non-heterosexual families bearing and raising children.

This chapter uses gender and social representations studies to develop a theoretical-conceptual framework for investigating the transitions, challenges and continuities of the family in the Anthropocene era, focusing on Mexico, especially following technological advances and legislative changes that have polarised the public sphere in terms of emerging forces and traditional resistances.

7.1.4 The Context of the Study: Social and Gender Violent Dynamics in Mexico

This chapter discusses violent gender social representations in relation to the family as institution, given the different levels of family that are part of the context where social representations originate, are accepted or challenged, are put in practice and are transformed. The first and deepest axis of violence is formed by the violent social representations that stem from the hegemonic gender system and that are expressed and reproduced and are very gradually being challenged in everyday life by cis and trans women and men, by the way the family has been historically constructed within the patriarchal order, and by the way these representations impact on and shape the modern and late modern age. This was discussed in the previous section and will be further explored throughout the text. Violent gender social representations are embodied representations, structuring social institutions and relations, meaning that it is very hard to challenge them, since they operate in a hegemonic dimension. Gender violence, power and domination, and their relation to the family as institution, have been widely addressed by the social sciences, although they have scarcely been addressed in social representations literature (Serrano 2013). Gender violence refers to different kinds of knowledge and practices that are based on the sex-gender system, including the following types of gender violence: physical, psychological, legal, political, economic, sexual, patrimonial, family or domestic, social, community, work, school, institutional, obstetric, against reproductive rights, moral, in mass media representations, femicides, etc.

A second axis of violence, whose identification was derived from feminism, comprises intersectional forms of structural violence. These are other forms of violence that intersect with gender-based violence, such as violence linked to social class, race, ethnicity, religion, age, ability, education, sexuality, etc. For example, although I refer to the Anthropocene era, there are references to modernity and late modernity, given the complex and contradictory characteristics and dynamics that shape everyday life in Mexico, where one finds contexts that are 'pre-modern' at

best. Structural inequalities entail violence expressed as sharp differences in terms
of who has access to which kinds of experiences, in a context where the most
developed and the most precarious and marginalised quotidian dimensions coexist
hand in hand. To name but a few structural indicators to make intersectional types
of violence that impact on family groups and individuals in the Mexican context
more explicit: according to the OECD *Economic Outlook for Latin America 2018*,
seven out of every ten Mexicans are living in poverty or vulnerability, while the
wealthiest twenty per cent of the population earns ten times as much as the poorest
twenty per cent; 57% of workers are employed under informal agreements, and
52% of Mexicans live below the poverty line in urban areas and 40% in rural areas
(Gurría, 13 January 2018). The proportion of Mexicans that can meet their
socio-economic needs above the well-being line is only 19.8%. According to
Oxfam data for 2014, ten per cent of the population holds 64% of the wealth, whilst
one per cent of the population holds 21% of the wealth. The gender pay gap in
Mexico is between thirty and forty per cent (Serrano 2016: 9). In terms of generation change, prospects are dim. According to the OECD, Mexico has the highest
rates in the world for sexual abuse of, physical violence towards and homicides of
children under 14 years of age, seven out of ten suffer some form of violence, and
only two per cent of cases are formally reported (Cámara de Diputados, 17 May
2018). In terms of extreme forms of gender violence, femicides have made the
country famous since the murders in Ciudad Juárez in the 1990s. The 'femicide
pandemic' has spread throughout the country. The annual number of femicides rose
from 1,485 in 1985 to 12,811 cases in 2017 of women's deaths that are presumed
femicides (ONU/INMUJERES 2017; ADN 40 28 December 2017).

Lastly, over the past decade at least, the country has been undergoing a process
of political and social violence derived from the so-called "war on drugs and
drug-trafficking", "strategy versus drugs" or "fight against organized crime",
including homicides estimated at between 60,000 and 150,000 deaths, in addition to
forced migrants, 'disappeared' and displaced people. This is in addition to the
challenges posed by the Anthropocene, and exacerbates them at the local level. It
has severely impacted on everyday life and family dynamics in Mexico, since most
families are related to at least one victim. According to Clara Jusidman (Centro
Tepoztlán 25 November 2017) there have been 310,000 homicides, 300,000 displaced persons, and 30,000 disappearances in Mexico, as well as half a million
people directly or indirectly linked to illegal activities. Linked to gender violence,
femicides in Mexico have more than doubled since 2007, with at least seven women
victims of gender-relating killings every day in 2016 (UN 29 November 2017). This
takes place in a country with a weak institutional structure, with high levels of
corruption and impunity, and where the rule of law is highly inefficient and needs to
be strengthened, meaning that both societal violence as well as ways of resisting
and alternatives to this violence are nurtured and reproduced in everyday life by lay
people.

Given the scope of this chapter, it is not possible to go into the discussion of
structural and gender violence in Mexico in depth. That would require a few books
by itself. The aim of this chapter is to develop a theoretical and methodological

model, based on Social Representations Theory and Gender, in order to study the transitions in the family as social institution in the light of homoparenting and human-assisted reproductive technologies. Nevertheless, it is important to provide at least a very general overview of the contextual violent dynamics where these gendered social representations are put into practice, resisted and transformed in everyday life. The core of the argument and contribution will be developed in the ensuing sections.

As noted above, this chapter is organised in three main sections, giving it contextual, theoretical and thematic coherence. Following this introduction, a theoretical-methodological section developing the model follows, and the chapter closes with a brief thematic zoom looking at homoparenting and human-*assisted reproductive technologies* (ARTs) in Mexico. Although the chapter does not provide research results, its contribution is to provide a social research framework that will prove useful for studies of the family and of gender and will be a solid basis for looking at diverse social aspects of the current era of human history called the Anthropocene.

7.2   The Family, Social Representations and Gender

It is following social representations that individuals and groups are situated in the social field (Duveen/Moscovici 2000), since that is where the dialectic relation between emerging and existing social representations takes place (Marková 1982, 2003). Just as new knowledge and practices are integrated in the repertoire of existing social representations (SRs), in the same way, existing knowledge is transformed in the light of its interaction with novelty. It is in this symbolic and relational space that gender socialisation processes take place (Duveen 1993, 1997, 2001; Duveen/Lloyd, 1986, 1990; Lloyd/Duveen 1992), as well as social comparison and esteem (Howarth 2002a, b). Besides, late modernity with its reflexivity (Giddens 1991), its liquidity (Bauman 2000, 2006), its risks (Beck 1998), and its unprecedented level of communicational and technological exchanges, forms the political field in which these processes come into contact. It is in this context, the so-called detraditionalised public sphere (Jovchelovitch 2007), that the processes of 'sliding' of representations and 'revolution of the social imaginary' (Arruda 2002) take place. According to Arruda, late modernity constitutes one such moment of rupture, in which changes at macro-systemic level, revitalising social life, generating contradictions and where exchanges with others imply such differences in identity terms, that greater questioning and dissent emerge, enabling us to see more clearly the gradual transformation of identities and social representations at their core. It is thus that this approach is rooted in the socio-genetic tradition of social representations.

Gender is a social category, an analytical concept, as well as a relational research perspective and a methodological tool (Velázquez 2014) derived from feminist political theory, which sees gender as a major axis of inequality. It implies a research stance as well as a political agenda (Phillips 1996) that enables us to make visible, analyse, historicise, dismantle, deconstruct and transform the hetero-cis-patriarchal hegemonic gender system in favour of substantive equality and social justice for all human beings (Lamas 1996, 2002; Lagarde 1990, 2001; Velázquez 2014; Squire 2000; Facio 1999).

The hegemonic gender system and its structural violence, also called patriarchal order or patriarchal ideology, refers to the coherent system of social representations and behavioural expectations based on the social construction of sexual difference structuring world outlooks, as well as prescriptions in terms of identities, ways of relation and association between genders (for example, establishing kinship networks, family and marriage, bonds, etc.). For Lagarde (1990: 15) it means "the harmful, destructive, oppressive and alienating aspects that result from the social organisation based in inequality, injustice and the political ranking of people based in gender". It can be made up of explicit norms and sociocultural constructions as well as an 'invisible web' (Walters et al. 1996) of expectations and sanctions.

The sex-gender system is "the set of arrangements by which a society transforms biological sexuality into products of human activity and in which these transformed sexual needs are satisfied" (Rubin 1998 [1975]: 37). As a social and cultural construction, the sex-gender system is a product of sexuality and not the other way around, it is not a natural sexuality that inevitably and ahistorically establishes identities and relations. According to Witting (1980), gender and sex are sociocultural constructions that have no existence before the social. They are historical, political, economic, cultural, relational, juridical, and sexual categories that are imposed in the subjective, practical and material realms on individual and collective subjects and are anchored as binaries in a logic of natural sexuality.

Gender identities, "feminine and masculine are not natural or given by biology, but must be constructed and should be understood therefore as cultural achievements" (Moore 1994: 42). From the standpoint of social representations theory with a historical, cultural and transverse gender lens I speak of the hetero-cis-patriarchal sex-gender system. Nevertheless, "facts are stubborn" according to Françoise Héritier, "'observation of the difference between the sexes underlies all thought, traditional as well as scientific': 'sexual difference and the different role of the sexes in reproduction' this is the 'ultimate limit of thought'" (Héritier, cited in Fassin 2005: 62–3).

The term patriarchy comes from the word 'patriarch', in Greek *patriárchees*, which is at the origin of terms such as parent, patrimony, patron, parricide, etc.; and *archee*, being the first and thus power, commanding others, such as the ruling father. Since the eleventh century it has also directly referred to a territory and to the government of the patriarch, called patriarchy. As a social system linked to kinship, patriarchy is defined as "a primitive social organisation in which authority [unique and absolute] is exercised by the male head of the family, spreading this power even over distant relatives of a same lineage" (Alonso 1982: 3177, in Lagarde 1990: 87).

In the formation process of the modern state in the seventeenth century, according to contractual theory, the 'patriarchal pact' systematically excluded women—the second sex, the feminine sex—from the political realm, from citizenship and ownership (Pateman 1991) subjugating women to the guardianship of the male household head, father, brother or spouse. As a social and ideological gender system, patriarchy implies the subordinate condition of the feminine to the masculine. In a patriarchal world, women are oppressed precisely because they are women, that is their gender condition under patriarchy. This corresponds to the historical construction of the feminine as the reproductive, emotional, undesirable, weak, passive, etc. that reflects nature and corresponds to the private realm. In contrast, the masculine is constructed as productive, rational, stable, strong, active, etc., which corresponds to culture and the public realm (Ortner 1974). Thus, through essentialism, dichotomisation and hierarchies masculine domination is enabled and legitimised as natural (Bourdieu 2002).

Currently, patriarchy, together with capitalism, has some central characteristics affecting women in their condition as gender subjects: (i) patriarchy or 'fraternity', according to Pateman (1991), subjects women to the authority of men in the domestic realm as housekeepers as well as in the public sphere as political subordinates; and (ii) capitalism subjects women and men as labour force to the owners of the means of production, communication and capital, predominantly hegemonic men (Burr 1998).

Dictionaries define 'family' as (amongst other things) "group of individuals that are related and live together", "usually under one head", "group of persons of common ancestry", "group of ascendants, descendants, collaterals who are part of a lineage", "sons or descendants" (RAE 2016; Merriam–Webster 2018). The family constitutes the basic institution of social life; its structure and modes of relation have been formed and transformed throughout history and geography. It is in the family that cultural transmission, material reproduction and differentiation take place. It is especially relevant for gender social representations given that, following Lévi-Strauss (1956), although the family is common and part of everyday life, human organisation in families following the universal prohibition of incest together with the sexual division of labour are two aspects that enable moving from the naturalist and animal outlook towards culture that is distinctively human.

It is in the family where institutionalisation and naturalisation of the models of femininity and masculinity take place, rooting sexual differentiation, gendered relationships (marriage), kinship and parentage, transfer of capital, extended exchanges inside and between groups, as well as the sexual division of work sexuality and reproduction. "The sexual division of labour is nothing more than a device to institute a reciprocal state of dependency between the sexes". The same could be said of the sexual aspect of family life. Although "the family cannot be explained on sexual grounds … sexual life and the family are by no means as closely connected as our moral norms would make them, there is a negative aspect that is much more important: the structure of the family, always and everywhere, makes certain types of sexual connections impossible, or at least wrong" (Lévi-Strauss 1956, in Ksenych/Liu 2001: 322–23).

Despite the importance of the work of Lévi-Strauss in questioning the hegemonic family as part of the natural order, championed by pioneer feminists such as Simone de Beauvoir, the anthropologist Gayle Rubin produces a strong critique since she considers that the jump to 'culture' in the sex-gender system disempowers and oppresses women, making the sex division natural and strengthening the dichotomisation of feminine and masculine identities, at the same time as it institutes compulsory heterosexuality and imposes a taboo on homosexuality, reinforcing the relations of patriarchal exchange of women—the relations of marriage and kinship between groups—rooting and justifying the practices of control and traffic of women in a system of political economy of sex. According to Rubin (1975: 179):

> gender is a socially imposed division of the sexes. It is a product of the social relations of sexuality. Kinship systems rest upon marriage. They therefore transform males and females into 'men' and 'women', each an incomplete half which can only find wholeness when united with the other. Men and women are, of course, different. But they are not as different as day and night, earth and sky, ying and yang, life and death. In fact, from the standpoint of nature, men and women are closer to each other than either is to anything else—for instance mountains, kangaroos, or coconut palms … Far from being an expression of natural differences, exclusive gender identity is the suppression of natural similarities. It requires repression: in men, of whatever is the local version of 'feminine' traits; in women, of the local definition of 'masculine' traits. The division of the sexes has the effect of repressing some of the personality characteristics of virtually everyone, men and women.

In Élisabeth Roudinesco's (2003) history of the family in the West and in the history of Western marriage by Stephanie Coontz (2006), it is clear that the models of gender-sex relation and of family organisation are mutually reinforcing binomials with an exclusionary logic that tend to reduce reality to a radical opposition based on which ideological[1] structures—of knowledge and practice—are constituted, reproducing inequalities based on gender despite exhibiting diverse manifestations, even if the contents and forms of organisation of gender-sex relations and family organisation have changed enormously throughout history. Currently, it is the state, at national and international level through multilateral treaties, that regulates marriage and the family institution, and the market organises production, distribution, and access to goods and services, a function that was fulfilled by the institution of marriage. Ways of knowledge and different types of practices are and have historically been beyond the regulation of the state or recognised social institutions. Civil society organises in order to put forward agendas to be considered by the state and the market. In the present, in the face of changes, it is useful to consider that "from different standpoints, feminism poses the need to revise the heterosexual matrix implied by gender and the effects of the naturalisation of masculinity and femininity in people's ontologies, as well as the regulation that is implied by its form of association, namely the family" (Mogrovejo 2015: 150).

In order to have a more systemic and complete outlook, it is useful to consider the classical distinction that disciplines such as Anthropology on the one hand and

---

[1]Ideology is not understood as false consciousness, but rather as "a coherent system of beliefs that guides peoples to concrete ways of understanding and valuing the world, providing a base for the evaluation of conducts and other social phenomena, and suggesting appropriate behaviours and responses" (Facio 1999: 3).

Sociology, Psychoanalysis and History on the other make of the family as an object of study (Roudinesco 2003, Chap. 1). Sociology, Psychoanalysis and History privilege the *vertical* study of the family, addressing topics such as filiation, generation and genealogy, continuities and distortions, transmission of material and symbolic goods among generations), whereas Anthropology privileges the *horizontal* study of kinship, for example diverse cultural ways of constituting a family, relations of exchange, mutual recognition and circulation of women and goods based on marriage, family structures and alliances, and prohibitions, as well as the conformation and transmission of cultural traditions and changes.

The theoretical and methodological perspective of Social Representations is useful in exploring knowledge and practices, as well as the structures and transformations of everyday life and common sense. Thus, researching the family from this standpoint is useful, given that as a social institution, the family is put into practice following common sense. Gender identities and relations are naturalised, legitimised and reproduced by the hetero-cis-patriarchal gender-sex system given common sense and in the quotidian dimension of the family. Social Representations Theory has linked to diverse theories and referents, each with very distinct methods, research interests and repercussions. Over time, it has evolved into diverse approximations (anthropological, interpretative, dialogical, structural and standpoint) and three main schools: the structural, the sociological or socio-structural and the socio-genetic school of social representations. This paper is situated in the socio-genetic or so-called 'French' school of Social Representations, following the lead of the work by Moscovici and Jodelet that focuses on looking at social representation both as content and as process.[2] By formulating the current model from the Socio-genetic School, it is possible to respond to the three questions—dimensions of representations—put forward by Jodelet (1989): (i) who knows and from where does one know?; (ii) how and what does one know?; (iii) about whom and with which effect does one know?, as well as including Jovchelovitch's (2007; 2001) (iv) why and what for? In this way, it is possible to relate together the current modern and late-modern transformation of common sense and everyday institutions such as the family.

---

[2]Although the study of social representations of the family as institution in transition in Mexico could also be addressed following the 'Structural' or the 'Sociological' school, it is important to note that the present model, given its objectives, is rooted in the 'Socio-genetic' school, enabling us to look at social representations in terms of both content and process, looking at the process though which they are formed, transformed and circulate, without obliterating their internal structure or the societal conditions that give rise to them. This is because undertaking a 'Structural' analysis would use multivariate quantitative methods in order to look at the representation as a heterogeneous product, as something already constituted, in order to analyse its internal structure (its nucleus and periphery), its stability, defensive system, themata and coherence. It would seek to respond to Jodelet's (1989) question "how and what does one know?" On the other hand, a 'Sociological' analysis would emphasise the specific conditions that determine social representations, their production and circulation, as well as the interactions and group dynamics following them, using predominantly quantitative methods, centred on answering Jodelet's (1989) question "who knows and from where does one know?".

Besides, the conditions for the emergence of a social representation are the existence of three main components that are present in the case of the family as a shifting institution in the current historical period: (i) *information*, the quality and quantity of knowledge linked to a social object; (ii) the *representational field* which expresses the organisation and content of a representation, the richness of its internal components and their hierarchical organisation, and (iii) the *attitudes* expressing orientation towards a social object of representation, enabling both consensus and dissent, as well as a variety of postures in a complex and evolving social field.

Also, in order to understand changes in representations such as the ones relating to the family as institution, it is useful to highlight the fact that there are three types of representations: hegemonic, emancipated, and polemic. Citing the definitions provided by Ben-Asher (2003: 6.3–6.4): (i) hegemonic representations may be uniform or coercive, they are "shared to some extent by all members of a society and signify the societal identity, allowing very few degrees of freedom on the individual level"; (ii) emancipated representations linked to subgroups are "distinctively constructed information by small sections of a society, which are not yet incompatible with the hegemonic representation. These representations are constructed when members of a society are differentially exposed to new information and consequently reflect differences between individuals or subgroups within a broad identity group"; whereas (iii) polemic representations construct new action scenarios, emerging from social conflicts as means of resistance or acceptance; they are "formed by subgroups in the course of a dispute or social conflict when society as a whole or the social authorities do not necessarily share them. They express rivalry or incongruity between representations" (Ben-Asher 2003: 6.4). According to Rodríguez (2007: 178), a single social representation may have contents or meanings that are hegemonic, emancipated and polemic. This is the case with the social representation of the family in the current era of the Anthropocene. There are hegemonic contents that are collectively shared and legitimised, derived from religious and mainstream biological beliefs, naturalising the social institution, which are only very gradually starting to be questioned. New reproductive technologies with their potential for (so far) up to six parents for a single child, as well as homoparenting laws that question dualistic male-female heteronormative cis-gender parenting, are very gradually presenting new forms of knowledge and practices. These new forms of social knowledge and practices are more commonly accepted in certain social groups and communities (emancipated social representational contents), at the same time as they are resisted and opposed by more traditional groups. Lastly, the public arena has become a space for open tension, for questioning and challenging of the legitimacy of traditional and emerging social representations regarding the family as institution, generating doubts, critiques, standpoints, acceptance, defiance, negotiations, and overt and covert forms of violence, as well as relativisms that are also characteristic of the Anthropocene. As Rodríguez states, "in any modern society, social actors—whether they are individuals or groups—are exposed to an impressive amount of cultural contents that are contradictory, imprecise and which express visions of socially distinct groups. That is to say, people and groups not only know, accept and contribute to the preservation or transformation of the representations of the groups they belong to, but they also recognise and discuss alien social representations" (Rodríguez 2007: 180).

Lastly, there is a critical and deconstructive aspect linked to this "if SR theory explains the state of things, it may also from its genesis, support the process of deconstruction and reconstruction of gender meanings, going beyond the descriptive dimension and suggesting strategies to deconstruct reductionist binomials such as subject/object, society/culture, feminine/masculine" (Flores 1996: 196). Besides, it enables us to study the position, situation and condition of gender subjects relationally, giving voice to their agency, identifying gender historical and cultural constructions, as well as power relations they establish and justify at micro-, meso- and macro-level, incorporating systemic horizontal and vertical elements.

In the context of contemporary Mexico, framed by the family as institution that legitimises reproduction, it refers to the way in which the hegemonic gender order is constituted as a patriarchal system, that has as its axis cisgender identities and heterosexual modes of relation.

'Cisgender' refers to the situation where gender identity is congruent with the assigned sex, which is different from 'transgender', a situation where we find that gender identity differs from the assigned sex. With technological advances, binary viewpoints have been challenged, and multidisciplinary research points to a sex-gender fluidity rather than a rigid polarisation in the continua of gender and sex fluidity. Diverse and interrelated aspects of the person are considered, such as: (i) the *sex or biological continuum* with the range male–intersex–female, considering organic markers at chromosomic, hormonal and genital level, as well as genes and secondary sexual traits. Although biology is structuring, it also varies throughout the developmental process of the person throughout their life cycle and is affected by context. Most people are situated on some point of the continuum rather than at one of the extremes: male or female. There are authors who consider the pertinence of working with at least five sexes (Fausto-Sterling 1991; Konner 2015); (ii) the *gender identity continuum* or gender self-conception that is unique in the case of each person. That is to say, being a man or a woman means different things to different people, at different times in their life cycles, in diverse groups and historical contexts; (iii) the *gender social expression continuum,* frequently called 'gender roles'; this encompasses the ways in which the social and cultural gendered constructions are expressed in how people behave, how they act, their dress codes, postures, gestures, etc., that oscillate between masculine–androgynous–feminine and also vary across cultures, times, and places; (iv) the *continuum of erotic and sexual orientation* that goes from heterosexuality to homosexuality/lesbianism, including bisexuality, pansexuality or omnisexuality, asexuality, monosexuality, demisexuality, queer sexuality, etc. (Malpas 2016). Although the research potential of gender and social representations is vast, it is useful to consider that recent legislative reforms in Mexico have generated a public debate and mobilised groups of activists throughout the country in relation to homoparenting; and that human-*assisted reproductive technologies* (ARTs) constitute potential rupture points in terms of gender social representations and the knowledge and relations they produce and transform on an everyday basis. The following section briefly presents the thematic outlook of human-assisted reproductive technologies (ARTs) and 'homoparenting' in the context of Mexico.

7.3   Thematic Section: Human-Assisted Reproductive Technologies and 'Homoparenting' in Mexico

7.3.1   Human-Assisted Reproductive Technologies

Human-*assisted reproductive technologies* (ARTs), also known as fertility technologies, appeared only recently in human history. The British citizen Louise Jay Brown was conceived in a Petri dish and she is the first known person born through *in vitro* fertilisation (on 25 July 1978). Nevertheless, it is currently estimated that one in every six or seven couples is infertile (The Economist, 27 August 2016: 16). Assisted reproductive technologies thus become a set of very attractive alternatives for those who can afford them and have access to them. In the whole world, the thirty-four-year period between July 1978 and July 2012 saw five million births using ARTs (Bryner 2012).

In the first hegemonic discussions of radical feminism, ARTs were seen as emancipating, potentially giving women choice and 'liberating' them from reproduction taking place in their bodies. Other feminist currents argued that this was neither ethical, nor empowering, nor desirable (O'Riley 2010). Some insisted on the importance of motherhood for women as an embodied experience of femininity (ecofeminism, material feminism), or the widening of reproductive capacity as an increase in the range of choice-making for women (liberal feminism). However, there are other criticisms that have seen in the proliferation of ARTs new ways of essentialising the feminine mystique in its relation to maternal femininity based on oppression and gender inequalities, as well as making parenthood linked to the heteronormative family the ultimate goal of women and men, even within the LGBTTTIQ+ (*lesbian, gay, bisexual, transvestite, transgender, transsexual, intersexual, and queer/questioning, etc.*) collectives that have historically struggled against the oppression and privileges of the patriarchal, heterosexual and nuclear family (for example, queer feminisms; Park 2013).

Up to the present, human reproduction and gestation outside women's bodies has not been possible. The only exceptions have been the conception and development of an embryo for a limited period of days (developing an embryo outside the body of a woman for a period longer than fourteen days is a serious medical offence), as well as temporarily renting or borrowing the reproductive capacity of other women. Nevertheless, discussions have become increasingly complex, involving fourth and fifth generations of human rights. On the one hand, ARTs do effectively widen reproductive technologies for many people, but on the other hand they involve serious political, ethical, moral, economic, juridical and health dilemmas.[3]

The case of Mexico is interesting. Following restrictive legislation in Thailand and India, Mexico started to become a 'reproductive tourism paradise'. The context is complex. Structural inequalities, on the one hand, mean that there is a large population that lives in extreme poverty, with women and mothers willing to assist the reproduction of other women and families of richer areas (especially North America and Europe), in order to make a living and to support themselves and their families and children. On the other hand, the country offers top-level medical infrastructure and credibility in private reproductive health services that generate innovation and medical-scientific research for rich Mexicans and international citizens who can afford the price. According to the *National Institute of Statistics* (INEGI), in Mexico there are currently around 2.6 million cases of infertility and sterility; in 2012 alone, according to data provided by Serono,[4] 82,000 assisted reproduction treatments took place in the country (RA, consulted on 30 May 2018).

In Mexico there are seventy-five clinics certified by the *Federal Commission for the Protection against Sanitary Risk* (COFEPRIS), catering for a market valued at an estimated hundred million dollars, with 15 million potential clients (Expansión 13 February 2015). This contrasts sharply with the two fertility clinics that Mexico had in 1986 (RA, consulted on 30 May 2018), and is an indication of the importance and increasing acceptance of such technologies in everyday life. Besides this, many clandestine, unregistered and uncertified clinics operate without regulation, and corruption and impunity are commonplace. Among the certified clinics, Mexico has the Ingenes Institute, "with the greatest share in the Mexican market in solving highly complex cases", with a "success rate that exceeds the average of the fertility clinics in the United States" according to their own website.[5] It is certified by the *National Council of Science and Technology* (CONACyT), and is a leader in assisted fertility and genetics in Latin America, according to the *Latin American Network of Assisted Reproduction* (RedLara).

Regarding surrogacy, renting a womb in Mexico is three times cheaper than in the USA, and it is intermediaries, not women, who collect the highest share of earnings. According to Ivan Davydov of Care Surrogacy Center México (EFE 9 April 2016), costs range from 500,000 to 700,000 Mexican pesos (US $28,000–

---

[3]"Currently, human assisted reproductive technologies include, at least, homologous, heterologous and *intraperitoneal artificial insemination* (IPI), in vitro fertilization, fertilization through embryo transfer, *gametes intrauterine transfer* (GIUT), zygote intrafallopian transfer both transferring pre-embryos at the earliest stages of fertilization and transferring embryos in the Fallopian tubes in more advanced development stages (GIFT and ZIFT), *intra-cytoplasmic sperm injection* (ICSI), *pronuclear stage tubal transfer* (PROST), Oocyte, spermatozoa and embryo culture and transfer to the uterus, and surrogacy" (Martínez 2015: 360).

[4]Note that the study by the company Merk Serono conducted in 2012 and presented in 2013 is cited, because although it may seem outdated, it is the only study of the market for fertility that has been undertaken in Mexico.

[5]Ingenes Institute, "Who We Are"; at: www.ingenes.com (11 March 2018).

39,000), to which medical expenses and childbirth costs must be added. Any future health risk, complication or condition that is not immediately derived from the pregnancy and childbirth is not the responsibility of the person hiring the womb. The woman renting her womb gets between 150,000 and 270,000 pesos in total. The states of Tabasco and Sinaloa are popular for surrogacy. Both states recognise surrogacy legally, although legislation is incomplete. In Tabasco, legislation dates back to the 1990s, but is obsolete and incomplete. In Mexico, especially in Tabasco, provision of an array of services is common: examples include *in vitro* fertilisation, donation of egg cells, surrogacy, sperm donation, and gender selection. The state of Quintana Roo, with its top tourist destinations of Cancun and the Mayan Riviera, is seen as a potentially important site for reproductive tourism. Plans have been made to establish a medical city with first-class human-assisted reproduction facilities.

In this context, the Senate is amending the existing National Health Law as it applies to assisted reproduction. At the same time, the *Supreme Court of Justice* (SCJN) is revising an action of unconstitutionality brought by the *Mexican Attorney General's Office* (PGR) that contests the reform to the Legal Code of Tabasco (Juan Pablo Reyes, Excelsior, 5 March 2017). This is because, as has been documented by the NGO GIRE, there are babies existing in a juridical limbo who are the children of foreigners, born in Mexico, but who are refused birth certificates by the authorities (GIRE 2015).

7.3.2 'Homoparenting'

Families and 'homoparental' (gay and lesbian parenting) relations have been addressed through various research lenses outside Mexico since the 1970s. In Mexico it is a field of studies of more recent debate and construction. Homoparenting as institution and practice has two sides. It refers to same-sex parental figures who have descendants who are recognised by the state. They can be common descendants or the children of one of the members of the couple, for example from a previous relationship, or registered to a single parent, the result of a 'natural' birth, of access to reproductive technologies or the product of adoption. Thus, it includes adoption rights, ARTs, social security, and the transmission of legal prerogatives and legal rights by the couple. The other side encompasses couples who are not recognised by the state, who cohabit and have a parental and family function, of procreation or upbringing.

'Homoparenting' is the legal term used in Mexico. It has become increasingly relevant as a result of recent legislative changes. On 17 May 2016, Mexican President Enrique Peña Nieto signed the initiative of reform to Article 4 of the Constitution, which recognises as a human right the marriage of people without any kind of discrimination. His reform was based on resolution 43-2015 by the Supreme Court of Justice (SCJN 12 June 2015), which obliges judges to follow that criterion in any legal protection action, anywhere in the country, even in states that do not recognise same-sex marriages.

The reaction of people, groups and institutions has been very significant and there is a strong polarisation between those who sympathise with 'marriage without discrimination' and those who favour 'the natural family'. The divide deepens when it comes to homoparenting. This indicates a potential moment of displacement or re-functionalisation of gender social representations. According to the *National Council for the Prevention of Discrimination* (CONAPRED), there is an acceptance rate of seventy per cent for same-sex couples, but only thirty per cent support homoparenting (Aristegui, 29 October 2015). Mexico has the second-highest rate of homophobia in the world. Between 1996 and 2015 there were 1,218 homicides due to homophobia (Aristegui, 17 May 2016).

In academic debates, matters relating to LGBTTTIQ+ populations in the Global South have been addressed from the perspective of the struggle for sexual rights (Ettelbrick/Shapiro 2004: 475). Marriage and homosexual adoption are taken to be a conquest and a decentring of heterosexism. They recognise people's basic human rights and marriage is taken to imply guarantees such as credits, pensions, labour, bank, economic, welfare and social benefits. According to Fassin (2005: 71) "'homoparenting families' are part of a 'family of families' without there being the need or possibility, beyond the partial commonalities, to assume there must be a common denominator: one must stop trying to substitute the sex difference by another central core, in order to make place for this new reality".

Critics denounce the institution of the 'patriarchal economic regime' that operates as a "legal trap, a historical instrument for social disorganisation that steals freedom and the rights of individuals and the collective in the name of the family" (Gargallo 2012: 1). We speak of reinforcing essentialisms and naturalizing gender roles and the sexual division of work anew, inciting people to have offspring and marry as a model of gender realisation, re-editing social representations and roles. This threatens human rights, since these "are constitutive of the subject by virtue of being born. They are human, not institutional, rights, thus obtaining them does not depend on the celebration of any civil, military or religious contract. That is to say, the exercise of human rights does not depend on belonging to any institution. The quality of the subject of rights cannot be made a condition of adhering to one of the most questioned institutions of heterosexual society, or any other for that matter" (Mogrovejo 2015: 158).

Besides, control by the state over the private life of individuals and collectives is re-functionalised, reinvigorating existing ideologies such as patriarchy, classism, colonialism, and capitalism. The state hands over its responsibility of social welfare to the couple, transferring its responsibilities. The married couple with progeny, based on romantic love, becomes the basic unit of procreation, production, consumption and ideological reproduction in a logic of exclusive models that are institutionally sanctioned. The class struggle dilutes, inequalities are strengthened. In the context of capitalism, accepting homoparenting becomes functional to late capitalism and to the rainbow market according to the capacity of consumption that nuclearises, atomises, individualises and subjects people to common debts and obligations. Otherness is masked and exclusions are made invisible without being dismantled. In cases of adoption and participation in conception, the use of novel human-assisted reproductive technologies threatens the basic rights of infants who are ignorant of their genetic origin (Park 2013).

Advocates of a third way called 'sceptic equality marriage' (Kim 2010) criticise marriage at the same time as they favour same-sex marriages. They claim that people who marry will be able to question, deconstruct and modify patriarchy in the micro-social sphere of their intimate and private behaviours, although their impact is marginal in the transformation of patriarchal society and class oppression as the main mechanism of capital inequality (Ettelbrick 1989).

7.4 Closing Reflection

It is surrounding these conflicting and contradictory postures of knowledge and practices that a relevant public debate concerning the family as institution and its naturalised and legal characteristics has emerged. As in Mexico, this is an ongoing debate that shares commonalities in other parts of the world and is linked as part of the processes of globalisation, although it needs to be addressed in its specific local context, where it will be coherent. It affects the family as social institution, its legal recognition, the provision of government services surrounding the family, and the way in which people construct and make sense of their gendered identities and relations. Although the study of the family as institution undergoing transition is not new, looking at it from the standpoint of Social Representations Theory and addressing gender in the case of homoparenting and human-assisted reproductive technologies is unique, and thus it seems a pertinent theoretical and methodological research framework applicable to Mexico and other local realities in the context of the rapid changes in the Anthropocene era.

In this rapid process of the 'sliding' of representations and the 'revolution of the social imaginary' regarding the family as institution, ruptures at the macro-systemic level expressed in terms of technological advances and legal changes have deep impacts in terms of the contradictions of coexisting forms of knowledge and practice. There is a greater opportunity to exercise rights in a context of democratic diversity, although extreme representational polarisation has an impact in terms of societal and gendered violence in different guises.

Social practices are rooted in everyday knowledge, and the changes in the logic of common-sense knowledge will have an impact as one of the characteristics of this *époque* of rapid changes and transitions. The 'Anthropocene era' is characterised by humans being both the threat and the solution to their own survival and that of the planet and its species. Historically, the family has been the social institution in charge of human reproduction. Just as in the environmental debates of the Anthropocene, looking at how to change and resist gendered social representations of the family means looking at power relations in the public sphere and the types of knowledge and practices that will have to be addressed, negotiated and re-enacted in the short, middle and long term, always considering their diversity and sustainability. Social Representations Theory offers an important set of tools for addressing this issue from a social research perspective, linking scientific discourses with common-sense practices.

References

ADN 40, 28 December 2017: "En 2017 se registraron más de 12mil feminicidios en México"; at: http://www.adn40.mx/noticia/mexico/nota/2017-12-28-13-50/en-2017-se-registraron-mas-de-12-mil-feminicidios-en-mexico/ (1 June 2018).
Alonso, Martín, 1982: *Enciclopedia de Idioma* (Mexico, D.F.: Ediciones Aguilar).
Aristegui Noticias (David Ordaz), 17 May 2016: "México, segundo país con más violencia por homofobia"; at: http://aristeguinoticias.com/1705/mexico/mexico-segundo-pais-con-mas-violencia-por-homofobia/ (18 October 2017).
Aristegui Noticias (Redacción), 29 October 2015: "Aún asocian homosexualidad con perversión, pedofilia por ello hay rechazos irracionales: Lamas"; at: http://aristeguinoticias.com/2910/mexico/aun-asocian-homosexualidad-con-perversion-pedofilia-por-ello-hay-rechazos-irracionales-lamas/ (18 October 2017).
Arruda, Ángela (org.): *Representando a Alteridade*, Rio de Janeiro, Editora Vozes, 2002 (1998).
Bauman, Zygmund, 2000: *Modernidad Líquida* (Mexico, D.F.: FCE).
Bauman, Zygmund, 2006: *Amor Líquido: Acerca de la Fragilidad de los Vínculos Humanos* (Buenos Aires: FCE).
Beck, Ulrich, 1998: *La sociedad del riesgo. Hacia una nueva modernidad* (Barcelona: Paidós).
Ben-Asher, Samadar, 2003: "Hegemonic, Emancipated and Polemic Social Representations: Parental Dialogue Regarding Israeli Naval Commandos Training in Polluted Water", *Papers on Social Representations*, Vol. 12: 6.1–6.12.
Bourdieu, Pierre, 2002 [2001]: *La Dominación Masculina* (Barcelona: Anagrama).
Bryner, Jeanna, 3 July 2012: "5 Million Babies Born from ivf, Other Reproductive Technologies", *Live Sciences*; at: http://www.livescience.com/21355-5-million-babies-born-ivf-technologies.html (16 October 2017).
Burr, Viven, 1998: *Gender and Social Psychology* (London: Routledge).
Cámara de Diputados, 17 May 2018: "Sufren violencia siete de cada 10 menores; se debe garantizar vida libre de maltrato", *boletín*, No. 5465; at: http://www5.diputados.gob.mx/index.php/esl/Comunicacion/Boletines/2018/Mayo/17/5465-Sufren-violencia-siete-de-cada-10-menores-se-debe-garantizarles-vida-libre-de-maltrato (1 June 2018).
Centro Tepoztlán (Clara Jusidman), 25 November 2017: "La política social ante las nuevas realidades y el futuro de México"; at: http://centrotepoztlan.org/reunion-dialogo-25-11-2017-la-politica-social-ante-las-nuevas-realidades-y-el-futuro-de-mexico/ (20 January 2018).
Coontz, Stephanie, 2006: *Historia del Matrimonio: Cómo el amor conquistó el matrimonio*, (Barcelona: Gedisa).
Duveen, Gerard; Lloyd, Barbara (Eds.), 1990: *Social Representations and the Development of Knowledge* (Cambridge: Cambridge University Press).
Duveen, Gerard; Lloyd, Barbara, 1986: "The significance of social identities", in: *British Journal of Social Psychology,* 25: 219–230.
Duveen, Gerard; Moscovici, Serge (Eds.), 2000: *"Social Representations: Explorations in Social Psychology"* (London: Polity Press).
Duveen, Gerard, 1997: "Psychological Developmental as a Social Process", in: Smith, Leslie, Dockerell, Julie; Tomlinson, Peter (Eds.): *Piaget, Vygotsky and beyond* (London, Routledge).
Duveen, Gerard, 2001: "Representations, Identities and Resistance", in: Deaux, Kay; Philogène, Gina (Eds.): *Representations of the Social* (Oxford: Blackwell): 257–271.

Duveen, Gerard, 1993: "The Development of Social Representations of Gender", *Papers on Social Representations*, 2, 3: 1–117.

EFE Agencia de noticias, 9 April 2016: "México se resiste a convertirse en 'próximo paraíso' del turismo reproductivo"; at: https://www.efe.com/efe/america/mexico/mexico-se-resiste-a-convertirse-en-proximo-paraiso-del-turismo-reproductivo/50000545-2891821 (16 June 2017).

Ettelbrick, Paula; Shapiro, Julie, 2004: "Are We on the Path to Liberation Now? Same Sex Marriage at Home and Abroad", in: *Seattle Journal for Social Justice*, 2: 475–493.

Ettelbrick, Paula, 1989: "Since When is Marriage a Path to Liberation?", in: *Out/Look*, 2,9: 14–17.

Excelsior (Juan Pablo Reyes), 5 March 2017: "Corte atrae cas de vientres rentados"; at: http://www.excelsior.com.mx/nacional/2017/03/05/1150193 (15 July 2017).

Facio, Alda, 1999: "Feminismo, género y patriarcado", in: Facio, Alda; Fries, Lorena (Eds.): *Género y Derecho*, (Santiago: LOM): 1–37; (18 August 2017).

Falicov, Celia Jaes, 2006: "El ciclo de vida familiar: Un esquema para la psicoterapia de familia", in: Roizzblatt, A. (Ed.): *Terapia Familiar y de Pareja* (Santiago: Mediterráneo): ch. 9.

Fassin, Eric, 2005: "Usos de la ciencia y ciencia de los usos. A propósito de las familias homoparentales", *Debate Feminista*, 32: 52–73.

Fausto-Sterling, Anne, 1993: "The Five Sexes: Why Male and Female Are Not Enough", *The Sciences*, March/April: 19–25.

Flores Palacios, Fátima, 1996: "Representación social: género y salud mental", in: Calleja, Nazira; Gómez-Peresmitré, Gilda (Eds.): *Psicología social: investigación y aplicaciones en México* (Mexico, D.F.: Biblioteca de Psicología, Psiquiatría y Psicoanálisis): 194–225.

Gargallo, Francesca, 2012: "La propiedad privada es la base del matrimonio", Conference at the Festival Internacional por la Diversidad Sexual, Museo del Chopo, Mexico City, 22 June 2012; at: https://francescagargallo.wordpress.com/ensayos/feminismo/feminismo-genero/la-propiedad-privada-es-la-base-del-matrimonio/ (18 October 2017).

Giddens, Anthony, 1991, *Modernity and Self-Identity: Self and Society in the Late Modern Age* (London: Polity Press).

GIRE (Grupo de Información en Reproducción Elegida), 2015: *Niñas y Mujeres sin Justicia: Derechos Reproductivos en México* (Mexico D.F.: GIRE).

Gurría, Ángel (OECD Secretary General), 2018: "Global and Economic Economic Outlook 2018"; at: http://www.oecd.org/fr/mexique/global-and-mexico-economic-outlook-2018.htm (31 May 2018).

Howarth, Caroline, 2002a: "'So, you're from Brixton?' The Struggle for Recognition and Esteem in a Multicultural Community", in: *Ethnicities*, 2, 2: 237–260.

Howarth, Caroline, 2002b: "Identity in Whose Eyes? The Role of Representations in Identity Construction", in: *Journal of the Theory of Social Behaviour*, 32, 2: 145–162.

INEGI, 2017: "Estadísticas a propósito del … Día de la Familia Mexicana" (5 Marz); at: http://www.inegi.org.mx/saladeprensa/aproposito/2017/familia2017_Nal.pdf (29 March 2018).

Jodelet, Denise, 1989: "Représentations sociales: une domaine en expansion", in: Jodelet, D. (Ed.): *Les Représentations Sociales* (Paris, PUF): 31–61.

Jovchelovitch, Sandra, 2007 [2001]: "Social representations, public life and social construction", in: *LSE Research Online*, at: http://eprints.lse.ac.uk/2649/1/Socialrepspubliclife.pdf (31 May 2018).

Jovchelovitch, Sandra, 2007: *Knowledge in Context: Representations, Community and Culture* (London: Routledge).

Kim, Suzanne A., 2010: "Skeptical Marriage Equality", *Harvard Journal of Law and Gender*, 34, December: 37–80.

Konner, Melvin, 2015: *Women After All: Sex, Evolution and the End of Male Supremacy* (New York: W. W. Norton).

Ksenych, Ed; Liu, David, 2001, *Conflict, Order & Action: Readings in Sociology* (Toronto: Canadian Scholars' Press).

Lagarde y de los Ríos, Marcela, 2001 [1996]: *Género feminismo, desarrollo humano y democracia.* Cuadernos Inacabados No. 25 (Madrid: Editorial Horas y Horas).

Lagarde y de los Ríos, Marcela, 2004 [1990]: *Los cautiverios de las mujeres: madresposas, monjas, putas, presas y locas* (Mexico City: PUEG/UNAM).

Lamas, Marta, 2002: *Cuerpo: Diferencia Sexual y Género* (Mexico City: Taurus).

Lamas, Marta, 1996: *El género: la Construcción Cultural de la Diferencia Sexual* (Mexico City: Editorial PUEG-UNAM/ Porrúa).

Lévi-Strauss, Claude, 1956: "La Familia", in: Lévi-Strauss, Claude; Spiro, M. E.; Gough, K. (Eds.): *Polémica sobre el Origen y la Universalidad de la Familia* (Barcelona: Anagrama).

Malpas, Jean, 19–21 May 2016: "'De lo Homo a lo Trans': Diversidad Sexual y Terapia", *Practicum Clínico Crisol* (Mexico City: Instituto Crisol e Instituto Ackerman, Colegio de Ingenieros).

Marková, Ivana, 2017: "The Making of the Theory of Social Representations", in: *Cadernos de Pesquisa*, 47, 163, January–March: 358–374.

Marková, Ivana, 2003: *Dialogicality and Social Representations: The Dynamics of Mind*, (Cambridge: Cambridge University Press).

Marková, Ivana, 1982: *Paradigms, Thought and Language* (Chichester: Wiley).

Martínez Martínez; Verónica Lidia, 2015: "Maternidad Subrogada: Una Mirada a su regulación en México", in: *Dikaionk*, 29, 2, Colombia, December 2015: 353–382; at: http://www.redalyc.org/html/720/72045844007/ (3 June 2017).

Merriam Webster Dictionary, 2018; at: https://www.merriam-webster.com/dictionary/family (accessed 15 February 2018).

Mogrovejo, Norma, 2015: "Matrimonio gay,? familias reconfiguradas?", in: Medina Trejo, José Antonio: *Familias Homoparentales en México: Mitos, Realidades y Vida Cotidiana*" (Mexico City: PRD, Secretaria de Igualdad de Género): 147–162.

Moore, Henrietta L., 1994: *A Passion for Difference: Essays in Anthropology and Gender* (London: Polity Press).

O'Riley, Andrea (Ed.), 2010: *Encyclopedia of Motherhood* (Thousand Oaks, CA: SAGE).

ONU/INMUJERES, 2017: *La Violencia Feminicida en México, Aproximaciones y Tendencias, 1985–2016* (Mexico City: INMUJERES/ONU).

Ortner, Sherry B., 1974: "Is female to male as nature is to culture?", in: Rosaldo, M.; Lamphere, L. (Eds.): *Woman, Culture and Society* (Stanford: Stanford University Press): 67–88.

Park, Shelley M., 2013: *Mothering Queerly, Queering Motherhood: Resisting Mono-maternalism in Adoptive, Lesbian, Blended and Polygamous Families* (New York: Suny Press).

Pateman, Carole, 1991 [1989]: *The Sexual Contract* (London: Polity Press).

Phillips, Anne, 1996: *Género y Teoría Democrática*, Universidad Nacional Autónoma de, (Mexico City: México/UNAM).

RA (Reproducción Asistida): "Situación Actual de la reproducción asistida en México"; at: https://reproduccion-asistida.mx/reproduccion-asistida/situacion-actual-mexico/ (30 May 2018).

RAE (Diccionario de la Real Academia Española), 2017; at: http://www.rae.es/ (19 September 2017).

Roudinesco, Élisabeth, 2003: *La Familia en Desorden* (Buenos Aires: Fondo de Cultura Económica).

Rodríguez Salazar, Tania, 2007: "Sobre el estudio cualitativo de la estructura de las representaciones sociales", in: Rodríguez Salazar, Tania; García Curiel, Ma. De Lourdes (Eds.): *Representaciones sociales. Teoría e investigación* (Guadalajara: Universidad de Guadalajara): 157–188.

Rubin, Gayle, 1975: "The Traffic in Women: Notes on the Political Economy of Sex", in: Reiter, Rayna R. (Ed.): *Toward an Anthropology of Women* (New York – London: Monthly Review Press): 157–210.

Satz, Debra, 2017: "Feminist Perspectives on Reproduction and the Family", in: Zalta, Edward N. (Ed.): *The Stanford Encyclopedia of Philosophy*; at: https://plato.stanford.edu/archives/sum2017/entries/feminism-family/ (17 March 2018).

Scott, Joan W., 1986: "Gender: A Useful Category for Historical Analysis", in: *The American Historical Review*, 91, 5 (December): 1053–1075.

Serrano Oswald, Serena Eréndira, 2016: "Reconceptualised security: core Mexican challenges", in: *Regions*, 301 (Spring): 18–21.

Serrano Oswald, Serena Eréndira, 2013: "The potential of Social Representations Theory (SRT) for gender equitable research", in: *Acta Colombiana de Psicología*, 16, 2: 63–70; at: http://editorial.ucatolica.edu.co/ojsucatolica/revistas_ucatolica/index.php/acta-colombiana-psicologia/article/view/180/221 (1 June 2018).

Squire, Corinne, 2000, *Culture in Psychology* (London: Routledge).

The Economist, 2016: "The Empty Crib: Demography and Desire", 27 August: 14–16.

United Nations, 1948: *The Universal Declaration of Human Rights*; at: http://www.un.org/en/universal-declaration-human-rights/ (25 March 2018).

UN, 29 November 2017: "The long road to justice, prosecuting femicide in Mexico"; at: http://www.unwomen.org/en/news/stories/2017/11/feature-prosecuting-femicide-in-mexico (2 June 2018).

Velázquez, Margarita, 2014: "Género, desarrollo y ambiente", keynote speech, 19 Encuentro Nacional de Desarrollo Regional en México, AMECIDER-Universidad de Guadalajara, Zapopan, 13 November.

Walters, Marianne; Carter, Betty; Papp, Peggy; Silverstein, Olga, 1996: *La red invisible: Pautas vinculadas al género en las relaciones familiares* (Barcelona: Paidós).

Wagner, Wolfgang; Hayes, Nicky, 2005: *Everyday Discourse and Common Sense: The Theory of Social Representations* (London: Palgrave Macmillan).

Wagner, Wolfgang; Duveen, Gerard; Farr, Robert; Jovchelovitch, Sandra; Lorenzi-Cioldi, Fabio; Marková, Ivana; Rose, Diana, 1999: "Theory and Method of Social Representations", in: *Asian Journal of Social Psychology* (Blackwell), 2: 95–125.

Wittig, Monique, 1980: "On ne naît pas femme", in: *Nouvelles Questions Féministes & Questions Feministes,* 8 (May): 75–84.

Chapter 8
Sustainable Peace Through Sustainability Transition as Transformative Science: A Peace Ecology Perspective in the Anthropocene

Hans Günter Brauch

Abstract This essay contributes to a conceptual discussion on the need for bridge-building between the natural and social sciences, among different social science disciplines, and the research programmes in political science focusing on peace, security, development and environment ('sustainable development'), by introducing the two new linkage concepts of 'political geo-ecology' and 'peace ecology'. It focuses on the policy goal of a 'sustainable peace' understood as 'peace with nature' in the newly proposed epoch of earth history, the Anthropocene.

The key argument of this chapter is that this goal may be achieved by a process of 'sustainability transition' that addresses the economic causes of greenhouse gas concentration in the atmosphere, in which where concerned individuals, families, local communities, states and nations as well as international governmental organisations and non-governmental bodies and social movements may contribute to the transition.

This text suggests that the goal of a 'sustainable peace' may be addressed from a peace ecology perspective that integrates both peace and security studies and ecology or ecological approaches aiming at the realisation of the goal of a 'sustainable development'. It is argued that this requires a shift from disciplinary and multidisciplinary research methods towards inter- and transdisciplinary approaches by moving towards a 'transformative science' aiming at a 'sustainable peace' where the needed policy changes and the actors and processes of this change towards sustainability should become a part of the research design and action research process.

Hans Günter Brauch has been a co-convener of IPRA's Ecology and Peace Commission (2012–2018) and is chairman of *Peace Research and European Security Studies* (AFES-PRESS) and editor of several English language book series published by Springer Nature; Email: brauch@afes-press.de.

Several colleagues have critically commented on an earlier draft: Úrsula Oswald Spring (Mexico), Francisco Rojas (Chile), Luis Alberto Padilla Menendez (Guatemala), Simon Dalby (Canada), Czeslaw Mesjasz (Poland) and Andrew Collins (UK). Their critical comments and suggestions are reflected in this text but the author did not change the main focus of his argument and tried to make it more transparent. For careful language editing the author is grateful to Ms. Margaret Gamberton.

© Springer Nature Switzerland AG 2019

H. G. Brauch et al. (eds.), *Climate Change, Disasters, Sustainability Transition and Peace in the Anthropocene*, The Anthropocene: Politik—Economics—Society—Science 25, https://doi.org/10.1007/978-3-319-97562-7_8

This essay touches on the manifold fundamental conceptual, methodological, theoretical and action-oriented research needs of a 'peace ecology approach' that aims at contributing to the realisation of a 'sustainable peace' as 'peace with nature' in the 'Anthropocene', where the societal outcomes of the physical effects of global environmental and climate change can be countered and mitigated by policies of adaptation, mitigation and an increase of resilience by the affected people.

From the perspective of a Hobbesian policy approach of 'business-as-usual,' but also from traditional scientific worldviews, this goal may appear at present to be utopian and for sceptics, it is not achievable. It requires a fundamental change in the dominant 'worldview' of many scientists and of the neoliberal mindset of most policymakers, practitioners, but also of ordinary citizens towards an alternative sustainability approach.

For the natural and social sciences it requires a new 'scientific revolution towards sustainability' – similar to what Kuhn (1962) called the 'Copernican Revolution' or Schellnhuber (1999) outlined as a 'Second Copernican Revolution' – with a new scientific paradigm of a 'peace ecology' that still needs to be developed in the future.

This holistic approach of linking different scientific discourses with discussions in the political realm deliberately distances itself from the mainstream of political science contributions with often narrowly focused theoretical discussions that solely appeal to a scientific audience and are hardly noted in societal and political discussions.

Keywords Anthropocene · Climate change · Copernican revolution
Global environmental change · Historical time · Mindset · Peace ecology
Political geo-ecology · Scientific revolution · Sustainable peace
Sustainability transition · Ttransformative research · Transformative science
Worldview

8.1   Introduction

Previous texts (Brauch 2008, 2009b, 2016b) addressed six levels of historical time (cosmic, geological, technical, structural, conjunctural, and history of events) and multiple turning points in international order (Holsti 1991, 2016a, b; Buzan/Lawson 2015), in modern history (peace of Münster and Osnabrück of 1648, peace of Utrecht of 1713, Vienna Congress of 1815), and in the short 20th century (Versailles Treaty and League of Nations of 1919; Charter of the United Nations 1945; 1989 end of the Cold War with the fall of the Berlin Wall).[1] This author

---

[1]This thinking was influenced by natural scientists on the Anthropocene (Crutzen/Stoermer 2000; Crutzen 2002, 2011, 2015; Clark et al. 2004), by human and physical geographers on geopolitics and geo-ecology (Dalby 2013, 2014, 2015; Huggett 1995), by historians (Braudel 1949, 1969, 1972; Osterhammel 2014; Hobsbawm 1994), by social scientists on sustainability transition (Grin et al. 2010; WBGU 2011), and by transformative research (Schneidewind/Singer-Brodowski 2012, 2014; Schneidewind et al. 2016).

argued that a technical (industrial) revolution and the start of the nuclear age (1945) triggered the first intervention of humankind into nature that may result in a change in the periodisation of earth history (Biello 2015).

8.1.1 Social Construction of the Anthropocene

Paul J. Crutzen's diagnosis of late February 2000 at a conference of the *International Geophysical Biological Programme* (IGBP) in Cuernavaca in Mexico that "we are now in the Anthropocene" has already triggered a controversial global debate in human history, in philosophy, in the natural and social sciences and in the humanities, including religion, which is detached from the geological debate (Crutzen/Stoermer 2000). Within 18 years the concept has spread and developed many different features, which this author will conceptually map in a different text (Brauch 2019).

In 2008, the *Anthropocene Working Group* (AWG) of the *Subcommission on Quaternary Stratigraphy* (SQS) was set up. This is a constituent body of the *International Commission on Stratigraphy* (ICS), the largest scientific organisation within the *International Union of Geological Sciences* (IUGS). The AWG offered this definition of the Anthropocene on its website:

- The 'Anthropocene' is a term widely used since its coining by Paul Crutzen and Eugene Stoermer in 2000 to denote the present time interval, in which many geologically significant conditions and processes are profoundly altered by human activities. These include changes in:

  - *erosion and sediment transport* associated with a variety of anthropogenic processes, including colonisation, agriculture, urbanisation and global warming,
  - the *chemical composition of the atmosphere, oceans and soils*, with significant anthropogenic perturbations of the cycles of elements such as carbon, nitrogen, phosphorus and various metals,
  - *environmental conditions* generated by these perturbations; these include global warming, ocean acidification and spreading oceanic 'dead zones',
  - *the biosphere* both on land and in the sea, as a result of habitat loss, predation, species invasions and the physical and chemical changes noted above.

- The 'Anthropocene' is not a formally defined geological unit within the Geological Time Scale. A proposal to formalise the 'Anthropocene' is being developed by the *'Anthropocene' Working Group* for consideration by the International Commission on Stratigraphy, with a current target date of 2016. …
- The 'Anthropocene' is currently being considered by the Working Group as a potential geological epoch, i.e. at the same hierarchical level as the Pleistocene and Holocene epochs, with the implication that it is within the Quaternary

Period, but that the Holocene has terminated. It might, alternatively, also be considered at a lower (Age) hierarchical level; that would imply it is a subdivision of the ongoing Holocene Epoch.

- Broadly, to be accepted as a formal term the 'Anthropocene' needs to be (a) scientifically justified (i.e. the 'geological signal' currently being produced in strata now forming must be sufficiently large, clear and distinctive) and (b) useful as a formal term to the scientific community. In terms of (b), the currently informal term 'Anthropocene' has already proven to be very useful to the global change research community and thus will continue to be used, but it remains to be determined whether formalisation within the Geological Time Scale would make it more useful or broaden its usefulness to other scientific communities, such as the geological community.

- The beginning of the 'Anthropocene' is most generally considered to be at *c*. 1800 CE, around the beginning of the Industrial Revolution in Europe (Crutzen's original suggestion); other potential candidates for time boundaries have been suggested, at both earlier dates (within or even before the Holocene) or later (e.g. at the start of the nuclear age). A formal 'Anthropocene' might be defined either with reference to a particular point within a stratal section, that is, a *Global Stratigraphic Section and Point* (GSSP), colloquially known as a 'golden spike; or, by a designated time boundary (a Global Standard Stratigraphic Age).

- The 'Anthropocene' has emerged as a popular scientific term used by scientists, the scientifically engaged public and the media to designate the period of Earth's history during which humans have a decisive influence on the state, dynamics and future of the Earth system. It is widely agreed that the Earth is currently in this state.[2]

During the International Geological Congress in August 2016 in Cape Town, South Africa, the AWG unanimously recognised in a report that the Anthropocene is a reality, and voted 30-to-three (with two abstentions) for the transition to be officially registered. The report argued that the Anthropocene started between 1945 (nuclear age) and 1950 (great acceleration). Its members pointed to "concentrations in the air of carbon dioxide, methane and stratospheric ozone; surface temperatures, ocean acidification, marine fish harvesting, and tropical forest loss; population growth, construction of large dams, international tourism—all of them take off from about mid-century. One of the main culprits is global warming driven by the burning of fossil fuels." However, the Phys.org website mentioned in its report "that the working group is not allowed to take any of these measures into consideration unless they show up in the geological record. If it can't be measured in rocks, lake sediments, ice cores, or other such formations—the criteria used to determine

---

[2]See: Subcommission on Quaternary Stratigraphy, Working Group on the 'Anthropocene', at: http://quaternary.stratigraphy.org/workinggroups/anthropocene/; http://quaternary.stratigraphy.org/majordivisions/anthropocene/ and the book edited by Waters et al. (2014). See the major stratigraphic divisions at: http://quaternary.stratigraphy.org/majordivisions/.

dozens of distinct eons, era, periods and ages going back four billion years—it doesn't count".[3]

Zalasiewicz et al. (2017c), leading researches active in the AWG, argued that

> the Anthropocene concept arose within the Earth System Science (ESS) community, albeit explicitly as a geological (stratigraphical) time term. Its current analysis by the stratigraphical community, as a potential formal addition to the Geological Time Scale, necessitates comparison of the methodologies and patterns of enquiry of these two communities. One means of comparison is to consider some of the most widely used results of the ESS, the 'planetary boundaries' concept of Rockström and colleagues, and the 'Great Acceleration' graphs of Steffen and colleagues, in terms of their stratigraphical expression. This expression varies from virtually non-existent (stratospheric ozone depletion) to pronounced and many-faceted (primary energy use), while in some cases stratigraphical proxies may help constrain anthropogenic process (atmospheric aerosol loading). The Anthropocene concepts of the ESS and stratigraphy emerge as complementary, and effective stratigraphic definition should facilitate wider transdisciplinary communication.[4]

While the debate among geologists is still ongoing and the Anthropocene has not yet been formally accepted as a new epoch of earth history by the organisations of the geological community, the concept has spread rapidly since 2000 and a discussion has emerged in many disciplines of the natural and social sciences and in the humanities that increasingly addresses the implications of this most significant turn in human history since human civilisationisations in the fertile crescent, Mesoamerica, India and in China emerged after the end of the last glacial period, which also marked the starting point of the Holocene.[5]

Zalasiewicz et al. (2017a: 55–60) summarised the interim recommendations of the Anthropocene Working Group:

> The majority opinion within the AWG holds the Anthropocene to be stratigraphically real, and recommends formalisation at epoch/series rank based on a mid-20th century boundary. Work is proceeding towards a formal proposal based upon selection of an appropriate Global boundary Stratotype Section and Point (GSSP), as well as auxiliary stratotypes. Among the array of proxies that might be used as a primary marker, anthropogenic radionuclides associated with nuclear arms testing are the most promising; potential secondary markers include plastic, carbon isotope patterns and industrial fly ash. All these proxies have excellent global or near-global correlation potential in a wide variety of sedimentary bodies, both marine and non-marine (Zalasiewicz et al. 2017a: 55).

Both with the effects of a massive use of the atomic bomb (e.g. nuclear winter; Crutzen/Birks 1982) and with the physical and societal effects of the accumulation of greenhouse gases in the atmosphere since the industrial revolution (1780s) and

---

[3]See: "The Anthropocene is here: Scientists" (29 August 2016), at: https://phys.org/news/2016-08-anthropocene-scientists.html (3 April 2018).

[4]See at: https://theanthropocene.org/petrifying-earth-process-the-stratigraphic-imprint-of-key-earth-system-parameters-in-the-anthropocene/.

[5]The emergence of the Anthropocene concept, its concept history, and concept mapping (Brauch 2008), will be discussed by Brauch (2019).

especially since the end of WWII (1945) humankind could for the first time threaten the survival of its own species by military means (atomic bomb) and by the silent effects of its economic behaviour and consumption pattern due to six greenhouse gases accumulated in the atmosphere.[6]

8.1.2   The Anthropogenic Nature of Climate Change

In its five Assessment Reports the *International Panel on Climate Change* (IPCC 1990, 1995, 2001, 2007, 2014) agreed that climate change is human-induced or 'anthropogenic' as stated in its most recent Summary for Policymakers of its Synthesis Report (IPCC 2014):

- *Observed changes and their causes*: Human influence on the climate system is clear, and recent anthropogenic emissions of greenhouse gases are the highest in history. Recent climate changes have had widespread impacts on human and natural systems {1}.
- *Observed changes in the climate system*: Warming of the climate system is unequivocal, and since the 1950s, many of the observed changes are unprecedented over decades to millennia. The atmosphere and ocean have warmed, the amounts of snow and ice have diminished, and sea level has risen {1.1}.
- *Causes of climate change*: Anthropogenic greenhouse gas emissions have increased since the pre-industrial era, driven largely by economic and population growth, and are now higher than ever. This has led to atmospheric concentrations of carbon dioxide, methane and nitrous oxide that are unprecedented in at least the last 800,000 years. Their effects, together with those of other anthropogenic drivers, have been detected throughout the climate system and are extremely likely to have been the dominant cause of the observed warming since the mid-20th century {1.2, 1.3.1}.
- *Impacts of climate change*: In recent decades, changes in climate have caused impacts on natural and human systems on all continents and across the oceans. Impacts are due to observed climate change, irrespective of its cause, indicating the sensitivity of natural and human systems to changing climate (IPCC 2014: 2–4).

This scientific diagnosis based on the peer-reviewed literature of the global scientific climate change community has been noted by the heads of states and governments of all but one country and by all major global (UN, UNEP, UNDP, UNFCC) and regional international organisations (EU, ASEAN et al.) in the post-Cold War era that signed the *United Nations Framework Agreement on Climate Change* (UNFCCC of 1992), which was ratified by 197 countries among

---

[6]The likelihood of human extinction by wholly natural scenarios is very low but anthropogenic extinction may result from global nuclear annihilation, biological warfare, a pandemic-causing agent, ecological collapse, and global warming (Anthropogenic ecocide).

then all member states of the United Nations and which entered into force on 21 March 1994. *The Kyoto Protocol* had 192 state parties in 2013.

On 12 December 2015 the Paris Climate Change Agreement was adopted by consensus by 196 states and by February 2018, "194 states and the European Union have signed the Agreement. 174 states and the EU, representing more than 88% of global greenhouse gas emissions, have ratified or acceded to the Agreement, including China, the United States ... and India, the countries with three of the four largest greenhouse gas emissions of the UNFCC members total (about 42% together). All 197 UNFCCC members have either signed or acceded to the Paris Agreement.

On 1 June 2017, US President Donald Trump announced his intention "to withdraw from the Paris Accord".

In accordance with Article 28 of the Paris Agreement the earliest possible effective with-drawal date for the United States is 4 November 2020. If it chooses to withdraw by way of withdrawing from the UNFCCC, notice could be given immediately (the UNFCCC entered into force for the US in 1994), and be effective one year later. On August 4, 2017, the Trump Administration delivered an official notice to the United Nations that the US intends to withdraw from the Paris Agreement as soon as it is legally eligible to do so. The formal notice of withdrawal cannot be submitted until the agreement is in force for 3 years for the US, in 2019.[7]

In March 2017 Scott Pruitt,[8] former head of the U.S. *Environmental Protection Agency* (EPA) in the Trump administration, denied "that carbon dioxide emissions arc a primary cause of global warming",[9] thus challenging the EPA's previous view that CO_2 is the "primary greenhouse gas that is contributing to recent climate change",[10] and NASA's findings that it is "the most important long-lived 'forcing' of climate change".[11]

Global environmental and climate change have become contested issues in American domestic politics, whcre ideology and belief systems have partly replaced the peer-reviewed empirical evidence of the "world of science".

8.1.3 Political Changes and Change in Geological Time

Since 1989 humankind experienced multiple political changes: (a) with the first peaceful change in international order; (b) a more cooperative international political

---

[7]See at: https://en.wikipedia.org/wiki/List_of_parties_to_the_Paris_Agreement (3 April 2018).

[8]Chris Mooney; Brady Dennis: "On climate change, Scott Pruitt causes an uproar—and contradicts the EPA's own website", in: *Washington Post*, 9 March 2017.

[9]Steven Mufson: "Rick Perry just denied that humans are the main cause of climate change", in: *Washington Post*, 19 June 2017.

[10]EPA: "Climate Change", at: https://19january2017snapshot.epa.gov/climatechange.html (26 August 2017).

[11]NASA: "Global Climate Change: A blanket around the Earth", at: https://climate.nasa.gov/causes/ (26 August 2017).

context on issues of global environmental and climate change; but (c) since the
mid-1990s disarmament initiatives have been blocked, military expenditures have
increased again, and new wars have taken place; and (d) regression has occurred in
disarmament, global environments, and climate change.

What were the manifold causes of this dual global change during the early 1990s
towards both global cooperation and a retreat to old patterns in a new global
disorder? In the 21st century populist movements, parties and politicians have
emerged, and nationalist policy orientations, authoritarian tendencies, and repres-
sive systems of rule are returning in many parts of the world.

Totally unrelated to these 'volatile' policy changes, a silent transition in earth
history from the 'Holocene' to the 'Anthropocene' occurred due to human-induced
changes in climate change. This transition is also a result of overconsumption
through over-exploitation of the natural resource base. However, these multiple
'anthropogenic' changes have been challenged by climate sceptics, economic
lobbyists, ideologues and policy-makers, most prominently by President Trump and
his administration, who are totally ignoring the result of the peer-reviewed natural
science literature on the causes of global environmental change and their physical
effects.

Crutzen/Stoermer (2000; Crutzen/Brauch 2016) claim that "we are now in the
'Anthropocene'" implies that we—as part of humankind—have increasingly
become the major threat to our own survival as a species and to biological diversity
through our way of life and our economic model based on cheap hydrocarbon energy
sources. Our burning of coal, oil, and gas has exponentially increased the global
concentration of CO_2 in the atmosphere from an average of 260–280 parts per
million (ppm) since the end of the glacial period 12.000 years ago, or 315 ppm (in
1958) to 405 ppm by end of 2016[12] and to 408.35 ppm in February 2018 and to
409.50 ppm on 2 April 2018 – so far the highest was 412.63 ppm on 26 April 2017 –
according to official NOAA data provided by this US government research agency.[13]

8.1.4   An Emerging Dispute: Scientific Evidence Versus Economic Interests, Political Ideologies, Belief Systems, and Ignorance

However, these objective scientific measurements of a leading US governmental
scientific institution have hardly any impact on the present U.S. administration and
its policies. With the election of the Trump administration in 2016, anti-intellectual
tendencies and powerful economic interests prevail in the most powerful country
with the biggest global economy and the second largest greenhouse gas emissions.

---

[12]See at: https://www.esrl.noaa.gov/gmd/aggi/aggi.fig2.png.

[13]See at: https://www.esrl.noaa.gov/gmd/ccgg/trends/ (3 April 2018); https://www.co2.earth/daily-co2 (3 April 2018).

Scientific evidence is being replaced by 'alternative facts' based on contested claims of climate change sceptics and deniers (Klein 2011). Donald Trump twittered in November 2012: "The concept of global warming was created by and for the Chinese in order to make U.S. manufacturing non-competitive"; on 1 June 2017, he announced his decision to "withdraw from the *Paris Climate Accord*."

Thus, a major new dispute has emerged since Donald Trump has been sworn in on 20 January 2017 as President of the most powerful country in the world, a country with a strong foundation in scientific research, whose scientists have provided much of the scientific evidence and knowledge on which the debate on Global Environmental Change (GEC) and the Anthropocene is based. Since 2017, major US funding for climate change research and for global climate change activities of international organisations has been cut to supress research results that conflict with the economic interests, ideological belief system, and ignorance of the world's most powerful policy maker. Don J. Frost, Jr. and Henry C. Eisenberg of Skadden Arps Slate Meagher & Flom LLP have summarised on 23 January 2018 this recent development, in which the "Trump Administration Rolls Back Climate Change Initiatives":

> The Trump administration has proposed reducing the EPA's 2018 budget by over 30 percent, including a proposed staffing cut of 25 percent. The administration has specifically targeted for elimination the EPA's Global Climate Change Research Program and various climate-related partnerships with outside groups, such as the EPA's state and local climate and energy programs. The justification for these proposed cuts is that climate change and sustainability are not among the EPA's core statutory obligations to protect air, water and land. The administration also has proposed substantial cuts to the Department of Energy's Office of Energy Efficiency and Renewable Energy; cuts to NASA earth science missions, including missions to track the distribution of carbon dioxide emissions and to better understand climate change; a reduction in support for climate science at the Department of the Interior; a reduction in funding for the U.S. Geological Survey's carbon sequestration research; and cuts to climate change programs at the U.S. Agency for International Development and the State Department.[14]

On 13 February 2018, Scott Waldman reviewed in *Scientific American* the proposed budget cuts for science and especially for climate science in the US Federal Government Budget Proposal for FY 2019 that focus specifically "on climate science, renewable energy research and climate mitigation efforts across a variety of federal agencies, including NASA, NOAA, U.S. EPA and the departments of the Interior and Energy", where at NOAA, "the budget would be trimmed 20 per cent, by about $1 billion, to $4.6 billion in 2019", and the National Science Foundation would face a 30 percent cut.[15]

The impact of these proposed budget cuts has already resulted in self-censorship, in which many US climate scientists have replaced "climate change" with "global

---

[14]See: Skadden Arps Slate Meagher & Flom LLP: "Trump Administration Rolls Back Climate Change Initiatives", 23 January 2018; Lexicology, at: https://www.lexology.com/library/detail.aspx?g=ebb4ab65-2664-46aa-abb7-8c9bf597e927.

[15]Scott Waldman: "Trump Budget Would Slash Science across Agencies", in: *Scientific American, 13 February 2018*.

change" or "extreme weather" in their funding proposals[16] and several US climate scientists have already considered continuing their research in other countries, responding to offers by French President Macron[17] and to job offers in other European countries.[18] However, many scientists did not remain silent and took their protest to mass demonstrations in the US and in many other countries as they saw the integrity of scientists challenged. On Earth Day, on 22 April 2017, more than one million scientists and citizens demonstrated in the US and in many countries abroad to emphasise "that science upholds the common good and to call for evidence-based policy in the public's best interest". A second 'March for Science' day was held on 14 April 2018 across the United States and in 6 other countries around the globe.[19]

The ideologically motivated climate scepticism, the efforts to cut science funding on themes that contest this ideology, and the related economic interests addressed core issues of the integrity, freedom and independence of evidence-based science and its contribution to addressing and solving fundamental challenges facing humankind and its survival as a species. The climate debate in the US is just one aspect of the growing challenges that have recently undermined cooperative policy approaches to addressing the risks posed by many human interventions into the earth system in a forward-looking and proactive way.

8.1.5 New Poisons or 'Ugly Responses to Globalisation' in Europe

Since 2017, the U.S. and parts of Europe have faced an ideology- and profit-driven ignorance that has resulted in global regression and in nightmares triggered by 'ugly responses to globalisation':

- *democratic regression,* shown in the authoritarian tendencies of Presidents Putin (Russia) to use force in support of territorial expansion and Erdoğan (Turkey) who has repressed a large minority in his home country, purged the government and the universities, put hundreds of journalists in prison, and invaded a neighbouring country;

---

[16]Rebecca Hersher, "Climate Scientists Watch Their Words, Hoping To Stave Off Funding Cuts", 29 November 2017; at: https://www.npr.org/sections/thetwo-way/2017/11/29/564043596/climate-scientists-watch-their-words-hoping-to-stave-off-funding-cuts (3 April 2018).

[17]"Macron awards US scientists grants to move to France in defiance of Trump", in: *The Guardian,* 11 December 2017; at: https://www.theguardian.com/environment/2017/dec/11/macron-awards-grants-to-us-scientists-to-move-to-france-in-defiance-of-trump (3 April 2018).

[18]"As America quits, Europe tries to lead on climate change", in: *The Economist,* 6 July 2017; at: https://www.economist.com/news/europe/21724834-g20-will-test-whether-world-can-implement-paris-emission-accords-america-quits (3 April 2018).

[19]"March for Science"; at: https://en.wikipedia.org/wiki/March_for_Science (3 April 2018).

- *democratic principles* have been challenged in several East European EU counties, especially by policies pursued by the Hungarian prime minister Orbán, by the government of Romania and by the efforts of Polish party leader J. Kaczyński and his party PiS to undermine the independence of the courts and judicial system, which has triggered countermeasures by the European Commission and Parliament;
- *nationalist and protectionist regression* by Trump, by the proponents of Brexit in the UK, by populist, nationalist, xenophobic and anti-Islamic parties in France (*Front National*), Germany (*Action for Germany* [AfD]), Italy, the Netherlands (G. Wilders' *Partij voor de Vrijheid*), in Belgium (*Vlamse Block*) and in Scandinavian countries but also in China and India and in many other regions;
- *scientific regression,* illustrated by a policy-driven purge of universities in *Turkey* where tens of thousands of professors, scholars and researchers have lost their jobs, and in the USA by announcements by the *Trump* administration to cut funding and constrain the academic freedom of climate scientists working for the US government;
- *possible environmental regression*: the growing respectability of climate change sceptics, ideologues and lobbyists of the carbon and car industries who have also increased their pressure on the European Commission to soften its commitment to its key climate change policy targets to be reached by 2030.

Some of these political tendencies were triggered or reinforced by the massive migration from war-torn countries in the wider Middle East (Syria, Iraq, Afghanistan) and economic migration triggered by the pull factors of the North, and the desperation in many countries of the global south. These alternative agendas have made forming a new government or progressive coalitions more difficult in many countries (e.g. in the Netherlands, Germany, and Italy).

8.1.6 Competing Mindsets and Worldviews

In research and in policies for dealing with global environmental and climate change in the Anthropocene, three mindsets among policymakers and two worldviews among scientists have been distinguished (Oswald Spring/Brauch 2011):

- *Business-as-usual* has been the dominant position, framed by the
  - *Hobbesian obsession*[20]: We have the power and the resources and must not learn!

---

[20]I am grateful for a critical comment on this reference to Amb. Luis Alberto Padilla Menendez (Guatemala).

- *Neoliberal Washington Consensus*: We control the policies, means and resources!
- *Adherence to the Western way of life* that may not be challenged: We will not change!

• *Proponents of sustainable development and sustainability transition* are often on the defensive and argue that

- *We are the major threat for the survival of humankind (climate change and conflicts);*
- *We can and must be the solution, e.g. by initiating strategies of sustainability transition;*
- *We need strategies and policies for environmental conflict prevention and sustainability transition.*

• A third mindset among extreme policymakers in several countries is spreading, and is partly driven by what was introduced above as the *ugly responses to globalisation*:

- *Neo-authoritarianism (Russia, Turkey and strong tendencies in Hungary, Poland, Romania);*
- *Populism (in the US and several West European countries);*
- *Nationalism (represented by political parties and xenophobic and anti-Islamic movements);*
- *Protectionism (at present pursued especially by the Trump Administration);*
- *Religious fundamentalism and intolerance (in parts of the Middle East, Asia, Europe and U.S.);*
- *Scientific scepticism and alternative facts (e.g. stressed by the Trump Administration).*

8.1.7 Key Questions of This Chapter

From the vantage point of the alternative worldview of *proponents of sustainable development and sustainability transition* this chapter will address the following questions:

• Which changes have occurred since 1989 that could be directly observed (fall of the Berlin Wall in 1989), retrospectively analysed (environmental regression), and which were only recently socially constructed with the concept of the Anthropocene? (part 2)
• How did policymakers, countries and international organisations react to the missed opportunities and populist and fundamentalist challenges? (part 3)
• How did the social sciences, political science, international relations and peace studies react to these challenges? (part 4)

- How should peace researchers address these challenges: reformulating goals, perspectives, strategies and policies? (part 5)
- What can a new scientific approach from a 'peace ecology perspective' contribute to the scientific diagnosis of present trends and challenges? (part 6)
- What does the goal of a 'sustainable peace' imply and how may a process of sustainability transition contribute to the realisation of this goal? (part 7)
- What may a new approach of a 'transformative science' contribute to the analysis of a process of sustainability transition? (part 8)
- What are possible pillars and components of a peace ecology perspective for the Anthropocene? (part 9)
- In the concluding part the author will respond to these questions (part 10) and summarise the results of his deliberations.

Some readers may disagree with the policy analysis of present and emerging future trends, for some scholars these questions are too broad to be discussed, and for some policy makers or advisers the analysis may appear to be too theoretical to appeal to practitioners. A threefold holistic bridge-building effort to break out of the narrow framing of most contemporary scientific contributions that appeal to only a few specialists is the goal of this contribution:

- to address the difficulties and obstacles for moving from knowledge to action;
- to build bridges for the discourse between natural and social scientists working on global environmental issues and problems of global environmental change; and
- to develop this author's emerging thinking on a peace ecology further by linking the knowledge developed both in environmental studies and in peace research.

The motive of this broad approach is to contribute to the framing of an action oriented research guided by principles of a peace ecology and aiming at the development of proactive policies for a sustainability transition that aims at a gradual decarbonisation of the economy towards what some scientists have called a 'good Anthropocene'. 'Peace with nature' has also been an aim for many indigenous people and cultures (e.g. Pacha Mama) who have lived in 'peace with nature' for generations (Oswald Spring et al. 2016a).

However, a world with 10 billion people by end of this century that aims at realising the goals of a sustainable development and of a sustainable peace requires far more complex political coping and transformation strategies than encompassed by the traditional knowledge of indigenous peoples.

For those 'realists' that are influenced by a *business-as-usual* mindset and who combine a *Hobbesian* approach to international relations with a *Neoliberal Washington Consensus* on the economy and globalisation and the defence of the *Western way of life*, the goals of these needed transformations strategies may appear utopian. However, if humankind wants to avoid the civilisational dangers the *business-as-usual* approach may face by not acting or acting too late, an action-oriented thinking is needed to cope with the ongoing political challenges and to face the new global environmental problems in a proactive way to avoid climate conflicts in the future. Building new and higher walls or fences against massive migration streams will not solve any of these problems, as has been evident since the Berlin Wall fell peacefully in November 1989.

Starting the argument from an alternative sustainability paradigm that aims at developing strategies and policies for sustainability transition for most modern economic sectors, a new 'scientific revolution (Kuhn 1962) for sustainability' (Clark et al. 2004), a 'second Copernican Revolution' (Schellnhuber 1999) or a new 'contract for sustainability' (WBGU 2011) are needed. These natural scientists have suggested a fundamental change in thinking, or a new scientific worldview on sustainability that must gradually develop from an intellectual 'niche' into the 'mainstream' in society, in the business community, and in the political realm. A paradigm shift to sustainability thinking is needed – they argue – similar to the Copernican worldview that challenged the prevailing orthodoxies of the Catholic Church and its inquisition practices in the 16th century in Europe.

Such a fundamental change in thinking and in 'Politik' was possible in the past – at the end of the medieval period in the 16th century after a period of turbulence and violence – and is possible and necessary in the future as human beings are the only ones that can bring about this fundamental transformation; we have been the 'threat' that took us into the 'bad Anthropocene'[21] therefore we must become and be the solution, as scientists, engineers, citizens, and consumers by moving towards 'a good Anthropocene'.[22] Elena Bennett of Future Earth and Albert Norström of the Stockholm Resilience Centre outlined this approach arguing:

> This is a suite of research activities that aim to solicit, explore, and develop a suite of alternative, plausible visions of "Good Anthropocenes" – positive visions of futures that are socially and ecologically desirable, just, and sustainable. Popular and scientific forecasts of the future are dominated by dystopian visions of environmental degradation and social inequality. Scientific assessments have demonstrated that more positive, desirable trajectories and futures appear to be possible, however thus far, the global community's efforts to imagine positive futures has led to visions that tend to be utopian, not well articulated, and too much like the world we already live in, and the steps to achieve these worlds remain unclear. This initiative aims to initiate wider global discussions of the kinds of positive social-ecological futures people would like to create and to expand discussions beyond efforts focused on avoiding negative futures or taking incremental steps forward.

> A future "Good Anthropocene" will probably be radically different from the world in which we are currently living. It will require fundamental changes in values, worldviews, relationships among people, and between people and nature. This initiative aims to scope out

---

[21]See Dalby (2016); at: http://journals.sagepub.com/doi/abs/10.1177/2053019615618681; Kunnas (2017); "at: http://journals.sagepub.com/doi/abs/10.1177/2053019617725538?journalCode=anra (7 April 2018).

[22]See the debates at: https://goodanthropocenes.net/definitions-of-a-good-anthropocene/ (7 April 2018); Andrew C. Revkin: "Building a 'Good' Anthropocene From the Bottom Up", 6 October 2016; at: https://dotearth.blogs.nytimes.com/2016/10/06/building-a-good-anthropocene-from-the-bottom-up/ (7 April 2018); "Bright spots: seeds of a good Anthropocene"; at: http://www.futureearth.org/bright-spots-seeds-good-anthropocene (7 April 2018).

some of these radical changes that go beyond incremental improvements (e.g., reducing pollution or increasing the environmental efficiency of agricultural production) that are the focus of much of today's sustainability dialogue. The seeds of these futures already occur in many places around the world. Identifying where these elements of a Good Anthropocene currently exist, and understanding how and why they occur, can help us envision how we might build on those examples to create new, positive futures for the Earth and humanity.

This initiative is co-led by ecoSERVICES[23] and PECS,[24] and will bring together researchers from multiple disciplines and a broad geographical distribution to (a) discuss the meaning and characteristics of a 'good' Anthropocene, (b) develop criteria by which one might assess seeds of a 'good' Anthropocene, (c) design workshops and other means to collect ideas of seeds from around the world and (d) identify broader networks that could be targeted and involved in this process. These discussions will be anchored in a diverse global set of regional studies of human-nature interaction, in places such as South Africa, the Arctic and Sweden. This regional variation will be linked to dialogue with a broad range of global leaders to facilitate wide-reaching discussions and engagement with society at large via social media channels and our existing scientific networks.

One goal of the following tour through many ongoing and often unconnected scientific debates is to contribute to a framing of such a needed comprehensive transformation in the context of a new 'ecological peace policy' aiming at a 'good Anthropocene'.

During the Cold War thinking on and framing a 'peace policy' implied thinking beyond the orthodoxies of the Cold War based on deterrence theories and policies driven by concepts of *mutually assured destruction* (MAD) on such issues as peaceful change, confidence building measures, and confidence building defence. The end of the Cold War in 1989 became possible not because of Western military and economic superiority, as the US conservatives claimed (Anderson 1990), but by jumping out of the dilemmas of nuclear deterrence and of the constraints of a militarised society by aiming at 'Glasnost' and 'Perestroika' (Lebow 1994; Lebow/Risse-Kappen 1996). Gorbachev did not only coin these terms but was also one of the key political conceptualisers of environmental security (1988). These reforms failed to 'save' the Soviet Union as a second superpower but this different worldview made peaceful transition possible all over central and eastern Europe, resulting in reunification of Germany but also reunification of Europe within the EU.

The rapid and often chaotic transformation from Socialism to Capitalism, from a planning to a market economy and society, produced in Russia and in many

[23]On EcoSERVICES see at: http://www.futureearth.org/projects/ecoservices and at: http://www.futureearth.org/ecoservices/ (7 April 2018).
[24]The Programme on Ecosystem Change and Society (PES) is directed by Albert Norström at the Stockholm Resilience Centre; see at: albert.norstrom@stockholmresilience.su.se and at: http://www.stockholmresilience.org/research/research-programmes-and-projects/2016-03-09-programme-on-ecosystem-change-and-society-pecs.html (7 April 2018).

post-communist countries in central and eastern Europe many losers from this transformation and from globalisation, who see in the 'poisons' of the 19th and early 20th century a solution for the future. This backward focus of political ideologies avoids and increasingly makes it more difficult for policy makers to form coalition governments that have the courage and political will to address the challenges we face in a proactive way, or to move from a 'bad' to a 'good Anthropocene'.

Such a broad holistic conceptual and political thinking is needed to develop politically relevant action, change, and transformation-oriented strategies and policies. If citizens fail to contain, control, and overcome the resurgent poisons of the past, those supporting the ugly responses to globalisation may make an agenda for the needed strategies for a sustainability transition impossible.

Thus, new policies for the Anthropocene require bold new thinking on achieving a strategy of sustainability transition with the goal of a sustainable peace without falling back into the nationalist traps of the past that contributed to two world wars during the short 20th century (1914–1989).

8.2   Contextual Change: We are in the Anthropocene!

Since 1989 humankind has experienced a manifold global contextual change that has affected policymaking and scientific analysis:

- through the first peaceful change in international order, with the fall of the Berlin Wall (1989), the dissolution of the Warsaw Treaty Organisation (1991) and the implosion of the Soviet Union (1991). This visible political turn has reduced—at least temporarily—the probability of a major nuclear war and made several disarmament treaties possible between 1990–1996. However, the unique historical opportunity was lost to build a new lasting cooperative global peace order based on a collective security system as envisioned in the UN Charter of 1945.
- In the immediate aftermath of this peaceful change and during a more cooperative international environment, issues of global environmental and climate change were put on the international agenda (of the UN GA and SC, G7 and G8) since 1988 and in June 1992 in Rio de Janeiro at the *UN Conference on the Environment and Development* (UNCED) that resulted in the setting up of three environmental regimes on global climate change (UNFCCC 1992), on biodiversity (CBD 1992) and on desertification and drought (UNCDD 1994).

These positive global changes towards disarmament of nuclear, chemical and conventional weapons, the reduction of troops in Europe, and the global decline in military expenditures between 1990 and 1995 and the progressive emergence of new global environmental regimes were gradually challenged by retrograde contrary policy developments:

- A gradual shift from military procurement spending towards military research and development (Brauch 1994) with an increase in arms exports to partly compensate for the decline in the defence spending in NATO and former WTO countries, an increase in nationally driven, ethno-religious wars in the Balkans and in new asymmetric wars (Kaldor 1999, 2002; Münkler 2004) 'in Europe as well as in the global South', and increasing fundamentalist terrorist attacks in North America, Europe and Asia.[25]
- Since the mid-1990s, regressions have occurred both with regard to disarmament (with the blocking of the Comprehensive Treaty on Nuclear Testing) and climate change (with the inability to ratify the Kyoto Protocol in the U.S. Senate), with the failure of the Copenhagen UNFCCC Conference (2009) to agree on a new climate change agreement and in June 2012 with the failure of the Rio+20 summit to adopt a legally binding regime on issues of governance of global environmental change.

What were the reasons for the changes that occurred during the 1990s in both the security realm and on environmental preferences and governance? The peaceful change since the fall of 1989 was unexpected for most national foreign and defence establishments, international organisations, civil and social movements, non-governmental organisations, for scientific experts, policy advisers, and the media. While there were policy and media discussions on a 'peace dividend' there was no integrated longer-term planning for a new cooperative global order and governance for peace, security, development, and the environment as had existed in the US State Department from 1939 to 1945 (Notter 1949; Layne 2006).

When Milosevic played the nationalist card in Yugoslavia (1989), Slovenia and Croatia seceded (1991), and ethno-religious conflict erupted in Bosnia-Herzegovina (1992–1995), neither the European Union, NATO, nor the national foreign and security establishments had any plan for dealing with this instrumentalisation of nationalist tendencies for acquiring power in the post-Communist and Yugoslav spaces. Short-term policy responses were driven by historical memories of World War II alliances (e.g. with Yugoslavia among leading EU members) and by traditional military mindsets (by the Clinton Administration, NATO), resulting in the use of outside military force and intervention in Bosnia-Herzegovina (1995/1996) and against Serbia (1999) over Kosovo's independence without an endorsement by the UN Security Council.

This implied a shift from a gradual strengthening of the *Conference on Security and Cooperation in Europe* (CSCE) that became the *Organisation on Security and*

---

[25]These changes have been documented in many publications http://www.pcr.uu.se/research/ publications/ of the Department of Peace and Conflict Research of Uppsala University and through its *Uppsala Conflict Data Program* (UCDP) at: http://ucdp.uu.se/?id=1 and in the "Conflict Barometer" (from 1997 to present) of the Heidelberg Institute for International Conflict Research; at: https://www.hiik.de/en/konfliktbarometer/.

Cooperation in Europe (OSCE) in 1994, which could be seen as a regional arrangement or agency under Chapter VIII of the UN Charter as the core of a regional collective security system. Although both CSCE and OSCE had legitimacy under the UN Charter, they lacked the political, economic and military resources to act. Rather, supported by traditional security and military mindsets, the operations gradually moved from Vienna (CSCE, OSCE) to Brussels (NATO) from the diplomats to military planners and establishments.

It took some time for the EU and its member states to develop powerful diplomatic tools to diffuse the nationalist tendencies in the citizenship laws in the Baltic republics (Max van der Stoel, OSCE High Commissioner for Ethnic Minorities in the mid-1990s) and with regards to the Hungarian minorities in neighbouring countries (Slovakia, Romania, Serbia) with its Balladur Plan (OSCE Stability Pact proposed by former French Prime Minister Eduard Balladur) by successfully using the 'carrot' of EU membership against nationalist politicians and policies.

While both initiatives in the OSCE context – supported by the EU – helped to diffuse and avoid additional violence in the Baltics and against the Hungarian minorities in the Balkans, the manifold turns backward between 1992 and 1997 both in U.S. domestic politics and its foreign and defence policies during the Clinton Administration (triggered by the Republican controlled Congress from 1995) prevented the ratification of new disarmament treaties (Comprehensive Nuclear-Test-Ban Treaty of 1996) and environmental conventions (Convention on Biological Diversity of 1992, Kyoto Protocol of 1997).

A decade later, following the global financial and economic crisis that gradually emerged from the U.S. sub-prime mortgage debacle (2007), the collapse of the investment bank Lehman Brothers (2008), and the great recession (2009–), environmental issues were in retreat. These challenges resulted in massive bail-outs of financial institutions by the Obama administration and many European governments as well. In Europe, a debt and banking crisis in several European countries using the Euro (Ireland, Spain, Portugal, Greece, Cyprus, Italy, France et al.) dominated the agenda and contributed to:

- the failure of the UNFCC's COP 15 in *Copenhagen* in 2009—this was partly due to the tactics of the Obama administration, which faced increasing opposition from important segments of U.S. industry, supported by an increasingly climate-sceptic U.S. Congress and a growing tendency towards climate denial in U.S. media, such as by Fox News;
- the failure of the second *Rio Conference* (Rio+20) in 2012 to adopt a legally binding final document entitled "The Future We Want", which included many suggestions made by the *United Nations Conference on Sustainable Development* (UNCSD) both before and during the Rio+20 meeting;
- The *Paris Agreement* of 2015, which combined 'voluntary commitments' by the parties into a legally binding international treaty. It entered into force on 4 November 2016 after the U.S. and China jointly announced their ratification on 3 September 2016.

Both the final documents of Rio de Janeiro in 2012 and the Paris Agreement of 2015 watered down the urgency for global action because U.S. domestic politics meant that a Republican-controlled U.S. Congress made it impossible for the Obama administration to ratify either document in the US Senate. This was a result of fundamental changes in the Republican Party since Presidents Ronald Reagan (1981–1989) and George Bush (1989–1993). In 1988 Reagan put climate change on the agenda of the G-7. Bush Sr. supported the negotiations of the climate and biodiversity conventions and signed and ratified the first, but only signed (and did not ratify) the second. During Clinton's presidency, the Republican opposition prevented the ratification of the *Comprehensive Test Ban Treaty* (CBD 1992) and of the Kyoto Protocol (1997).

The Obama administration decided in summer 2016 to join the *Paris Climate Accord* without Senate ratification, while the Republican presidential candidate Donald Trump opposed the Paris Agreement during his campaign and on 31 May 2017 the Trump Administration announced its withdrawal. Thus, the international community faces a severe dilemma:

- Not acting now to implement the voluntary national commitments to reduce the projected global average temperature increase to 2 °C or even 1.5 °C above pre-industrial levels by the end of this century will increase the projected physical consequences of climate change: (a) rise in global average temperature, (b) increase in the variability of precipitation, (c) rise in sea level, and (d) increase in the number and intensity of climate-induced natural hazards.
- Future generations will face the societal, political and economic consequences and their costs. The probability will increase that the physical effects of climate change may result in massive climate-induced migration and violence at the regional and national levels (Schellnhuber et al. 2006, 2012, 2016; Scheffran et al. 2012).

The alternative course has been addressed by the agreements between the heads of states of the G-8 in 2007 to aim for 80–95% reduction in CO_2 by 2050, and by those of the G-7 in 2015 to move towards a decarbonised economy by the end of the twenty-first century. The European Commission has taken up these goals in its long-term roadmaps for energy and transport leading to a *competitive low carbon economy* by 2050 (Brauch 2016a).

These two opposite trends in peace and security and environmental policies during the 1990s require a wide scope of historical and policy analysis that combines both political issue areas and scientific research programmes and combined scientific programmes of peace and ecology studies or a more comprehensive, holistic and consilient (Wilson 1998) peace ecology approach.

Independent from these policy trends and occasional observations in the social sciences, a more stringent change remained unnoticed in the natural sciences for a long time: the 'silent transition' in the interactions between humankind and its natural, global, geophysical environment for which Nobel Laureate Crutzen (2002) coined the term 'Anthropocene', suggesting that we have entered a new phase of earth history. That this change has been human-induced was confirmed in five assessment reports of the *International Panel on Climate Change* (IPCC 1990, 1995, 2001, 2007, 2013/2014).[26]

This silent transition from the Holocene to the Anthropocene (Steffen et al. 2007, 2011; Zalasiewicz et al. 2008, 2010) was brought about by a complex interaction among five major drivers – initially in West and Central Europe and later in the U.S. between the *first phase* (invention of the steam engine) and *third phase* (use of the nuclear bomb and start of the nuclear era) of the *industrial revolution*:

- A major *cultural change* with the new worldview of 'enlightenment' (*Aufklärung*);
- The emergence of new *political ideologies* and a major *political change* with the American (1776), French (1789) and Russian (1917) Revolutions;
- A fundamental change in the *'capitalist' economic system* in the United Kingdom where a new *financial bourgeoisie* provided financial resources and capital for *entrepreneurs* (Polanyi 1944);
- A *scientific revolution* in the natural sciences where new basic knowledge triggered a period of practical inventions resulting in *technological innovations* (engineering) that facilitated a process of fundamental *economic change* in which the UK, Germany and later the U.S. were among the key pioneers of change in the natural sciences and in technology development (Jochum 2010).
- A fundamental *economic change* (Landes 1969; Ziegler 2010) that became possible as a result of the complex interaction of different drivers with a change in the energy system relying on cheap fossil energy sources (coal, oil and natural gas).

However, fossil fuel, this cheap facilitator that fostered the 'American' or 'Western way of life' spread globally during the 'American century' and with the U.S. political, economic and military dominance during the Cold War, resulted in a massive human interference in the earth system with ozone layer depletion and the accumulation of greenhouse gases in the atmosphere. The consequences were gradually analysed, interpreted, and understood by the expanding research on global environmental change and climate change during the past 50 years.

8.3 Policy Responses to the Contextual Changes

There have been several policy responses to the contextual changes in the early 1990s and to the opposite developments in the second part of the 1990s both in the political-military and in the economic and environmental areas that have represented different interests:

---

[26]IPCC: *Fifth Assessment Report* (2014); United States Global Change Research Program (2009); Oreskes (2004: 1686).

- At the Paris summit in November 1990 the mobilisation of the world for a war against Iraq had become a major concern of the U.S. Administration of George Bush (1989–1993).
- After the failure of the CSCE and OSCE initiatives of the early 1990s in the wars between Yugoslavia (Serbia) and Slovenia and Croatia, in the next two wars in the post-Yugoslav space (in Bosnia, Hercegovina [1995] and Serbia [1999]) the Clinton Administration returned to a military strategy in the NATO context without an endorsement by the UN Security Council.
- On environmental issues the Clinton Administration continued the cooperative approach of its predecessor partly in its effort to widen its national security approach to increasingly incorporate environmental security concerns and later climate security threats for legitimating its national security strategy (Brauch 2011).
- The major political turning point occurred when the U.S. Republican Party controlled both houses of Congress after the elections in November 1994, 1996 and 1998 and used their majority in the U.S. Senate to block the ratification of the Comprehensive Test Ban Treaty (1996) and of the Kyoto Protocol (1997), thus preventing two cooperative global policy agreements.
- During the George W. Bush Administration (2001–2009) military strength and unilateralism in the 'war on terror' in Afghanistan (1991–) and Iraq (2003–) were central national strategy goals while climate change issues were internally challenged and downgraded.
- A third major turn in U.S. domestic politics on climate change occurred between 2008 and 2012. While Barack Obama won the Presidential election of 2008 on a progressive climate change policy, he failed to implement it in the U.S. Congress despite a Democratic majority (2009–2012) in the Senate. Public opinion in the U.S. on climate change issues had significantly changed, partly due to a massive economically-fuelled, ideologically-driven campaign supported by conservative media (Klein 2011).
- The increasing strength of the opposition on climate change issues in the U.S. media, in the U.S. Congress and in public opinion constrained the leverage of U.S. international policies on global environmental and especially global climate change policies, one among several reasons for the paralysis of international climate change governance since 2009 that had partly come to an end with the signing of the Paris Climate Change Treaty of 2015.

In retrospect, after the initial cooperative initiatives in the early 1990s both on disarmament and global environmental issues, international policies returned to short-term reactive crisis management partly driven by the dominant Hobbesian national security mindset and a growing scepticism towards and downgrading of the urgency of global environmental challenges.

However, in many EU countries and increasingly in China, India, South Africa, Brazil, and Mexico, energy efficiency has been rising and renewable energy sources (solar and wind power) are increasingly used for electricity generation. Processes for a sustainability transition were launched in the industrial, energy, housing and transport sectors, and in urban planning in many countries. Employment in these new industrial areas has been rising both in industrialised and in industrialising countries. Thus, a new worldview aiming at sustainability and a process of sustainability transition has been spreading since the early 21st century.

While the Trump Administration has stressed the importance of the carbon sector in its economic and climate policy, the EU, China, and many other countries in North and South are increasingly stressing sustainability goals and economic initiatives. In the security realm, national security strategies and military tools in dealing with persistent terrorist and fundamentalist threats have been strengthened by the Trump Administration.

The assessment that we are now in the Anthropocene has also been attacked by many climate sceptics and deniers from a mindset of *business-as-usual* or by more radical 'ugly responses to globalisation'. The Anthropocene concept has been challenged by some social scientists and it has been attacked by ideology-driven propaganda institutes in the USA. On 24 January 2017, Ian Angus wrote that "a new conservative campaign aims to discredit efforts to define the new and dangerous stage of planetary history, by driving a wedge between social scientists and the Anthropocene Working Group" (Angus 2015, 2016, 2017). This was partly inspired by "anti-green, pro-nuclear and pro-capitalist ideologues at the *Breakthrough Institute* (BTI)" that was founded by Nordhaus and Shellenberger (2004) who deny any environmental crisis, call for more technology, expand capitalism, and give up trying to harmonise society with nature. They partly rely on Erle Ellis, the sole dissenter within the *Anthropocene Working Group* (AWG).

Thus, social scientists face a dual challenge in putting the Anthropocene on the research agendas of their disciplines:

- from within by the so-called 'ecomodernists' who praised the death of environmentalism (Shellenberger/Nordhaus 2004; Nordhaus et al. 2015a, b);
- from the political realm by the Trump Administration and its climate sceptical allies and partners in different parts of the world.

On 6 November 2012, Donald Trump tweeted: "The concept of global warming was created by and for the Chinese in order to make U.S. manufacturing non-competitive."[27] On 1 June 2017, the Trump Administration announced that it would withdraw from the Paris Treaty on Climate Change.[28] After receiving an internal governmental Climate Science Special Report in early August 2017 that concluded that global climate change was real, on 21 August the Trump

---

[27]Tim Marcin: "What Has Trump Said About Global Warming? Eight Quotes on Climate Change as He Announces Paris Agreement Decision", in: *Newsweek*, 1 June 2017.
[28]Michael D. Shear: "Trump Will Withdraw U.S. From Paris Climate Agreement", in: *The New York Times*, 1 June 2017, Ari Natter: "Donald Trump Notifies UN of Paris Exit While Keeping Option to Return", in: *Time*, 5 August 2017.

administration announced the non-renewal of the Advisory Committee for the Sustained National Climate Assessment.[29]

The scientific and political attacks on the anthropogenic causes of global environmental and climate change and on the concept of the Anthropocene, the cut in funding for global climate change research, and other decisions have all pointed to a fundamental conflict between peer-reviewed scientific evidence-based research results on the one hand and economically driven interests and ideological belief systems on the other.

How should social scientists, international relations experts, and peace and environmental specialists respond to these regressive trends in international relations and many national policies? Are our scientific concepts, theories and methods sufficient for examining and interpreting these manifold old threats, challenges, and new vulnerabilities and risks?

Due to space constraints, this text will focus below only on new global environmental vulnerabilities and risks, the question of whether the potential processes of sustainability transition may contribute to avoiding new and additional military threats, and their potentials for achieving the goal of peace with nature in the context of a wider goal of a sustainable peace.

8.4 Contributions of the Social Sciences and Peace and Ecology Studies for this New Global Agenda

The present prevailing scientific trends are not very conducive for such a value-oriented wider conceptualisation of a *sustainable peace*. The dominant trend of professionalisation, important for future careers, focuses on highly specialised, often esoteric methodological and theoretical debates that avoid wider, holistic, inter- and trans-disciplinary and politically critical approaches, resulting in a depoliticised, uncritical social science that is primarily oriented at a narrow, specialised scientific community and much less on the enlightenment of a wider political and societal audience (Flinders/John 2013).

Engaging with a wider audience has been partly filled by specialised policy consultants and policy advisers who try to satisfy their clients, customers or funding agencies. They sometimes end up tasked to legitimise the policy interests and preferences of their funders and not to initiate politically relevant and critical scientific discourses and policy debates. This role in initiating scientific and policy debate has sometimes been taken over by political activists, social movements, and nongovernmental organisations.

---

[29]Juliet Eilperin: "The Trump administration just disbanded a federal advisory committee on climate change", in: *Washington Post*, 20 August 2017; at: https://www.washingtonpost.com/news/energy-environment/wp/2017/08/20/the-trump-administration-just-disbanded-a-federal-advisory-committee-on-climate-change/?utm_term=.fe019cabd981.

The impact of the work of peace scholars and ecologists is influenced by their primary audience: the *scientific community* that communicates increasingly through highly specialised peer-reviewed scientific journals. Some peace researchers try to serve political activists, societal movements, and nongovernmental organisations, with the goal of influencing policymakers in parties, governments and international organisations.

For young social scientists specialising in peace and ecological issues and problems in academia, the scientific community has become and remains the key audience and source of funding and the *contemporary requirements* for academic careers have discouraged wider, holistic, inter- and transdisciplinary approaches that challenge the prevailing specialisation trends. These new professional requirements do not encourage new approaches that are needed for addressing, examining, understanding, and communicating the complexity of the interactions between peace and security issues on the one hand and environmental problems and ecological concerns on the other.

Only a very few social and political scientists have been 'public intellectuals' who have addressed both audiences; (a) their own discipline by innovative theoretical, methodological and empirical publications in prestigious journals and widely noted and cited books, and (b) the public at large by writing articles in political journals and columns and op-ed pieces in high quality dailies, or addressing a policy-oriented audience via the new social media.

France has long had a tradition of 'public intellectuals'; Emile Zola, Jean-Paul Sartre, Pierre Bourdieu, Michel Foucault, and Jean-François Lyotard, to name but a few, who often intervened in public debates. This position of 'public intellectual' has been held in Germany for decades by Carl Friedrich von Weizsäcker, Jürgen Habermas, Hans-Magnus Enzensberger, and Ulrich Beck. In the US a few scientists, like Noam Chomsky, George Lakoff, Garry Wills, Francis Fukuyama, Stanley Hoffmann or Samuel Huntington, are widely noticed and cited outside academia.

However, on issues of peace research and global environmental change and on the Anthropocene, most 'public intellectuals' did not address these global environmental challenges. Exceptionally, Johan Galtung's early publications have been widely cited by authors in many scientific disciplines since the 1960s and Kenneth Boulding's work transgressed the boundaries of economics stimulating debates both in peace research and in ecology.

The recent debates on the 'Anthropocene', on the need for a 'second Copernican Revolution' for sustainability, or on new 'global boundaries' were triggered by the Dutch Atmospheric Chemist and Nobel Laureate, Paul J. Crutzen, by the German physicist Jochen Schellnhuber, director of the *Potsdam Institute on Climate Impact Research* (PIK), and by the Swedish ecologist Johan Rockström, director of the *Stockholm Resilience Centre* (SRC). While these three natural scientists are not 'public intellectuals' in the traditional sense, they introduced major new concepts that are being discussed by a wider audience outside their respective scientific disciplines but they have not yet triggered any broad debate within major societal groups, NGOs, or social movements nor have their concepts been taken up by high-level policymakers.

While the concept of 'sustainable development' that was first politicised by the Brundtland Commissions Report (1987) is widely accepted in academia, in society, in the business community, and in the political realm, the concept of 'sustainable peace' has been discussed only by small epistemic communities working on conflict prevention, peacebuilding, peace psychology or on 'peace with nature' (Brauch 2016a); it has been used in a few high-level UN debates (e.g. in 2017, 2018).[30]

The discussion on 'sustainability transition' has occurred so far primarily in the US (since 1976) and in Europe (since 2005) in the Netherlands, Belgium, UK, Germany, Switzerland, Denmark, and Sweden within the framework of the Sustainability Transition Research Network (STRN). The Handbook on *Sustainability Transition and Sustainable Peace* (Brauch et al. 2016b) has tried to bring these different conceptual discussions together. However, so far contributions by the social sciences and peace and ecology studies to this new global agenda have been limited and primarily addressed to within each practitioner's own scientific discipline or respective epistemic community.[31]

8.5 Scientific Tasks of Peace Research in the 21st Century

Conceptually and analytically linking peace and security studies with ecological analysis as suggested by Boulding (1966, 1978; Stephenson 2016) fifty years ago has remained an exception. The academic survival of young peace researchers and ecologists has become a major constraint and impediment for a more active societal and political role that reaches out to society at large. Overcoming these professional

---

[30]See: Regional Dialogues on Sustaining Peace: Shaping UN Strategies for 2018 and Beyond An Informal Planning/Scoping Discussion at Columbia University Law School; at: https://www.stimson.org/; see: UNGA, UNSC: "Peacebuilding and sustaining peace - Report of the Secretary-General, 18 January 2018, A/72/707–S/2018/43; UNGA High-level Meeting on Sustaining Peace, 24–25 April 2018; at: http://sdg.iisd.org/events/unga-high-level-meeting-on-sustaining-peace/: This meeting will "strengthen the UN's work on peacebuilding and sustaining peace. As specific objectives, the meeting will reflect on how to: (1) Adjust to the new UN approach to peace with the emphasis on conflict prevention; (2) Strengthen operational and policy coherence within the United Nations system towards peacebuilding and sustaining peace; (3) Increase, restructure and better prioritize funding to United Nations peacebuilding activities; (4) Strengthen partnerships between the UN and key stakeholders in the field; (5) Address the root causes of conflict to sustain peace; (6) Address the role of women and youth in peacebuilding."

[31]See Holzner (1968); Holzner/Marx (1979); Haas (1992). Holzner/Marx (1979) defined them as "knowledge-oriented work communities in which cultural standards and social arrangements interpenetrate around a primary commitment to epistemic criteria in knowledge production and application". Haas (1992) defined an epistemic community as "a network of professionals with recognised expertise and competence in a particular domain and an authoritative claim to policy relevant knowledge within that domain or issue-area."

constraints has remained difficult. Thus, combined potential contributions of peace and ecology studies for this global agenda are scarce. Since 1989, there have been two conceptual and empirical approaches that have addressed possible security impacts of global environmental change:

- *Environmental security* in the context of different referent objects (a) the nation state (national security), (b) international organisations (international security) or (c) of affected victims (human security) that have been used by the hazard, disaster or humanitarian communities (Brauch et al. 2008, 2009, 2011a, b)
- *Climate security* focusing on possible societal outcomes of the physical effects of global environmental and climate change in the Anthropocene (UN 2009; UNEP 2011; UNSC 2011; Scheffran et al. 2012) in the context of the same referent objects.

These analyses have been of interest for different organisations and actors: militaries in search of new humanitarian missions, development and humanitarian organisations, and the national and international disaster response community, striving to be better prepared for addressing, facing, responding and coping with extreme weather events (storms, floods, landslides, droughts, forest fires, famines etc.).

From a peace research perspective, the focus should go beyond the analysis of the question of coping with the consequences of global environmental change to addressing the issue area of "peace with nature", with strategies of sustainable development to reduce the probability of severe societal impacts of climate change with anticipatory or preventive measures.

This theme was addressed in the *Handbook on Sustainability Transition and Sustainable Peace* (Brauch et al. 2016b) where the "Scientific and Policy Context, Scientific Concepts and Dimensions" were outlined (Brauch/Oswald Spring 2016) and "Key Messages and Scientific Outlook" for a "Sustainability Transition with Sustainable Peace" were summarised. This author explored the concept of "Sustainable Peace in the Anthropocene" and the possible contribution of two multi-, trans- and interdisciplinary approaches we had introduced earlier as "political geo-ecology" for the need of linking global environmental research in the natural and social sciences (Brauch et al. 2011a, b) and to combine two research programmes of peace studies and ecology as "peace ecology" (Oswald Spring et al. 2014).

While this author examined in a previous text "Building Sustainable Peace by Moving Towards Sustainability Transition" (Brauch 2016a), this chapter tries to carry the argument in the next sections a step further by analysing from a peace ecology perspective how the goal of a sustainable peace through a process of sustainability transition could be furthered by a new scientific approach of a transformative science.

8.6 Scientific Perspective: The Peace Ecology Perspective

Conca (1994: 20) suggested an "environmental agenda for peace studies" and a discussion on whether "ecologically desirable futures include concerns for peace and justice". He argued that it is not enough "to place 'sustainable development' and 'ecological security' alongside peace or social justice as 'world-order values'" but that scholars must ask whether "not only their formal definitions, but also their metaphorical and institutional associations, further the purposes of peace, justice, and community". Later Conca (2002: 9) fundamentally challenged a core premise of the debate on environmental (in)security and conflict by asking "whether environmental cooperation can trigger broader forms of peace defined as a continuum ranging from the absence of violent conflict to the inconceivability of violent conflict" by also addressing "problems of structural violence and social inequality" and by "building an imagined security community" based on peaceful conflict resolution.

The concept "peace ecology" was first proposed by Kyrou (2007) of American University in three conference papers. In carrying the debate on "environmental peacemaking" (Conca/Dabelko 2002) further, Amster (2014) developed the peace ecology concept from discussions on the war economy, the commons, community resilience, resource conflicts and transborder cooperation. For Amster (2015)

> peace ecology is more than merely a conceptual synthesis of peace and ecology. In essence, it contemplates the ways in which the same environmental processes that often drive conflict—e.g. resource depletion, anthropogenic climate change, food and water shortages—can also become opportunities for peaceful engagement. ... People around the world who strive to manage scarce essential resources ... often find that their mutual reliance on the resource transcends even profound cultural and political differences—and in some cases even warring parties have found ways to work together positively on such issues. ... Peace ecology is concerned equally with the human–human and human–environmental interfaces as they impact the search for peace at all levels (Amster 2015: 8–9).

Amster (2015: 143ff.) reviewed different concepts of and approaches to environmental cooperation, of *environmental dispute resolution* (Caplan 2010), *environmental peacemaking* (Conca/Dabelko 2002) and *environmental peacebuilding*. He did not discuss issues of sustainability transition or of the long-term global transition and transformation of the national and global economies that needs to be achieved through strategies aiming at a gradual decarbonisation and dematerialisation of the economy.

Oswald Spring, Brauch and Tidball (2014: 18–19) introduced peace ecology as a linkage concept bridging the research programmes on peace and ecology within the framework of six conceptual pillars: peace, security, equity, sustainability, culture, and gender, where *negative peace* (non-war) is defined by the linkages between peace and security, while for the relationship between peace and equity the concept of *positive peace* is defined by peace with social justice and global equity; for interactions between peace, gender and environment, the concept of *cultural peace* was proposed, and for the relations between peace, equity, and gender the concept of an *engendered peace* was suggested.

These five pillars of peace ecology point to different conceptual features of peace. The classic relationship between 'international peace and security' in the UN Charter refers only to a narrow *negative peace* without war and violent conflict. It aims for the prevention, containment and resolution of conflicts and violence and the absence of 'direct violence' in wars and repression. To achieve peace with equity, or *positive peace,* requires the absence of 'structural violence', which is achieved by overcoming social inequality, discrimination, marginalisation and poverty, where there is no access to adequate food, water, health, or educational opportunities.

This author (Brauch 2016c) discussed peace ecology primarily in the context of his research on global environmental change and the Anthropocene. Policies aiming for 'sustainability transition' are part of a positive strategy that addresses possible new causes of instability, crises, conflicts, and in the worst case, even war. These causes may be either the scarcity of fossil energy sources or the possible security consequences of anthropogenic global environmental and climate change, either of which may be triggered by linear trends as well as by chaotic tipping points (Lenton et al. 2006). Thus, a peace ecology perspective of global environmental change should be developed, including

- a new scientific agenda that should formulate a goal and outline a process;
- new methods in coping with and communicating complexity;
- the issue areas of peace, security and the environment combined as major themes of human survival that require extraordinary measures.

Addressing 'peace with nature' differs from 'ahimsa' and its Hinduist and Jainist contexts and should address the causes of *global environmental change* (GEC) and *sustainability transition* (ST) as the process of a fundamental change aiming at a progressive decarbonisation and dematerialisation of the economy.

8.7 Sustainable Peace Through Sustainability Transition

Possible consequences of policies of non-action (Stern 2006) driven by a worldview or mindset of 'business-as-usual' have been discussed within the debate on *environmental* and *climate security*: environmentally-induced or forced migration, crises, conflicts (Brauch 2002) and in the very worst case as 'climate wars'.[32] Climate sceptics and deniers have fuelled the propaganda of populist politicians,

---

[32]See: Williams, Nathan, film, 2008: *The Climate Wars:* BBC Earth: Episode 2; at: http://freedocumentaries.org/documentary/fightback-bbc-earth-the-climate-wars-episode-2-of-3; Dyer (2011); Welzer (2012); Butler (2017).

and parties, among them U.S. President Trump and his Administration.[33] Some have also challenged the argument on the security consequences of climate change. The Trump Administration called for cuts of public spending on climate change research and for a stop to research on climate change impacts for security.[34]

From a peace ecology perspective, the policy problem and the related scientific research question should be defined as how these negative consequences could be avoided by forward-looking, anticipatory, and preventive activities. This requires a totally different scientific worldview, aiming at the development and implementation of strategies, policies, and measures focusing on the dual goals of a 'sustainable development' and 'sustainable peace'.

In the policy context, the first concept was introduced by the Brundtland Commission three decades ago (1987) and has been discussed since then in the UN context by global bodies (UN Commission on Sustainable Development, UNEP, UNDP et al.), at conferences (World Summit on Sustainable Development [WSSD] in 2002 in Johannesburg and at the UN Conference on Sustainable Development [UNCSD] in 2012 in Rio de Janeiro), and as the UN Sustainable Development Goals (SDC) that were adopted by the UN General Assembly in September 2015.

On 24 January 2017, *H.E. Peter Thomson, President of the UN General Assembly,* convened a high-level dialogue on *"Building Sustainable Peace for All: Synergies between the 2030 Agenda for Sustainable Development and the Sustaining Peace Agenda"*[35] in line with an agreement of the United Nations' Member States in April 2016 calling for

> a new organisational approach to the maintenance of international peace and security, through the concept of 'sustaining peace'… that covers the restoration of peace after conflict, as well as ensuring that the conditions for sustainable peace are in place – particularly by addressing the root causes of conflict. Sustaining peace is based on the premise that it will not be possible to achieve lasting peace in the long-term without sustainable development, equitable economic opportunity, and human rights protections for all. This central tenant was also recognized by world leaders in September 2015, when they adopted the 2030 Agenda for Sustainable Development.

---

[33]See: Capstick/Pidgeon (2014); Runciman, David, 2017: "How climate scepticism turned into something more dangerous", in: *The Guardian*, 7 July; at: https://www.theguardian.com/environment/2017/jul/07/climate-change-denial-scepticism-cynicism-politics (5 April 2018); see for an overview with many scientific references on: "Climate change denial"; at: https://en.wikipedia.org/wiki/Climate_change_denial (5 April 2018).

[34]Plautz, Jason, 2015: "CIA Shuts Down Climate Research Program", in: *The Atlantic*, 21 May; at: https://www.theatlantic.com/politics/archive/2015/05/cia-shuts-down-climate-research-program/452502/ (5 April 2018); Hill, Alice C., 2017: "Trump's environmental order jeopardizes our national security", in: *The Hill*, 28 March; at: http://thehill.com/blogs/pundits-blog/energy-environment/326221-trumps-climate-change-order-jeopardizes-our-national; Scholl, Ellen; Livingston, David, 2018: "Intelligence Community Continues to See Threat from Climate Change", 15 February; at: http://www.atlanticcouncil.org/blogs/new-atlanticist/intelligence-community-continues-to-see-threat-from-climate-change (5 April 2018).

[35]Peter Thomson: "Building Sustainable Peace for All"; at: http://www.un.org/pga/71/2017/01/20/building-sustainable-peace-for-all/, (20 January 2017).

> The 2030 Agenda provides humanity with a universal masterplan to transform our world, by eliminating extreme poverty, ensuring access to quality education, empowering women and girls, combating climate change, and protecting our natural environment. Critically, the 2030 Agenda recognises the importance of fostering peaceful, just and inclusive societies that are free from fear and violence, if all 17 Sustainable Development Goals are to be realised. Implementing the 2030 Agenda and the concept of sustaining peace are therefore priority tasks for the United Nations.[36]

The high-level dialogue was to address these objectives:

- enhance greater understanding on how Sustaining Peace relates to the 2030 Agenda in its entirety by highlighting country examples and good practice;
- discuss the linkages between Sustaining Peace and the 2030 Agenda across SDG targets, including those related to the promotion of peaceful and inclusive societies, the fostering of responsive, inclusive, participatory and representative decision-making at all levels, the building of effective, transparent and accountable institutions and the tackling of the root causes of conflict;
- consider the role that women and youth can play in the effort to implement the 2030 Agenda and to sustain peace.

Not surprisingly, both the high-level dialogue and the outcome document by the President of the UN GA remained within the traditional policy rhetoric and debate of the UN and its Peacebuilding Commission. The debate did not touch on how the likely consequences of GEC and climate change could be avoided and how the *business-as-usual* mind-set could be overcome, nor did it refer to the debate on sustainability transition. A Google search of the first 50 entries on 'sustainable peace' address primarily issues of peace, security, and peace education and avoid the new security challenges posed by our economic behaviour and consumption pattern in the Anthropocene.

This is the point of departure of the argument in the remaining part of this chapter. With the adoption of the concept of the Anthropocene and the recognition that we are now living in a new era of earth history where anthropogenic interferences into the geophysical and chemical processes of nature are increasing, a new level of threat to the survival of humankind is emerging: If 'we are the threat' to our own survival 'only we can also be the remedy' by dealing with the causes of this threat that has resulted in an exponential increase of greenhouse gases in the atmosphere, most particularly carbon dioxide.

Thus, addressing this threat and developing coping strategies for adaptation and mitigation becomes part of a global survival strategy and a major task for a 'sustainable peace' policy. This requires thinking on 'sustainable peace' beyond the

---

[36]See the schedule and all relevant documents; at: http://www.un.org/pga/71/event-latest/building-sustainable-peace-for-all-synergies-between-the-2030-agenda-for-sustainable-development-and-sustaining-peace/ and a summary by the president of the GA, HR Peter Thomson, 29 March 2017, at: http://www.un.org/pga/71/wp-content/uploads/sites/40/2015/08/Summary-of-the-High-level-Dialogue-on-Building-Sustainable-Peace-for-All.pdf.

narrow and traditional agendas of the UN High-level Dialogue on Sustainable Peace of January 2017 and of the United Nation's Peacebuilding Commission.

A 'sustainable peace' concept that links the two policy fields of peace and security and sustainable development must overcome the constraints of the present scientific discourse and of the UN policy debates. If the problems of global environmental change and climate change are caused by the accumulation of greenhouse gases in the atmosphere due to the burning of carbon energy sources, then we are the cause due to our energy use and consumption patterns and thus we pose a threat that fundamentally differs from traditional security threats posed by other countries or in the new wars by asymmetric non-state actors.

To cope with these threats, greenhouse gas emissions must decline through energy efficiency improvements and a shift to non-fossil renewable energy sources must occur. This goal cannot be achieved by international disarmament or arms control agreements but only by unilateral or joint individual, local, regional, and national initiatives in the areas of the environment, economy (energy, industrial, agricultural), urban planning, housing, and transportation policies (shift to electric mobility relying on renewables only).

The debate in the social sciences on sustainability transition in these different economic sectors and in the area of consumption, initially to a large extent in the energy sector (*Energiewende*), refers both to the *supply side* which can be framed and influenced by government norms and regulations but also by the *demand side* by the individual daily choices of citizens and their consumptive behaviour (basic food staples, food culture and mobility preferences). If 'we are the new threat' in the Anthropocene impacting directly on GEC and climate change, individual citizens can directly respond to this threat by their consumptive choices.

Adaptation and mitigation of GEC and CC directly involves the individual, the family (economic decisions), and the local community (town and city planning, e.g. by the role of public transportation) besides the nation state and international organisations. The influence of non-state actors and processes of *subpolitics* (Beck 1997) is growing in regard to this new threat. However, this requires media who are not controlled or manipulated by climate change sceptics and deniers and who do not falsify scientific evidence.[37] Therefore, public awareness and an active participation in local policy debates are needed. The new social media can act both as a tool of deliberate disinformation and propaganda but also of public control (Anderson 2017; Segerberg 2017; Tandoc/Eng 2018).

---

[37]See in the US on the role of Fox News, Breitbart News and the Sinclair Media Group on Climate Change; at: http://www.foxnews.com/category/us/environment/climate-change.html (5 April 2018); http://www.breitbart.com/environment/ (5 April 2018); Graves, Lucia, 2017: "This is Sinclair, 'the most dangerous US company you've never heard of'", 17 August; at: https://www.theguardian.com/media/2017/aug/17/sinclair-news-media-fox-trump-white-house-circa-breitbart-news (5 April 2018); Ward, Bob, 201x: "President Trump's fake news about climate change, in: LSE, *Grantham Research Institute on Climate Change and the Environment;* at: http://www.lse.ac.uk/GranthamInstitute/news/president-trump-fake-news-climate-change/ (5 April 2018).

Thus, aiming at a sustainable peace or a 'peace with nature' in the Anthropocene can be directly influenced by the individual and the family in their daily decisions and does not solely rely on the policies resulting from local, state and national decisions and international treaties. Aiming at a 'sustainable peace in the Anthropocene' requires scientific agenda setting, formulating the dual goal of our research and of political action towards a sustainable peace by developing a scientific and policy agenda for achieving a sustainability transition.

This process requires scientists, academics, practitioners, politicians and technocrats to identify, mobilise and involve the needed actors of such a transformation process: (a) political parties and social movements; (b) the government and the political bureaucracy; and (c) scientists, technological actors and industry (companies) who can develop, produce, market and distribute the new products and services if there is a public awareness and a demand by customers. Such a broad goal of a sustainable peace that includes 'peace with nature' does not primarily involve diplomats, development specialists, and the military but those involved in science and technology development and in the industrial, energy, transportation and housing sectors.

The strategies, policies, and measures of such a sustainable peace do not require the consent of the Security Council nor can the five permanent members block it with their veto. Implementation does not depend on the political leadership of major countries but on thousands of decisions at several levels in many sectors. A president of a major country cannot stop these alternative measures with its policies but it can delay the decisions by removing economic incentives. Among the G-8 (now G-7) the EU member countries (Germany, France, UK, Italy) have been among the key promoters of major reductions of GHG or CO_2 emissions by 80–95% and of a decarbonisation of the economy between 2050 and 2100. The success of such initiatives will depend to a large extent on the economic attractiveness for the rapidly growing new emitters in the new economic threshold states, such as China and increasingly India and many other developing countries in the global sunbelt.

The debate on 'sustainability transition' emerged first in the U.S. in the 1970s and was taken up in a report by the U.S. Academy of Science (NRC 1999; Johnston 2016; Raskin et al. 2002). It has looked forward to the processes of a long-term system transformation necessary to contain and reduce the effects of the dominant *business-as-usual* paradigm and to reduce GHG emissions through both *multilateral* quantitative emission reduction obligations and *unilateral* transformations.

Since 2005 a specific 'sustainability transition' research paradigm emerged from the work of the *Dutch Knowledge Network on Systems Innovation and Transition* (KSI)[38] and from the Amsterdam Conference in 2009 where the *Sustainability Transition Research Network* (STRN)[39] was founded. This approach was reviewed

---

[38]For KSI see for details: http://www.ksinetwork.nl/home.

[39]For STRN, see: http://www.transitionsnetwork.org/; selected results are published in: *Journal on Environmental Innovation and Sustainability Transition*; at: http://www.journals.elsevier.com/environmental-innovation-and-societal-transitions/.

elsewhere (Brauch/Oswald Spring 2016), together with long-term transformations in culture, values, behaviour and lifestyle. Grin et al. (2010) combined "three perspectives on transitions to a sustainable society: complexity theory, innovation theory, and governance theory". The authors

> seek to understand transitions dynamics, and how and to what extent they may be influenced. ... They do so from the conviction that only through drastic system innovations and transitions it becomes possible to bring about a turn to a sustainable society to satisfy their own needs, as inevitable for solving a number of structural problems on our planet, such as the environment, the climate, the food supply, and the social and economic crisis. ... The transition to sustainability has to compete with other developments, and it is uncertain which development will gain the upper hand. ... The ... authors address the need for transitions, as well as their dynamics and design (Grin et al. 2010: xvii–xix).

This research focus influenced a policy report by the *German Advisory Council on Global Change* (WBGU 2011) on a 'Social Contract for Sustainability' (2011), which argued that the transformation to a low-carbon society requires that we

> not just accelerate the pace of innovation; we must also cease to obstruct it. ... We must also take into account the external costs of high-carbon (fossil energy-based) economic growth to set price signals, and thereby to provide incentives for low-carbon enterprises. Climate protection is ... a vital fundamental condition for sustainable development on a global level.

To achieve a major transformation towards a low-carbon economy and society, the WBGU proposed specific measures for the energy sector, land-use changes and global urbanisation that could accelerate and extend the transition to sustainability. The WBGU Report suggested that "research and education are tasked with developing sustainable visions, in co-operation with policy-makers and citizens; identifying suitable development pathways, and realizing low-carbon and sustainable innovations". The WBGU argued "that transformation costs can be lowered significantly if joint decarbonisation strategies are implemented in Europe." (WBGU 2011).

A report by the *International Social Science Council* (ISSC) on *Transformative Cornerstones of Social Science Research for Global Change* defined transformation as "a process of altering the fundamental attributes of a system, including in this case structures and institutions, infrastructures, regulatory systems, financial regimes, as well as attitudes and practices, lifestyles, policies and power relations" (ISSC 2012: 16).[40] It argued that the necessary additional social science research will contribute to producing change by calling "for an additional response to global change and to climate change; additional to and building on the enduring focus in this field on adaptation and mitigation; ... a critical questioning of the systems and paradigms that have created climate change and on which climate change rests". But the STRN did so far hardly discuss its relevance for or impact on a 'sustainable

---

[40]International Social Science Council, 2012: "Transformative Cornerstones of Social Science Research for Global Change" (Paris: ISSC); at: http://www.worldsocialscience.org/documents/transformative-cornerstones.pdf (2 May 2018).

peace'. In August 2017 the ISSC launched a website for its 'Transformations to Sustainability Programme' that

> supports research to help advance transformations to more sustainable and equitable societies around the globe. By generating knowledge that produces a broader and deeper understanding of the conditions, processes, outcomes and impacts of transformative social change in the context of global environmental change, the programme is intended to: a) help craft more effective, durable and equitable solutions to the problems of environmental change and sustainability, in a context of social and cultural diversity; [and] b) promote the habitual use of the best knowledge about social transformations by researchers, educators, policy makers, practitioners, the private sector and citizens.[41]

Since 2014 the ISSC awarded 38 seed grants and three major grants for 'Transformative Knowledge Networks' (TKNs) to conduct empirical research projects for (a) the 'Academic-Activist Co-Produced Knowledge for Environmental Justice' network, (b) the 'Transformative Pathways to Sustainability' network and (c) the 'Transgressive Learning' network.[42] Within this ISSC and SIDA-sponsored programme the Pathways Network:

> explores how creative methods informed by social learning can be combined with research to respond to socio-ecological problems in six sites around the world, focusing on: a) sustainable urban water and waste; b) low carbon energy transitions for the poor and c) sustainable agricultural and food systems for healthy livelihoods. Transformative pathways call for new social science approaches that directly address global environmental and social imperatives, requiring context-sensitive critical engagement and practical responses.

In contributing to the construction of transformative pathways to sustainability this network applies: "transdisciplinary approaches to understand and catalyse change in diverse historical, political and cultural contexts and to communicate lessons learnt to wider research and user communities".[43]

Between the STRN (since 2009) and ISSC's TKNs (since 2016), where primarily SIDA-sponsored scholars and activists cooperate, there hardly exists any scientific exchange. Both had so far no impact on the UN's High-level Dialogue on sustaining peace nor was this theme addressed by both networks. In 2017, the ISSC trained African scholars in partnership with ICSU and the Network of African Science Academies (NASA) on "Transdisciplinary research … for Agenda 2030" addressing complex sustainability challenges in Africa and to increase participation of the African scientific community in global research programmes with the goal to "enable researchers to build meaningful inter- and trans-disciplinary projects".[44]

---

[41]See: "Transformations to Sustainability"; at: http://www.worldsocialscience.org/activities/transformations/, and at: https://transformationstosustainability.org/.

[42]See: Dougsiyeh, Hibaq, 2017: "Media coverage of the Transformative Knowledge Networks", 17 May; at: https://transformationstosustainability.org/magazine/transformative-knowledge-networks/ (5 April 2018).

[43]See: "Transformative pathways to sustainability: learning across disciplines, contexts and cultures"; at: https://transformationstosustainability.org/research/pathways/.

[44]See: "Transformative pathways to sustainability …", *Ibid.*

Göpel (2016), the new Secretary-General of the WBGU, proposed in her book on: *The Great Mindshift* to combine both worlds of economic thinking and practice on *How a New Economic Paradigm and Sustainability Transformations go Hand in Hand*.

> This book describes the path ahead. It combines system transformation research with political economy and change leadership insights when discussing the need for a *great mindshift* in how human wellbeing, economic prosperity and healthy ecosystems are understood if the Great Transformations ahead are to lead to more sustainability. It shows that history is made by purposefully acting humans and introduces transformative literacy as a key skill in leading the radical incremental change.[45]

However, the STRN, ISSC and ICSU and the mainstream innovative thinking on 'sustainability transition' have so far neither framed these issues in the context of 'sustainable peace' in the Anthropocene nor of a need to move towards a 'transformative science' from a 'peace ecology' perspective.

8.8 Method of Analysis: Transformative Science

The linkages between sustainability transition and sustainable peace require a bridge-building among scientific disciplines in the natural and social sciences between environmental and development studies focusing on sustainable development, and on peace and security studies. This implies a shift from disciplinary approaches to *multi- and interdisciplinary perspectives* as well as *transdisciplinary* and *transformative research designs* and *policy proposals*.

Each discipline has its specific epistemology, its premises and its methods of generating new knowledge. As the problems and issues that need to be examined scientifically become more complex, multidisciplinarity offers a first step in analysing complex problems from different disciplinary perspectives. These multidisciplinary studies rely on the methodologies of their respective disciplines.

The Swiss scholar Jean Piaget pioneered a new transdisciplinary scientific approach, when he proposed in the 1960s the term 'interdisciplinary' to integrate knowledge from different disciplines. Given the complexity of the Anthropocene, of global environmental change, and of resource scarcity, several research centres proposed transdisciplinarity as a new scientific approach to overcome the narrow disciplinary boundaries of specialised subfields and epistemic schools of knowledge creation. For Hadorn et al. (2008), transdisciplinarity refers to "the cause of the present problems and their future development (*system knowledge*)"; to the "values and norms … [to] be used to form goals of the problem-solving process (*target knowledge*)"; and to "how a problematic situation can be transformed and improved (*transformation knowledge*)". While "*multidisciplinarity* draws on knowledge from

---

[45]See Göpel (2016); at: http://www.afes-press-books.de/html/APESS_02.htm and at: http://www.springer.com/de/book/9783319437651.

different disciplines but stays within their boundaries" (Choi/Pak 2006), transdisciplinary research[46] is defined as research efforts conducted by investigators from different disciplines working jointly to create new conceptual, theoretical, methodological, and translational innovations that integrate and move beyond discipline-specific approaches to address a common problem. *Interdisciplinary Research* is any study or group of studies undertaken by scholars from two or more distinct scientific disciplines. The research is based upon a conceptual model that links or integrates theoretical frameworks from those disciplines, uses study design and methodology that is not limited to any one field, and requires the use of perspectives and skills of the involved disciplines throughout multiple phases of the research process.

In short, *transdisciplinarity* refers to a research strategy that establishes a common research objective that crosses disciplinary boundaries. The goal is to create a holistic approach by addressing complex problems that require close cooperation between several disciplines, e.g. of issues of global environmental change. Funtowicz and Ravetz (1993, 2003) argued that "transdisciplinarity can help determine the most relevant problems and research questions involved".

Holistic system analysis also contributed to *transdisciplinary* research, which includes all possible aspects and focuses on the interaction among different elements. *Transdisciplinarity* takes a structural approach and distinguishes between different levels of analysis. The surrounding conditions facilitate dynamic adjustment of undesirable disturbers. Of particular interest is a systemic dissipative and self-regulating approach, based originally on Ilya Prigogine (*Prigogine/Stengers* 1997: *184*) and Haken's (1983) *Synergetics*.

Luhmann (1991) applied dynamic system analysis to sociology and used the term 'autopoiesis', with which he refers to the complexity of dynamic systems which interact with the complexity of the environment. Luhmann insisted on the radical nature of the concept and assessed five key characteristics: autonomy, emergency, operative closure, self-structuration and autopoietic reproduction. These elements are essential for the analysis of new risks and uncertainties caused by changes in the environment and social behaviour in the Anthropocene.

Schneidewind et al. (2016) proposed moving from a 'transdisciplinary' approach to a 'transformative science', while Swilling (2016) suggested an 'anticipatory science'. The concept of 'transformative research' or 'science' has been used since the 2000s for a new approach that cuts across the dominant scientific paradigms.

The U.S. National Science Board (2007) adopted the following working definition of 'transformative research': "[it] involves ideas, discoveries, or tools that radically change our understanding of an important existing scientific or engineering concept or educational practice or leads to the creation of a new paradigm or field of science, engineering, or education. Such research challenges current understanding or provides pathways to new frontiers".

---

[46]"Definitions"; at: http://www.hsph.harvard.edu/trec/about-us/definitions/ (4 February 2016).

Trevors et al. (2012) reviewed *different definitions of "transformative research" that have been used in the scientific literature and argued that*

> Transformative discoveries leading to paradigm shifts can transform at many levels including scientific, personal and sociological. They may change cultural values and transform a society. When this occurs, there are sociological 'stages' of resistance preceding transformation: first denial, then anger, and finally acceptance. Arthur Schopenhauer summarized this progression as follows: 'All truth …. self-evident.'

Building on this approach, in *World in Transition—A Social Contract for Sustainability*, the *German Advisory Council on Global Change* (WBGU 2011: 21–23, 321–356) referred to "four transformative pillars of the knowledge society": 'transformation research' and 'transformation education', as well as 'transformative research' and 'transformative education'.

The WBGU (2011: 21) proposed that 'transformation research' should "specifically addresses the future challenge of transformation realisation" by exploring "transitory processes in order to come to conclusions on the factors and causal relations of transformation processes" and should "draw conclusions for the transformation to sustainability based on an understanding of the decisive dynamics of such processes, their conditions and interdependencies. … Transformative research supports transformation processes with specific innovations in the relevant sectors and it should encompass, for example, "new business models such as the shared use of resource-intensive infrastructures, and research for technological innovations like efficiency technologies" by aiming at a "wider transformative impact". Schneidewind and Singer-Brodowski (2013) and Göpel (2016) have developed this transformative approach further for climate policy and sustainability transition.

The International Social Science Council (ISSC 2012: 21–22) in its report on the *Transformative Cornerstones of Social Science Research for Global Change* identified six cornerstones: (1) historical and contextual complexities; (2) consequences; (3) conditions and visions for change; (4) interpretation and subjective sense-making; (5) responsibilities; and (6) governance and decision-making. The report concluded that

> the transformative cornerstones framework speaks to the full spectrum of social science disciplines, interests and approaches—theoretical and empirical, basic and applied, quantitative and qualitative. By not fashioning a global change research agenda around a substantive focus on concrete topics—water, food, energy, migration, development, and the like—the cornerstones are not only inclusive of many social science voices but … show that climate change and broader processes of global environmental change are organic to the social sciences, integral to social science preoccupations, domains par excellence of social science disciplines. … The transformative cornerstones of social science function not only as a framework for understanding what the social sciences can and must contribute to global change research. They function as a charter for the social sciences, a common understanding of what it is that the social sciences can and must do to take the lead in developing a new integrated, transformative science of global change.

Various initiatives by the U.S. National Science Board (2007), the WBGU (2011), the ISCC (2012), and the STRN (2016) have called for a new scientific paradigm in research into both global environmental change and sustainability transitions. The policy dimension should be included in the research design, by moving from knowledge creation to action, to policy initiatives, development, and implementation.[47]

These efforts are still highly dependent on the top-down activities of governments and multinational enterprises. In the South, meanwhile, many excluded stakeholders have, for decades, put into practice transformative education, e.g. by implementing the pedagogy of liberation inspired by Freire (1975).

Understanding or redefining 'peace research' and 'ecology' or the combined approach of a 'peace ecology' aiming at the goal of a 'sustainable peace' as a 'transformative science' would imply a broader holistic approach that would benefit from Wilson's (1998) proposed 'consilience' that refers to the interlocking of causal explanations across disciplines, in which the "interfaces between disciplines become as important as the disciplines themselves" that would "touch the borders of the social sciences and humanities". In addressing and coping with the complexity of human-nature interactions, such a new approach fundamentally challenges the present tendency of a narrow professionalisation of peace studies and ecology with progressing segmentalisation and overspecialisation of scientific analysis as a precondition for career and publication strategies of young peace and ecology scholars.

A "transformative science" approach to complex new research questions and problem areas that requires the knowledge of peace research and ecological studies and also the relevant knowledge from the natural sciences cannot be achieved by one scholar; it would require a research team that combines expertise from the natural and social sciences as well as from peace and ecology programmes to address questions of peace and security in the Anthropocene triggered by multiple anthropogenic impacts due to the burning of hydrocarbon energy sources (coal, oil, gas) and the related overconsumption of scarce resources that were fostered by lifestyles that resulted in an 'economy of waste'.

Trans- and interdisciplinary research programmes in the natural sciences, such as *Earth Systems Studies* (ESS) or Analysis (ESA), and in geography, such as *Geoecology* (Huggett 1995) have left social science approaches, especially the political dimension, uncovered (Brauch et al. 2011a, b).

A 'transformative research design' that includes the transformation of the prevailing economic system and its impacts on 'international peace and security' in the Anthropocene as a key research goal requires such a much wider holistic research approach that transcends the present research practice and career patterns of

---

[47]The theme of the 7th International Sustainability Transition Conference (IST) conference in Wuppertal (Germany) was "Exploring Transition Research as Transformative Science"; at: https://ist2016.org/ and https://transitionsnetwork.org/past-ist-conferences/ (2 May 2018).

scholars. Such an approach requires courage and readiness to take risks of failure by major research funding agencies.

In the United States, the *National Science Foundation* (NSF) and the *National Institutes of Health* (NIH) decided to fund projects with a "high-risk, high-reward", which they defined as "research with an inherent high degree of uncertainty and the capability to produce a major impact on important problems in biomedical/ behavioural research", what has been called by the *European Research Council*, as 'frontier research'.

Not surprisingly the *U.S. National Science Board* defined transformative research as "research that has the capacity to revolutionize existing fields, create new sub-fields, cause paradigm shifts, support discovery, and lead to radically new technologies".[48] Peer reviewers of NSF funded projects are requested "to include an emphasis on potentially transformative research".[49]

Since the NSF has been under attack by the Trump Administration and its future budget may face major cuts, the European Union and the European Research Council may be the sole funding source for such ambitious transformative research projects and research centres to address the challenges humankind in general and European citizens in particular will face during the Anthropocene.

However, individual researchers can and should lay the conceptual ground-work for a "Peace Ecology Perspective in the Anthropocene". But the first three independent explorations on peace ecology by Kyrou (2007), Amster (2014), Oswald Spring et al. (2014) have so far not been taken up neither by peace nor by ecological researchers.

8.9   Pillars and Components of a Peace Ecology Perspective for the Anthropocene

8.9.1   Overcoming Narrow Disciplinary Approaches and Projects

Earlier approaches by Kenneth Boulding (Scott 2015; Stephenson 2016) since the 1960s and by Elise Boulding in the 1980s to link peace studies with ecological concerns have not been taken up, neither by peace researchers nor by ecologists or environmental studies.

---

[48]This paragraph relies on: "transformative research"; at: https://en.wikipedia.org/wiki/Transformative_research (5 April 2018).

[49]In 2012, the NSF hosted a brainstorming workshop on "Transformative Research: Ethical and Societal Dimensions" that explored the history and alternative conceptions that play an important role in policy debates, and in public discourse on the future of science in society. Key points of the discussion were summarised by Michael E. Gorman, a Professor in the department of science, technology, and society (STS) at the University of Virginia, at: http://www.cccblog.org/?s= Gorman (5 April 2018).

The environmental security debate (since 1989) and on the climate-security nexus (since 1988) did not address classical peace issues (Brauch et al. 2009; Scheffran et al. 2012). Human security approaches to environmental or ecological security were developed primarily by peace researchers without an effort to bring these two separate modes of thinking together.

IPRA's Ecology and Peace Commission has been, since IPRA's 24th conference in Mie (Japan), one of the very few scientific voices that has aimed at a bridge-building trying to bring authors from both groups of thought into a dialogue and to expand the 'peace ecology' concept.[50] However, similar proposals by Conca (1994) and Amster (2015) have not yet been widely cited and taken up by both research programmes.[51]

8.9.2   Five Pillars of a Wider Sustainable Peace Concept

The peace concept of the UN Charter of June 1945 is confined to its 'international' dimension and mostly used with security as "international peace and security".[52] 'Peace' is noted as the UN Charter's key mission in Art. 1,1: "to maintain international peace and security", and "to take effective collective measures for the prevention and the removal of the threats to the peace, and for the suppression of acts of aggression or other breaches of the peace", as well as peaceful conflict settlements.

In interpreting the peace concept used in the UN Charter, Wolfrum (1994: 50) pointed to both narrow and wide interpretations of peace. In Art. 1(2) and 1(3) the Charter uses a wider and more positive peace concept when it calls for the development of "friendly relations among nations" and for "international cooperation in solving international problems of an economic, social, cultural, or humanitarian character." In Chapter VI on the Pacific Settlement of Disputes, Art. 33 uses a 'negative' concept of peace that is "ensured through prohibitions of intervention and the use of force" (Tomuschat 1994: 508). In Chapter VII of the UN Charter dealing with "Action with Respect to Threats to the Peace, Breaches of the Peace, and Acts of Aggression", in Art. 39, a 'negative' concept of peace dominates, with a reference to "the absence of the organized use of force between states". In the framework of

---

[50]See the three previous volumes edited by Oswald Spring et al. (2014, 2016b); Brauch et al. (2016a).

[51]The proposal by Conca (1994) on "environmental peacemaking" has achieved a total of 322 citations according to 'google scholar' from 2002 until April 2018; Amster's book on peace ecology has gained only 26 citations since 2015. The book by Oswald Spring, Brauch, Tidballs achieved since December 2013 until February 2018 a total of 4799 chapter downloads and the introductory chapter that specifically touches peace ecology received 546 chapter downloads; see at: http://www.bookmetrix.com/detail_full/book/b795abb3-6400-4db9-8073-c03a6ec82f5d#downloads (5 April 2018).

[52]This section relies on Brauch (2016c), where the peace concept has been presented in more detail. The text of the UN Charter is at: http://www.un.org/en/sections/un-charter/un-charter-full-text/ (6 April 2018).

Chapter IX on "International Economic and Social Cooperation", Art. 55(3) refers to "universal respect for, and observance of, human rights and fundamental freedoms." It has been suggested that "the right of self-determination, to peace, development, and to a sound environment" (Partsch 1994: 779) should be incorporated as "human rights of the third generation" (Vasak 1984: 837).

Thus, in the UN Charter of June 1945, a narrow or 'negative' concept of peace has been at the centre with a few direct references to 'positive' aspects to be achieved by 'friendly relations among nations', and by 'international cooperation'. The 'positive peace' concept indicates peaceful social and cultural beliefs and norms, the presence of economic, social and political justice and a democratic use of power including non-violent mechanisms of conflict resolution. 'Sustainable peace' or 'peace with nature' was added later to the debate in the UN.

The concept of peace has been widely used in different cultures, religions, and languages with different meanings. The English word 'peace' originates from the Latin 'pax'. The German word 'Frieden' indicates a 'condition of quietness, harmony, resolution of warlike conflicts', and in Russian 'mir' means both 'peace' and 'the world'. While the Greek *eirene*, the Hebrew *shalom*, and the Arab *salām* all imply 'peace with justice', the Hindi *ahimsa* adds the ecological dimension.[53]

In the European tradition, influenced by its Greek and Roman origins, the peace concept has been widely used in philosophy, history, religion, and international law in relation with other concepts, especially with violence and war. In the social and political sciences and in the peace research programme, the concept of peace has gradually widened since the 1960s. Galtung (1967, 1968, 1969; Galtung/Fischer 2013) distinguished between 'negative peace' (absence of physical or personal violence) and 'positive peace' (absence of structural violence, repression, and injustice). He distinguished negative, indifferent and positive relations that often result in *negative peace* (absence of violence, cease-fire, indifferent relations) or *positive peace* (harmony).[54] Later, the concepts of 'cultural peace' (Galtung 2003) and 'engendered peace'[55] and 'sustainable peace'[56] were added to bring in culture, gender, and sustainability. So far, most texts on sustainable peace have ignored

---

[53]For a discussion of the "meanings of peace" that includes the philosophical debate in China (including Lao Tzu, Confucius, Mo Tzu), and in Buddhist, Hindu, and Judaeo-Christian thinking with the goal of "achieving positive peace", see Webel (2008) and Oswald Spring (2008).

[54]Johan Galtung commented on an earlier draft: "To me both sustainable peace and environment are guided by two basic deep structures: diversity and symbiosis; hence very dynamic. To impose one culture, Western, one structure, capitalism will lead to the collapse of both; cultures-structures in partnership and species, abiota and biota in symbiosis will lead to higher complexity, evolution —that then has to be watched but is promising. What I read seems to me very compatible with these simple propositions."

[55]See e.g. Peck (1998); Ekiyor (2006); International Fellowship of Reconciliation (2010); Barnes (2010); Ellerby (2011); Oswald Spring et al. (2014): "Engendered and Sustainable Peace with Resilience Building", presentation at the 25th IPRA conference in Istanbul; at: http://www.afes-press.de/html/pdf/2014_UOS/text_16.pdf (6 April 2018).

[56]See the discussion by Brauch (2016c: 210–211) on "Developing Sustainable Peace Further" based on the then available literature.

environmental challenges and their possible consequences for new types of conflict in the Anthropocene. Therefore, conceptual, theoretical, and empirical research is needed to develop the concept of a sustainable peace in the Anthropocene.

Sustainable peace refers to manifold links among peace, security, and the environment, where humankind and environment as two key parts of global Earth face the consequences of destruction, extraction, and pollution. The sustainable peace concept includes also processes of recovering from environmental destruction, reducing the human footprint in nature through a less carbon-intensive and – in the long-term – possibly carbon-free and dematerialised production processes so that future generations may still be able to decide on their own resources and development strategies.

Oswald Spring et al. (2014: 18–19) referred to these five pillars of a wider peace concept, on which peace may be grounded as: security, development (equity), environment (sustainability), culture, and gender conceptualised as a framework for a conceptualisation of peace ecology (Fig. 5.1).

These five pillars of peace ecology refer to different conceptual features of peace. The classic relationship between 'international peace and security' in the UN Charter refers to a narrow political agenda of *negative peace* without war and violent conflict aiming at the prevention, containment and resolution of conflicts and violence or the absence of 'direct violence' in wars and repression. To achieve peace with equity or *positive peace* points to the absence of 'structural violence' due to overcoming social inequality, discrimination, marginalisation and poverty with a lack of access to sufficient food, water, health and educational opportunities. Oswald Spring, Brauch, Tidballs argued that

> Peace ecology in the Anthropocene era of earth and human history may be conceptualised within the framework of five conceptual pillars we introduce here as the 'peace ecology quintet' consisting of peace, security, equity, sustainability and gender. To conceptualise the linkages between peace and security we refer to 'negative peace' and for the relationship between peace and equity we use the 'positive peace' concept, for interactions between peace, gender and environment we suggest the 'cultural peace' concept and finally for the relations between peace, equity and gender we propose the concept of an 'engendered peace' (Oswald Spring et al. 2014: 18).

However, more conceptual work and reflection is needed to develop this proposed peace ecology perspective in the Anthropocene further. Below several components for such a perspective for the Anthropocene epoch of earth and human history will be outlined.

8.9.3 Components of a Peace Ecology Perspective in the Anthropocene

In the debate of the geologists on establishing the 'Anthropocene' as a new epoch of earth history referred to above in the introduction, the natural scientists must provide the evidence in the sediments that such a transition has occurred in the mid-20th century. Peace ecology as a proposed scientific approach grounded in the social sciences and trying to link the knowledge of peace and environmental studies requires a clear research programme that could combine the following components in terms of *time, impacts, space, values, issue areas,* and *research problems* aiming at a 'good Anthropocene':

In terms of *time,* peace ecology should focus on the *Anthropocene era of earth and human history,* addressing the manifold developments since the end of World War II due to the observed and well-documented human intervention into the earth system as a result of the global availability and massive increase and burning of cheap fossil energy sources (coal, oil, natural gas) in a globalised economy with an expansion of free trade, uncontrolled financial flows, and a massive movement of people as businessmen, tourists or migrants. These anthropogenic interventions into the earth system or 'nature' have affected the atmosphere (ozone layer depletion, greenhouse gas accumulation), but also the ground (composition of earth, decline of soil fertility), water (degradation, pollution, overuse, scarcity, stress), and biodiversity (decline of flora and fauna).

This period has been described as the period of 'great acceleration' (Steffen et al. 2015a). The IGBP defined the great acceleration as follows:

> The second half of the 20th Century is unique in the history of human existence. Many human activities reached take-off points sometime in the 20th Century and sharply accelerated towards the end of the century. The last 60 years have without doubt seen the most profound transformation of the human relationship with the natural world in the history of humankind.

> The effects of the accelerating human changes are now clearly discernible at the Earth system level. Many key indicators of the functioning of the Earth system are now showing responses that are, at least in part, driven by the changing human imprint on the planet. The human imprint influences all components of the global environment - oceans, coastal zone, atmosphere, and land. Dramatic though these human-driven impacts appear to be, their rates and magnitudes must be compared to the natural patterns of variability in the Earth system to begin to understand their significance.[57]

These human interventions into the earth system have already caused manifold societal impacts that can be analysed in a rapid increase in production, consumption, urbanisation, pollution, migration, crises and conflicts for whose analysis this author suggested a more complex pressure-response model he called a PEISOR model for the analysis of the linkages between *P*ressure-*E*ffect-*I*mpact-*S*ocietal *O*utcome-policy *R*esponse (Fig. 8.1).

- Peace ecology should address the *impacts* of the human intervention into the earth system that have partly already crossed the "planetary boundaries",[58] a

---

[57]See at: http://www.igbp.net/globalchange/greatacceleration.4.1b8ae20512db692f2a680001630. html.

[58]The global scientific debate on "planetary boundaries" was triggered by these two publications: Rockström et al. (2009); Steffen/Stafford Smith (2013); Steffen et al. (2015b); see also the special website of the Stockholm Resilience Centre: http://www.stockholmresilience.org/research/research-news/2015-01-15-planetary-boundaries—an-update.html (6 April 2018).

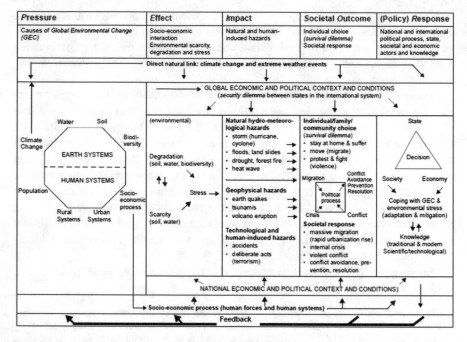

Fig. 8.1 The PEISOR model. *Source* Brauch (2005, 2009a); Brauch/Oswald Spring (2009)

concept that has been promoted by the Stockholm Resilience Centre (see Fig. 8.3). Its proponents referred to nine planetary boundaries: 1. Climate change; 2. Change in biosphere integrity (biodiversity loss and species extinction); 3. Stratospheric ozone depletion; 4. Ocean acidification; 5. Biogeochemical flows (phosphorus and nitrogen cycles); 6. Land-system change (for example deforestation); 7. Freshwater use; 8. Atmospheric aerosol loading (microscopic particles in the atmosphere that affect climate and living organisms); and 9. Introduction of novel entities (e.g. organic pollutants, radioactive materials, nanomaterials, and micro-plastics).

- In terms of *space,* the proposed peace ecology perspective should address issues that are of a global nature, as addressed in the debate on global environmental change, such as climate change, soil erosion, water degradation and diversity loss. However, these global processes have already brought about consequences at the regional level in the so-called 'environmental hotspots' (see Fig. 8.3) and at the local level where most of the extreme weather events occur and the societal impacts become evident and urgent (Fig. 8.2).

- As peace research, the proposed peace ecology perspective should be a value-oriented scientific approach aiming at a future where a 'peace with nature' becomes possible where new types of war, like resource conflicts over scarce natural resources, or the worst societal impacts of climate change in terms of violent conflicts can be avoided through a strategy of environmental conflict avoidance (Brauch 2002).

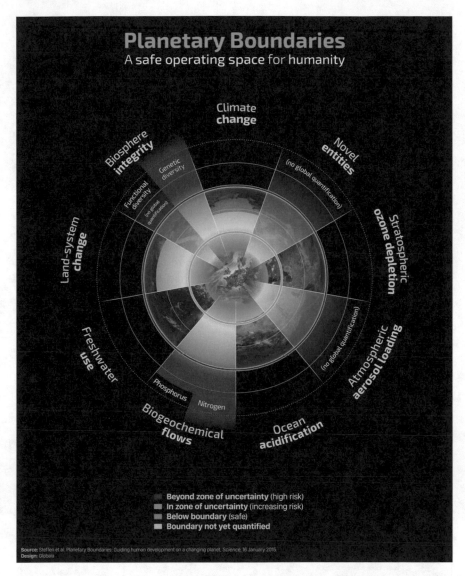

Fig. 8.2 An illustration of the concept of "Planetary Boundaries" developed by Rockström et al. since 2009–2015. *Source* Stockholm Resilience Centre; at: http://www.stockholmresilience.org/research/planetary-boundaries.html (6 April 2018), *Image source* F. Pharand-Deschênes/Globaïa. Permission was granted by Sturle Hauge Simonsen, Head of communications, Stockholm Resilience Centre on 9 April 2018

- A peace ecology approach should aim at the analysis of both anthropogenic causes and impacts of global environmental change, thus providing the best possible science-based assessment of the challenges and diagnosis of probable and likely societal impacts.
- Among its specific research problems should be the societal impact of anthropogenic and natural developments of global environmental change on domestic and international crises, conflicts and in the worst case even war. The WBGU (2007) addressed in its Report: *Security Risk Climate Change* four conflict constellations in selected hotspots (Fig. 8.3): (a) climate induced degradation of fresh water resources; (b) climate-induced decline in food production; (c) climate-induced increase in storm and flood disasters; and (d) environmentally-induced migration.
- The report pointed to the following regions as environmental hotspots: (1) Mexico, Central America and the Caribbean; (2) the Amazon region; (3) the Andes; (4) North Africa; (5) the Sahel zone; (6) southern Africa; (7) West Asia; (8) Central Asia; (9) South Asia; and (10) East Asia and Mongolia.
- As a 'transformative science' or 'transformative research' approach, a peace ecology perspective should not be limited to the causal analysis of potential

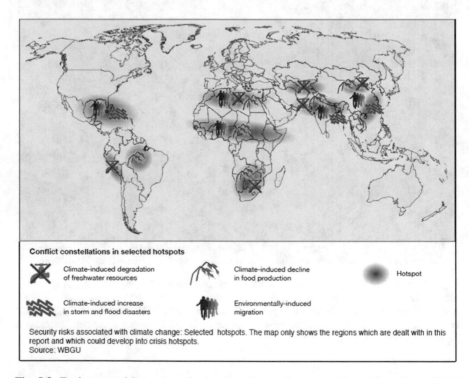

Conflict constellations in selected hotspots

Climate-induced degradation of freshwater resources

Climate-induced decline in food production

Hotspot

Climate-induced increase in storm and flood disasters

Environmentally-induced migration

Security risks associated with climate change: Selected hotspots. The map only shows the regions which are dealt with in this report and which could develop into crisis hotspots.
Source: WBGU

Fig. 8.3 Environmental hotspots and migration. *Source* German Advisory Council on Global Change (WBGU 2007: 8). Reprinted with permission. Permission was granted by Maja Göpel, Secretary General of the WBGU on 9 April 2018

conflict initiating factors of the so-called 'bad Anthropocene'; it should also develop policy scenarios, strategies, policies, and measures for how such factors can be avoided or contained, focusing on a 'good Anthropocene'.

- A peace ecology perspective should also contribute to a 'transformative education' aiming at public awareness on the potential causes of environmental and climate-induced conflicts in the media, in the population at large, among societal movements, parties and policy-makers.
- As a transformative science, a peace ecology perspective should provide both conceptual and empirical contributions for policy development aiming at peace with nature.
- As a transformative science, a peace ecology approach could analyse, assess, and interpret the implementation of policies to avoid negative societal outcomes, focusing specifically on political, economic, and societal obstacles and interest constellations that are opposed to a strategy of a sustainability transition and a transformation of the economy aiming at a gradual decarbonisation of productive and consumptive processes and of the way of life of citizens.

These ten components of a proposed peace ecology approach aim to contribute not only to the analysis of the planetary challenges in the Anthropocene but also as a 'transformative science'. They also aim to contribute to a change of the societal causal factors that may endanger the human species if policies of *business-as-usual* prevail, by reducing the costs of non-action to which the *Stern Review on the Economics of Climate Change* referred to in 1996.

8.10   Conclusions

This essay has touched on knowledge, concepts, discourses, and debates from peace studies and ecology, trying to contribute with a wide concept of a 'sustainable peace' and a combined approach of a 'peace ecology' to a redefinition of peace and ecology studies in the Anthropocene. Below the discussion in the previous sections will be briefly summarised.

8.10.1   A New Turn Backwards in Europe and in North America?

Since 1989, with the fall of the Berlin Wall (November 1989), the dissolution of the Warsaw Pact (March 1991) and the implosion of the Soviet Union (December 1991), the first peaceful change in international order has occurred. However, this 'global turn' neither resulted in a peace dividend, nor in a new European or global peace order. While during the early 1990s major disarmament agreements and environmental conventions were signed and ratified and new cooperative global regimes emerged, since the mid-1990s new disarmament and environmental agreements have been increasingly blocked (not only in the U.S. Senate), new wars occurred, and military expenditures have risen again in a new global disorder.

Against this background of contradictory global policy trends, Paul J. Crutzen argued that "we are in the Anthropocene!", that humankind has for the first time directly interfered into nature and thus provoked a change to a new era of earth history, the Anthropocene. Crutzen pointed to the most fundamental change in earth and human history that has occurred silently and that is gradually being socially constructed by the natural and social sciences but not yet fully understood by most policymakers.

How did policymakers, countries and international organisations react to these changes and new opportunities? Nearly all policymakers were surprised by the rapid dynamic of this peaceful change between 1989 and 1991 and tried to react and adapt to these changes and to new security challenges (ex-Yugoslav space, Iraq occupation of Kuwait) with their old Hobbesian mind-set and with military tools (military force and unilateral 'humanitarian' military interventions or with UN peacekeeping missions and operations). There was no post-Cold War planning by national governments and international organisations and very few general contributions or schemes from peace scholars or public intellectuals were tabled. Rather, since the incorporation the Crimea into Russia in 2014 and the Russian sponsored violent conflict in the primarily Russian-speaking Eastern Ukraine, with the active foreign involvement on behalf of Syrian President Assad, with the Skripal episode in the UK, elements of a New Cold War between the West and Russia have re-emerged.

The policies of the Trump Administration have contributed further to a deterioration of the relations between the West, represented by NATO and the EU, and Russia. A new trade war between the US and China and possibly also with the US neighbours and partners of NAFTA and European NATO allies is emerging that is replacing the traditional US Open Door policies since the late 19th century and the Bretton Woods system that was set up by the US since 1944/1945 and which resulted in GATT and later the WTO. A century after the End of World War I and nearly three decades since the Fall of the Berlin Wall, it is unclear whether the new protectionist policies of 'America First' will erode the foundations of the post-World War II global economic system and will result in new major violent conflicts and new global alliances that further reduce the possibility to develop and implement forward looking cooperative global environmental policies aiming at a decarbonisation of the world economy.

A new turning point in international order may emerge, where national self-interest, unilateral tariffs, and a trade war may fundamentally challenge the rules and treaties of international trade and economic relations.[59] The increase in

---

[59]This may require for the author to revise his more positive assessment he expressed during the 50th anniversary of IPRA during its 25th Conference in Istanbul in August 2014, one hundred years after the start of the first World War (see Brauch 2016b).

political, military and economic tensions may further reduce the potential for cooperative agreements in the environmental realm, especially regarding climate change and the implementation of the Paris Climate Change Treaty of December 2015.

The growth of populist movements and parties and the rapid rise to power of unpredictable politicians in some West European countries has already made it more difficult to form reform-oriented government coalitions in Belgium, The Netherlands, Austria, Germany, and Italy. The growth in authoritarian tendencies in Hungary, Poland, Romania, and xenophobic perspectives that are opposed to asylum seekers, especially from Muslim countries, in the former East Germany or in the new states of the Federal Republic of Germany, especially in the states of Saxonia, Saxonia-Anhalt, and Mecklenburg Vorpommern, but also in the Czech Republic and Slovakia have already weakened the European Union. Its political and economic influence may also decline after the British exit (Brexit) from the EU in 2019.

8.10.2   Reaction of the Social and Political Science to this Turn

How have the social sciences, political science, international relations and peace studies reacted to these challenges? In international relations, some scholars celebrated the victory of liberalism of the West (Fukuyama 1989, 1992), others claimed a return to old conflicts (Mearsheimer 1990) or pointed to new types of conflicts (Huntington's (1993, 1996)) such as a clash of civilisations.

The soft voices of European public intellectuals (Habermas, Beck, Giddens, Latour etc.) and of the tiny group of peace scholars were largely ignored and contributed little to a design of a new international order despite the intensive debate on components of a European peace order after the global turn of 1990. There has been a dual failure of social scientists, peace scholars and ecologists to foresee the turn and to contribute significantly to the goals and structures of a new peaceful international order.

8.10.3   Diagnosing Present Political Trends from a Peace Ecology Perspective

What may a proposed new scientific approach of a 'peace ecology perspective' contribute to the scientific diagnosis of present trends and challenges? A peace ecology perspective is still in its early stages of scientific development and its initial proponents have associated with it different goals, e.g. environmental peacemaking, post-conflict peace-building, coping with or avoiding the violent consequences of global environmental change. The major challenge is to think beyond the boundaries of each of the two research programmes of peace and ecology. In the framework of the UN's High-level Dialogue on 'sustaining peace' (2017) the conceptual boundaries between peace and sustainable development have not yet been overcome.

What does the goal of a 'sustainable peace' imply and may a process of sustainability transition contribute to the realisation of this goal? If sustainable peace is defined as 'peace with nature' the specific environmental causes of conflict, e.g. of environmental scarcity, degradation, abundance, and stress (of the old debate of the 1990s) but also of the violent outcomes of the physical impacts of global environmental change (new debate since the early 21st century) must be addressed by scientists and citizens alike. Sustainable peace in the Anthropocene (third debate) includes the processes (e.g. of sustainability transition) that address the economic and societal causes of the GHG accumulation in the atmosphere and the economic and societal countermeasures to prevent the worst impacts from occurring.

What may a new approach of a 'transformative science' contribute to the analysis of a process of sustainability transition? Disciplinary research within peace or ecology studies may be insufficient in addressing the complexity of the challenge we are facing. Multi-, inter- and transdisciplinary approaches require the expertise of scholars from several disciplines by distinguishing among different perspectives and integrating this segmentalised knowledge into a new approach.

8.10.4 Transformative Research and Peace Ecology

The idea of a transformative research was stimulated by Kuhn's (1962) notion of scientific revolutions that result in changing paradigms. The term has been used by U.S. Agencies (the National Science Foundation [NSF] and the National Institutes of Health [NIH]), where it often referred to "research with an inherent high degree of uncertainty and the capability to produce a major impact on important problems in biomedical/behavioural research", by the U.S. National Science Board as "research that has the capacity to revolutionise existing fields, create new sub-fields",[60] and by the European Research Council) as "frontier research".[61] For the Wuppertal Institute in Germany:

> [t]ransformative research contributes to solving societal problems and is characterized by an explicit aspiration to get involved: The aim is to catalyse processes of change and to actively involve stakeholders in the research process. In this way, transformative research generates 'socially robust' knowledge needed for sustainability transitions.

---

[60]Wikipedia: "Transformative research" (30 August 2017); at: https://en.wikipedia.org/wiki/Transformative_research; Trevors et al. (2012).

[61]European Research Council: "Frontier Research"; at: http://ec.europa.eu/research/participants/data/ref/h2020/other/guides_for_applicants/h2020-guide17-erc-adg_en.pdf.

Research at the Wuppertal Institute thus follows a transdisciplinary concept of knowledge: it does not only serve to generate 'systems knowledge' (e.g. technological or resource-oriented systems analysis), but also integrates stakeholders in the process of generating 'target knowledge' (visions and guiding principles) and 'transformation knowledge' in concrete settings of urban or sectoral transitions to sustainability. …

Doing transformative research for sustainability transitions, the Wuppertal Institute applies a comprehensive set of methods – ranging from scenario analysis, resource and energy systems modelling to policy analysis and evaluation – for the generation of systems, target and transformation knowledge. A specific type of transformative research is carried out in real-world laboratories where scientists and stakeholders do research together and work on solutions for real-world problems.[62]

Whether 'sustainable peace' can become a transformative concept and peace ecology can become a 'transformative research approach' remains to be seen. More fundamental rethinking on human interference in nature and the potential of countermeasures are needed to transform a widely used but still underdefined concept into a tool of innovative research that helps to initiate societal processes that fundamentally change political practice. For the time being this remains a utopia. Galtung's (1967, 1968, 1969) concept of a 'positive peace' and the extended peace concepts (Fig. 8.1) with societal justice, gender and global equity have been utopian as well; the goal of a 'sustainable peace' will remain for a long time if not forever.

References

Amster, Randall, 2014: *Peace Ecology* (Boulder, CO: Paradigm).
Anderson, Ashley A., 2017: "Effects of Social Media Use on Climate Change Opinion, Knowledge, and Behavior", in: *Oxford Research Encyclopedias, Climate Change*, March; at: http://climatescience.oxfordre.com/view/10.1093/acrefore/9780190228620.001.0001/acrefore-9780190228620-e-369 (5 April 2018).
Anderson, Martin, 1990: *Revolution: The Reagan Legacy* (Stanford: Hoover Institution Press).
Angus, Ian, 2015: "When did the Anthropocene begin and why does it matter?", jn: *Climate & Capitalism* (10 September); at: http://climateandcapitalism.com/2015/09/10/when-did-the-anthropocene-begin-and-why-does-it-matter/ (1 September 2017).
Angus, Ian, 2016: "Earth System: Anthropocene Working Group: Yes, a new epoch has begun", posted on 9 January 2016; at: http://climateandcapitalism.com/2016/01/09/anthropocene-working-group-yes-a-new-epoch-has-begun/ (1 September 2017).
Angus, Ian, 2017: "Protecting business as usual: Another attack on Anthropocene science", posted on 24 January 2017; at: http://climateandcapitalism.com/2017/01/24/another-attack-on-anthropocene-science/ (1 September 2017).
Barnes, Karen, 2010: *Engendering peace or a gendered peace? The UN and liberal peacebuilding in Sierra Leone, 2002–2007*, Ph.D. thesis, London School of Economics and Political Science (London, UK); at: http://etheses.lse.ac.uk/2375/ (6 April 2018).

---

[62]See for an illustration of the *transformative research concept* of the Wuppertal Institute; at: https://wupperinst.org/en/research/transformative-research/ (30 August 2017).

Beck, Ulrich: "Subpolitics. Ecology and the Disintegration of Institutional Power", in: *Organization & Environment*, 10(1) (March 1997): 52–65.

Biello, David, 2015: "Did the Anthropocene Begin in 1950 or 50,000 Years Ago?", in: *Scientific American*, 2 April.

Boulding, Kenneth E., 1966: "The Economics of the Coming Spaceship Earth", in: Jarrett, H. (Ed.): *Environmental Quality in a Growing Economy* (Baltimore: Johns Hopkins Press).

Boulding, Kenneth E., 1978: *Ecodynamics* (Beverly Hills, CA: Sage).

Brauch, Hans Günter, 1994: *Global Structural Change, System Transformation and no National Change. Armament and Disarmament Policy in American-Soviet Relations (1981–1992)* (Berlin, Free University, habilitation, unpublished).

Brauch, Hans Günter, 2002: "Climate Change, Environmental Stress and Conflict—AFES-PRESS Report for the Federal Ministry for the Environment, Nature Conservation and Nuclear Safety", in: Federal Ministry for the Environment, Nature Conservation and Nuclear Safety (Ed.): *Climate Change and Conflict. Can Climate Change Impacts Increase Conflict Potentials? What is the Relevance of This Issue for the International Process on Climate Change?* (Berlin: Federal Ministry for the Environment, Nature Conservation and Nuclear Safety, 2002): 9–112; at: http://www.afes-press.de/pdf/Brauch_ClimateChange_BMU.pdf.

Brauch, Hans Günter, 2005: *Threats, Challenges, Vulnerabilities and Risks in Environmental Human Security*. Source, 1/2005 (Bonn: UNU-EHS); at: http://collections.unu.edu/eserv/UNU:1868/pdf4040.pdf.

Brauch, Hans Günter, 2008: "Introduction: Globalisation and Environmental Challenges: Reconceptualising Security in the 21st Century", in: Brauch, Hans Günter; Oswald Spring, Ursula; Mesjasz, Czeslaw; Grin, John; Dunay, Pal; Behera, Navnita Chadha; Chourou, Béchir; Kameri-Mbote, Patricia; Liotta, P. H. (Eds.): *Globalization and Environmental Challenges: Reconceptualizing Security in the 21st Century* (Berlin–Heidelberg–New York: Springer): 27–43.

Brauch, Hans Günter, 2009a: "Securitizing Global Environmental Change", in: Brauch, Hans Günter; Oswald Spring, Ursula; Grin, John; Mesjasz, Czeslaw; Kameri-Mbote, Patricia; Behera, Navnita Chadha; Chourou, Béchir; Krummenacher, Heinz (Eds.): *Facing Global Environmental Change: Environmental, Human, Energy, Food, Health and Water Security Concepts*. Hexagon Series on Human and Environmental Security and Peace, vol. 4 (Berlin–Heidelberg–New York: Springer): 65–102.

Brauch, Hans Günter, 2009b: "Introduction: Facing Global Environmental Change and Sectorialization of Security", in: Brauch, Hans Günter; Oswald Spring, Ursula; Grin, John; Mesjasz, Czeslaw; Kameri-Mbote, Patricia; Behera, Navnita Chadha; Chourou, Béchir; Krummenacher, Heinz (Eds.): *Facing Global Environmental Change: Environmental, Human, Energy, Food, Health and Water Security Concepts*. Hexagon Series on Human and Environmental Security and Peace, vol. 4 (Berlin–Heidelberg–New York: Springer): 21–42.

Brauch, Hans Günter, 2011: "Security Threats, Challenges, Vulnerabilities and Risks in US National Security Documents (1990–2010)", in: Brauch, Hans Günter; Oswald Spring, Úrsula; Mesjasz, Czeslaw; Grin, John; Kameri-Mbote, Patricia; Chourou, Béchir; Dunay, Pal; Birkmann, Jörn (Eds.): *Coping with Global Environmental Change, Disasters and Security – Threats, Challenges, Vulnerabilities and Risks* (Berlin–Heidelberg–New York: Springer): 249–274.

Brauch, Hans Günter, 2016a: "Building Sustainable Peace by Moving Towards Sustainability Transition", in: Brauch, Hans Günter; Oswald Spring, Úrsula; Bennett, Juliet; Serrano Oswald, Serena Erendira (Eds.): *Addressing Global Environmental Challenges from a Peace Ecology Perspective* (Cham–Heidelberg–New York–Dordrecht–London: Springer International Publishers, 2016): 145–180.

Brauch, Hans Günter, 2016b: "Historical Times and Turning Points in a Turbulent Century: 1914, 1945, 1989 and 2014", in: Brauch, Hans Günter; Oswald Spring, Úrsula; Bennett, Juliet; Serrano Oswald, Serena Erendira (Eds.): *Addressing Global Environmental Challenges from a Peace Ecology Perspective* (Cham–Heidelberg–New York–Dordrecht–London: Springer International Publishers, 2016): 11–54.

Brauch, Hans Günter, 2016c: "Conceptualizing Sustainable Peace in the Anthropocene: A Challenge and Task for an Emerging Political Geoecology and Peace Ecology", in: Brauch, Hans Günter; Oswald Spring, Úrsula; Grin, John; Scheffran; Jürgen (Eds.), 2016: *Handbook on Sustainability Transition and Sustainable Peace* (Cham–Heidelberg–New York–Dordrecht–London: Springer): 187–236.

Brauch, Hans Günter, 2019: "Review of the Emerging Debates on the Anthropocene in the Natural and Social Sciences, in the Humanities, National and International Law and in Politics and Policies" in: Benner, Susanne; Lax, Gregor; Crutzen, Paul J.; Lelieveld, Jos; Pöschl, Ulrich (Eds.): *Paul J. Crutzen: The Anthropocene Fundamental Concepts, Perspectives and Implications* (Cham: Springer International Publishing, 2019).

Brauch, Hans Günter; Oswald Spring, Úrsula, 2009: *Securitizing the Ground - Grounding Security* (Bonn: UNCCD).

Brauch, Hans Günter; Oswald Spring, Úrsula, 2016: "Sustainability Transition and Sustainable Peace: Scientific and Policy Context, Scientific Concepts and Dimensions", in: Brauch, Hans Günter; Oswald Spring, Úrsula; Grin, John; Scheffran; Jürgen (Eds.), 2016: *Handbook on Sustainability Transition and Sustainable Peace* (Cham: Springer International Publishing): 3–66.

Brauch, Hans Günter; Oswald Spring, Úrsula; Mesjasz, Czeslaw; Grin, John; Dunay, Pal; Behera, Navnita Chadha; Chourou, Béchir; Kameri-Mbote, Patricia; Liotta, P. H. (Eds.), 2008: *Globalization and Environmental Challenges: Reconceptualizing Security in the 21st Century* (Berlin–Heidelberg–New York: Springer).

Brauch, Hans Günter; Oswald Spring, Úrsula; Grin, John; Mesjasz, Czeslaw; Kameri-Mbote, Patricia; Behera, Navnita Chadha; Chourou, Béchir; Krummenacher, Heinz (Eds.), 2009: *Facing Global Environmental Change: Environmental, Human, Energy, Food, Health and Water Security Concepts* (Berlin–Heidelberg–New York: Springer).

Brauch, Hans Günter; Dalby, Simon; Oswald Spring, Úrsula, 2011a: "Political Geoecology for the Anthropocene", in: Brauch, Hans Günter; Oswald Spring, Úrsula; Mesjasz, Czeslaw; Grin, John; Kameri-Mbote, Patricia; Chourou, Béchir; Dunay, Pal; Birkmann, Jörn (Eds.): *Coping with Global Environmental Change, Disasters and Security – Threats, Challenges, Vulnerabilities and Risks* (Berlin–Heidelberg–New York: Springer): 1453–1486.

Brauch, Hans Günter; Oswald Spring, Úrsula; Mesjasz, Czeslaw; Grin, John; Kameri-Mbote, Patricia; Chourou, Béchir; Dunay, Pal; Birkmann, Jörn (Eds.), 2011b: *Coping with Global Environmental Change, Disasters and Security—Threats, Challenges, Vulnerabilities and Risks* (Berlin–Heidelberg–New York: Springer).

Brauch, Hans Günter; Oswald Spring, Úrsula; Bennett, Juliet; Serrano Oswald, Serena Erendira, 2016a: *Addressing Global Environmental Challenges from a Peace Ecology Perspective* (Cham: Springer International Publishing).

Brauch, Hans Günter; Oswald Spring, Úrsula; Grin, John; Scheffran, Jürgen (Eds.), 2016b: *Handbook on Sustainability Transition and Sustainable Peace* (Cham–Heidelberg–New York–Dordrecht–London: Springer).

Braudel, Fernand, 1949: *La Méditerranée et le monde méditerranéen à l'époque de Philippe II* (Paris: Armand Colin).

Braudel, Fernand, 1969: "Histoire et science sociales. La longue durée", in: *Écrits Sur l'Histoire* (Paris: Flammarion): 41–84.

Braudel, Fernand, 1972: *The Mediterranean and the Mediterranean World in the Age of Philip II*, 2 volumes (New York: Harper & Row).

Brundtland Commission (World Commission on Environment and Development), 1987: *Our Common Future. The World Commission on Environment and Development* (Oxford–New York: Oxford University Press).

Butler, Mark, 2017: *Climate Wars* (Melbourne: Melbourne University Press).

Buzan, Barry; Lawson, George, 2015: *The Global Transformation: History, Modernity and the Making of International Relations* (Cambridge: Cambridge University Press).

Caplan, James A., 2010: *The Theory and Principles of Environmental Dispute Resolution* (San Bernardino, CA: edrusa.com).

Capstick, Stuart Bryce; Pidgeon, Nicholas Frank, 2014: "What *is* climate change scepticism? Examination of the concept using a mixed methods study of the UK public", in: *Global Environmental Change*, 24 (January): 389–401; at: https://www.sciencedirect.com/science/article/pii/S0959378013001477 (5 April 2018).

Choi, B. C.; Pak, A. W., 2006: "Multidisciplinarity, interdisciplinarity and transdisciplinarity in health research, services, education and policy: 1. Definitions, objectives, and evidence of effectiveness", in: *Clinical and Investigative Medicine*, 29(6) (December): 351–364; at: http://www.ncbi.nlm.nih.gov/pubmed/17330451.

Clark, William C.; Crutzen, Paul J.; Schellnhuber, Hans Joachim, 2004: "Science and Global Sustainability: Toward a New Paradigm", in: Schellnhuber, Hans Joachim; Crutzen, Paul J.; Clark, William C.; Claussen, Martin; Held, Hermann (Eds.): *Earth System Analysis for Sustainability* (Cambridge, MA; London: MIT Press): 1–28.

Conca, Ken, 1994: "In the Name of Sustainability: Peace Studies and Environmental Discourse", in: Kakonen, Jyrki (Ed.): *Green Security or Militarized Environment* (Dartmouth: Aldershot): 7–24.

Conca, Ken, 2002: "The Case for Environmental Peacemaking", in: Conca, Ken; Dabelko, Geoffrey (Eds.): *Environmental Peacemaking* (Baltimore: Johns Hopkins University Press): 1–22.

Conca, Ken; Dabelko, Geoffrey, 2002: "The Case for Environmental Peacemaking", in: Conca, Ken; Dabelko, Geoffrey (Eds.): *Environmental Peacemaking* (Baltimore: Johns Hopkins University Press): 1–22.

Crutzen, Paul J., 2002: "Geology of Mankind", in: *Nature*, 415(3) (January): 23.

Crutzen, Paul J., 2011: "The Anthropocene: A geology of mankind", in: Brauch, Hans Günter; Oswald Spring, Úrsula; Mesjasz, Czeslaw; Grin, John; Kameri-Mbote, Patricia; Chourou, Béchir; Dunay, Pal; Birkmann, Jörn (Eds.), 2011: *Coping with Global Environmental Change, Disasters and Security—Threats, Challenges, Vulnerabilities and Risks* (Berlin–Heidelberg–New York: Springer): 3–4.

Crutzen, Paul J.; Brauch, Hans Günter (Eds.), 2016: *Paul J. Crutzen: The Anthropocene: A New Phase of Earth History: Impacts for Science and Politics* (Cham–Heidelberg–New York–Dordrecht—London: Springer, 2015.

Crutzen, Paul J.; Stoermer, Eugene F., 2000: "The Anthropocene", in: *IGBP Newsletter*, 41: 17–18.

Crutzen, Paul J.; Birks, John W., 1982: "The atmosphere after a nuclear war: Twilight at noon", in: *Ambio*, 11: 114–125.

Dalby, Simon, 2013: "The Geopolitics of Climate Change", in: *Political Geography*, 37: 38–47.

Dalby, Simon, 2014: "Rethinking Geopolitics: Climate Security in the Anthropocene", in: *Global Policy*, 5(1): 1–9.

Dalby, Simon, 2015: "Climate Geopolitics: Securing the Global Economy", in: *International Politics,* 52(4): 426–444.

Dalby, Simon, 2016: "Framing the Anthropocene: The good, the bad and the ugly", in: *The Anthropocene Review*, 3(1); at: http://journals.sagepub.com/doi/abs/10.1177/2053019615618681.

Dyer, Gwynne, 2011: *Climate Wars: The Fight for Survival as the World Overheats* (London: Oneworld Publications).

Ekiyor, Thelma, 2006: *Engendering Peace - How the Peacebuilding Commission can live up to UN Security Council Resolution 1325*, FES Briefing Paper (New York: Friedrich-Ebert Stiftung, June); at: http://library.fes.de/pdf-files/iez/global/50420.pdf (6 April 2018).

Ellerby, Kara L., 2011: *Engendered Security: Norms, Gender and Peace Agreements*, Ph.D. thesis, University of Arizona (Tucson, Arizona, USA); at: http://citeseerx.ist.psu.edu/viewdoc/download?doi=10.1.1.466.9065&rep=rep1&type=pdf (6 April 2018).

Flinders, Matthew; John, Peter, 2013: "The Future of Political Science", in: *Political Studies Review*, 11: 222–227.

Freire, Paulo, 1975: *Conscientization* (Geneva, World Council of Churches).

Fukuyama, Francis, 1989: "The End of History?", in: *The National Interest* (Summer): 3–18.

Fukuyama, Francis, 1992: *The End of History and the Last Man (Free Press).*

Funtowicz, S.; Ravetz, J. R., 1993: "Science for the Post-Normal Age", in: *Future*, 25: 735–755.
Funtowicz, S.; Ravetz, J. R., 2003: "Post-Normal Science", in: International Society for Ecological Economics, *Internet Encyclopaedia of Ecological Economics*; at: http://isecoeco.org/pdf/pstnormsc.pdf (6 February 2016).
Galtung, Johan, 1967: "Peace Research: science, or politics in disguise?", in: *International Spectator*, 21(19): 1573–1603.
Galtung, Johan, 1968: "Peace", in: *International Encyclopedia of the Social Sciences* (London–New York; Macmillan): 487–496.
Galtung, Johan, 1969: "Violence, Peace and Peace Research", in: *Journal of Peace Research*, 3: 167–191.
Galtung, Johan, 2003: "Cultural Peace: Some Characteristics", 12 October; at: https://www.transcend.org/files/article121.html.
Göpel, Maja, 2016: *The Great Mindshift: How a New Economic Paradigm and Sustainability Transformations go Hand in Hand* (Cham: Springer International Publishing).
Grin, John; Rotmans, Jan; Schot, Johan, 2010: *Transitions to Sustainable Development. New Directions in the Study of Long Term Transformative Change* (New York, NY–London: Routledge).
Haas, Peter M. (Ed.), 1992: *Knowledge, Power and International Policy Coordination (International Organization*, 46(1)), (Cambridge, Mass: MIT Press).
Haken, Hermann, 1983: *Synergetics: An Introduction. Nonequilibrium Phase Transition and Self-Organization in Physics, Chemistry, and Biology* (Berlin–Heidelberg: Springer).
Hirsch Hadorn, Gertrude; Hoffmann-Riem, Holger; Biber-Klemm, Susette; Grossenbacher-Mansuy, Walter; Joye, Dominique; Pohl, Christian; Wiesmann, Urs; Zemp, Elisabeth (Eds.), 2008: *Handbook of Transdisciplinary Research* (Berlin–Heidelberg: Springer).
Hobsbawm, Eric, 1994: *The Age of Extremes: The Short Twentieth Century, 1914–1991* (London: Michael Joseph).
Holsti, Kalevi J., 1991: *Peace and War: Armed Conflicts and International Order 1648–1989* (Cambridge: Cambridge University Press).
Holsti, Kalevi J., 2016a: *Major Texts on War, the State, Peace, and International Order* (Cham: Springer International Publishing).
Holsti, Kalevi J., 2016b: *Kalevi Holsti: A Pioneer in International Relations Theory, Foreign Policy Analysis, History of International Order, and Security Studies* (Cham: Springer International Publishing).
Holsti, Kalevi, 1996: *The State War and the State of War* (Cambridge: Cambridge University Press).
Holzner, Burkart, 1968: *Reality Construction in Society* (Cambridge: Schenkman).
Holzner, Burkart; Marx, John, 1979: *Knowledge Application: The Knowledge System in Society* (Boston: Allyn and Bacon, Inc.).
Huggett, Richard John, 1995: *Geoecology. An Evolutionary Approach* (London–New York: Routledge).
Huntington, Samuel P., 1993: "The Clash of Civilizations", in: *Foreign Affairs*, 72(3) (Summer): 22–50.
Huntington, Samuel P., 1996: *The Clash of Civilizations and the Remaking of World Order* (New York: Simon & Schuster).
International Fellowship of Reconciliation, 2010: *Engendering Peace: Incorporating a Gender Perspective in Civilian Peace Teams* (Alkmar: International Fellowship of Reconciliation); at: http://www.peacewomen.org/assets/file/Resources/NGO/reconpb_engenderedpeaec_iforwpp_2010.pdf (6 April 2018).
IPCC, 1990: *Climate Change. The IPCC Impacts Assessment* (Geneva: WMO; UNEP; IPCC).
IPCC, 1995: *IPCC Second Assessment Climate Change 1995. A Report of the Intergovernmental Panel on Climate Change* (Geneva: WMO, UNEP).
IPCC, 2001: *Climate Change 2001: Synthesis Report. A Contribution of Working Groups I, II, and III to the Third Assessment Report of the Intergovermental Panel on Climate Change* (Cambridge: Cambridge University Press).

IPCC, 2007: *Climate Change 2007. Synthesis Report* (Geneva: IPCC); at: http://www.ipcc.ch/pdf/assessment-report/ar4/syr/ar4_syr.pdf.

IPCC, 2014: *Climate Change 2014—Synthesis Report, Summary for Policymakers* (Geneva: IPCC); at: http://www.ipcc.ch/report/ar5/syr/.

ISSC (International Social Science Council), 2012: *Transformative Cornerstones of Social Science Research for Global Change* (Paris: ISSC).

Jochum, Uwe, 2010: "Wissensrevolution", in: *WBG Weltgeschichte*, vol. V: *Entstehung der Moderne 1700 bis 1914* (Darmstadt: Wissenschaftliche Buchgesellschaft): 150–194.

Johnson, Twig, 2016: "Policy, Politics and the Impact of Transition Studies", in: Brauch, Hans Günter; Oswald Spring, Úrsula; Grin, John; Scheffran; Jürgen (Eds.), 2016: *Handbook on Sustainability Transition and Sustainable Peace* (Cham–Heidelberg–New York–Dordrecht–London: Springer): 481–492.

Kaldor, Mary, 1999: *New and Old Wars: Organized Violence in a Global Era* (Cambridge: Polity–Stanford: Stanford University Press).

Kaldor, Mary, 2002: *New and Old Wars: Organized Violence in a Global Era* (Cambridge: Polity).

Klein, Naomi: "Capitalism vs. the Climate—Denialists are dead wrong about the science. But they understand something the left still doesn't get about the revolutionary meaning of climate change", in: *The Nation*, 28 November 2011; at: http://www.thenation.com/article/164497/capitalism-vs-climate (25 January 2014).

Kuhn, Thomas, 1962: *The Structure of Scientific Revolutions* (Chicago: University of Chicago Press).

Kunnas, Jan, 2017: "Storytelling: From the early Anthropocene to the good or the bad Anthropocene", in: *The Anthropocene Review*, 4(2) (14 August); at: http://journals.sagepub.com/doi/full/10.1177/2053019617725538 (7 April 2018).

Kyrou, Christos N., 2007: "Peace Ecology: An Emerging Paradigm in Peace Studies", in: *The International Journal of Peace Studies*, 12(2) (Spring/Summer): 73–92.

Landes, David S., 1969: *The Unbound Prometheus: Technological Change and Industrial Development in Western Europe from 1750 to the Present* (Cambridge: Cambridge University Press).

Layne, Christopher: *The Peace of Illusions: American Grand Strategy from 1940 to the Present* (Ithaca-London: Cornell University Press, 2006): 39–50.

Lebow, Richard Ned, 1994: "The long peace, the end of the cold war, and the failure of realism", in: *International Organization*, 48(2) (Spring): 249–277.

Lebow, Richard Ned; Risse-Kappen, Thomas, 1996: *International Relations Theory and the End of the Cold War* (New York: Columbia University Press).

Lenton, Timothy; Held, Hermann; Kriegler, Elmar; Hall, Jim W.; Lucht, Wolfgang; Ramstorf, Stefan; Schellnhuber, Hans Joachim, 2008: "Tipping elements in the Earth's climate system", in: *Proceedings of the National Academy of Science* (PNAS), 105(6) (12 February): 1786–1793.

Luhmann, Niklas, 1991: *Soziale Systeme* (Frankfurt a. M: Suhrkamp).

Mearsheimer, John J., 1990: "Back to the Future. Instability in Europe after the Cold War", in: *International Security*, 15(1) (Summer): 5–56.

Münkler, Herfried, 2004: *Die neuen Kriege* (Reinbek: Rowohlt).

National Science Board, 2007: *Enhancing Support of Transformative Research at the National Science Foundation* (Washington DC: National Science Foundation); at: https://www.nsf.gov/about/transformative_research/definition.jsp (6 February 2016).

Nordhaus, Ted; Shellenberger, Michael, 2004–2007: *Break Through: From the Death of Environmentalism to the Politics of Possibility* (Boston: Houghton Mufflin Company).

Nordhaus, Ted; Shellenberger, Michael et al., 2015a: "An Ecomodernist Manifesto." (April).

Nordhaus, Ted; Shellenberger, Michael; Mukuno, Jenna, 2015b: "Ecomodernism and the Anthropocene," in: *Breakthrough Journal* (Summer).

Notter, Harley: *Postwar Foreign Policy Preparation, 1939–1945* (Washington, D.C.: US Government Printing Office, 1949).

NRC (National Research Council), 1999: *Global Environmental Change: Research Pathways for the Next Decade* (Washington DC: National Academy Press).

Oreskes, Naomi, 2004: "The Scientific Consensus on Climate Change", in: *Science*, 3 December 2004: 306(5702) (December): 1686; at: https://doi.org/10.1126/science.1103618.

Osterhammel, Jürgen, 2014: *The Transformation of the World: A Global History of the Nineteenth Century* (Princeton: Princeton University Press).

Oswald Spring, Ùrsula, 2008: "Oriental, European and Indigenous Thinking on Peace in Latin America", in: Brauch, Hans Günter; Oswald Spring, Úrsula; Mesjasz, Czeslaw; Grin, John; Dunay, Pal; Behera, Navnita Chadha; Chourou, Béchir; Kameri-Mbote, Patricia; Liotta, P.H. (Eds.): *Globalization and Environmental Challenges: Reconceptualizing Security in the 21st Century* (Berlin–Heidelberg–New York: Springer): 175–194.

Oswald Spring, Úrsula; Brauch, Hans Günter, 2011: "Coping with Global Environmental Change —Sustainability Revolution and Sustainable Peace", in: Brauch, Hans Günter; Oswald Spring, Úrsula; Mesjasz, Czeslaw; Grin, John; Kameri-Mbote, Patricia; Chourou, Béchir; Dunay, Pal; Birkmann, Jörn, 2010: *Coping with Global Environmental Change, Disasters and Security—Threats, Challenges, Vulnerabilities and Risks* (Berlin–Heidelberg–New York: Springer): 1487–1504.

Oswald Spring, Úrsula; Brauch, Hans Günter; Tidball, Keith G., 2014: "Expanding Peace Ecology: Peace, Security, Sustainability, Equity and Gender", in: Oswald Spring, Úrsula; Brauch, Hans Günter; Tidball, Keith G. (Eds.), 2014: *Expanding Peace Ecology: Peace, Security, Sustainability, Equity and Gender—Perspectives of IPRA's Ecology and Peace Commission* (Cham: Springer International Publishing): 1–32.

Oswald Spring, Úrsula; Brauch, Hans Günter; Scheffran, Jürgen, 2016a: "Sustainability Transition with Sustainable Peace: Key Messages and Scientific Outlook", in: Brauch, Hans Günter; Oswald Spring, Úrsula; Grin, John; Scheffran; Jürgen (Eds.), 2016: *Handbook on Sustainability Transition and Sustainable Peace* (Cham–Heidelberg–New York–Dordrecht–London: Springer): 887–928.

Oswald Spring, Úrsula; Brauch, Hans Günter; Serrano Oswald, Serena Erendira; Bennett, Juliet, 2016b: *Regional Ecological Challenges for Peace in Africa, the Middle East, Latin America and Asia Pacific* (Cham: Springer International Publishing).

Partsch, Karl-Josef, 1994: "Art. 55 (c)", in: Simma, Bruno (Ed.): *The Charter of the United Nations. A Commentary* (Oxford: Oxford University Press): 776–793.

Peck, Connie, 1998: *Sustainable Peace: The Role of the UN and Regional Organizations in Preventing Conflicts* (Lanham–Boulder–New York: Carnegie Commission on Preventing Deadly Conflict).

Polanyi, Karl, 1944: *Great Transformation: The Political and Economic Origins of our Time* (Boston, MA: Beacon Press).

Prigogine, Ilya; Stengers, Isabelle, 1997: *The End of Certainty* (Glencoe: The Free Press).

Raskin, P.; Banuri, T.; Gallopin, G.; Gutman, P.; Hammond, A.; Kates, A.; Swart, R., 2002: *Great Transition. The Promise and the Lure of the Times Ahead*. Report of the Global Scenario Group. Pole Start Series Report 10 (Stockholm: Stockholm Environment Institute).

Rockström, Johan; Steffen, Will; Noone, Kevin; Persson, Åsa; Chapin, III, F. Stuart; Lambin, Eric; Lenton, Timothy M.; Scheffer, Marten; Folke, Carl; Schellnhuber, Hans Joachim; Nykvist, Björn; De Wit, Cynthia A.; Hughes, Terry; Leeuw, Sander van der; Rodhe, Henning; Sörlin, Sverker; Snyder, Peter K.; Costanza, Robert; Svedin, Uno; Falkenmark, Malin; Karlberg, L.; Corell, Robert W.; Fabry, Victoria J.; Hansen, James; Walker, Brian; Liverman, Diana M.; Richardson, Katherine; Crutzen, Paul; Foley, Jonathan A., 2009: "Planetary boundaries: exploring the safe operating space for humanity", in: *Ecology and Society*, 14(2), 32; at: http://www.ecologyandsociety.org/vol14/iss2/art32/.

Scheffran, Jürgen; Brzoska, Michael; Brauch, Hans Günter; Link, Peter Michael; Schilling, Janpeter (Eds.), 2012: *Climate Change, Human Security and Violent Conflict: Challenges for Societal Stability* (Berlin–Heidelberg–New York: Springer).

Schellnhuber, Hans Joachim, 1999: "'Earth system' analysis and the second Copernican revolution", in: *Nature*: 402: C19–C23 (2 December); at: https://doi.org/10.1038/35011515.

Schellnhuber, Hans Joachim; Cramer, Wolfgang; Nakicenovic, Nebojsa; Wigley, Tom; Yohe, Gary (Eds.), 2006: *Avoiding Dangerous Climate Change* (Cambridge: Cambridge University Press).

Schellnhuber, Hans Joachim; Hare, William L.; Serdeczny, Olivia; Adams, Sophie; Coumou, Dim; Frieler, Katja; Marin, Maria; Otto, Ilona M.; Perrette, Mahé; Robinson, Alexander; Rocha, Marcia; Schaeffer, Michiel; Schewe, Jacob; Wang, Xiaoxi; Warszawski, Lila, 2012: *Turn Down the Heat: Why a 4 °C Warmer World Must Be Avoided* (Washington DC: The World Bank).

Schellnhuber, Hans Joachim; Serdeczny, Olivia Maria; Adams, Sophie; Köhler, Claudia; Otto, Ilona Magdalena; Schleussner, Carl-Friedrich, 2016: "The Challenge of a 4 °C World by 2100", in: Brauch, Hans Günter; Oswald Spring, Úrsula; Grin, John; Scheffran; Jürgen (Eds.), 2016: *Handbook on Sustainability Transition and Sustainable Peace* (Cham–Heidelberg–New York–Dordrecht–London: Springer, *forthcoming*).

Schneidewind, Uwe; Singer-Brodowski, Mandy, 2013, [2]2014: *Transformative Wissenschaft-Klimawandel im deutschen Wissenschafts- und Hochschulsystem* (Marburg: Metropolis).

Schneidewind, Uwe; Singer-Brodowski, Mandy; Augenstein, Karoline, 2016: "Transformative Science for Sustainable Transitions", in: Brauch, Hans Günter; Oswald Spring, Úrsula; Grin, John; Scheffran; Jürgen (Eds.), 2016: *Handbook on Sustainability Transition and Sustainable Peace* (Cham–Heidelberg–New York–Dordrecht–London: Springer).

Scott, Robert, 2015: *Kenneth Boulding: A Voice Crying in the Wilderness* (London: Palgrave Macmillan).

Segerberg, Alexandra, 2017: "Online and Social Media Campaigns for Climate Change Engagement", in: *Oxford Research Encyclopedias, Climate Change*, June; at: http://climatescience.oxfordre.com/view/10.1093/acrefore/9780190228620.001.0001/acrefore-9780190228620-e-398 (5 April 2018).

Shellenberger, Michael; Nordhaus, Ted, 2004: *The Death of Environmentalism: Global Warming Politics in a Post-Environmental World* (Oakland, Ca: The Breakthrough Institute); at: https://www.thebreakthrough.org/images/Death_of_Environmentalism.pdf.

Steffen, W.; Stafford Smith, M., 2013: "Planetary boundaries, equity and global sustainability: Why wealthy countries could benefit from more equity", in: *Current Opinion in Environmental Sustainability*, 5: 403–408.

Steffen, Will; McNeill, John; Crutzen, Paul J., 2007: "The Anthropocene: Are Humans Now Overwhelming the great forces of Nature?, in: *Ambio*, 36: 614–621.

Steffen, Will; Grinevald, Jacques; Crutzen, Paul J.; McNeill, John, 2011: "The Anthropocene: conceptual and historical perspectives", in: *Phil. Trans. R. Soc. A* 369: 843.

Steffen, Will; Broadgate, Wendy; Deutsch, Lisa, 2015a: "The trajectory of the Anthropocene: The Great Acceleration", in: *The Anthropocene Review*, 2(1) (15 January).

Steffen, Will; Richardson, Katherine; Rockström, Johan; Cornell, Sarah E.; Fetzer, Ingo; Bennett, Elena M.; Biggs, R.; Carpenter, Stephen R.; de Vries, Wim; de Wit, Cynthia A.; Folke, Carl; Gerten, Dieter; Heinke, Jens; Mace, Georgina M.; Persson, Linn M.; Ramanathan, Veerabhadran; Reyers, B.; Sörlin, Sverker, 2015b: "Planetary Boundaries: Guiding human development on a changing planet", in: *Science*, 347(6223).

Stephenson, Carolyn M., 2016: "Paradigm and Praxis Shifts: Transitions to Sustainable Environmental and Sustainable Peace Praxis", in: Brauch, Hans Günter; Oswald Spring, Úrsula; Grin, John; Scheffran; Jürgen (Eds.), 2016: *Handbook on Sustainability Transition and Sustainable Peace* (Cham–Heidelberg–New York–Dordrecht–London: Springer): 89–104.

Stern, Nicholas, 2006, 2007, [4]2008: *The Economics of Climate Change—The Stern Review* (Cambridge–New York: Cambridge University Press).

Swilling, Mark, 2016: "Preparing for Global Transition; Implications of the Work of the International Resource Panel", in: Brauch, Hans Günter; Oswald Spring, Úrsula; Grin, John; Scheffran; Jürgen (Eds.), 2016: *Handbook on Sustainability Transition and Sustainable Peace* (Cham–Heidelberg–New York–Dordrecht–London: Springer): 249–274.

Tandoc Jr., Edson C.; Eng, Nicholas, 2018: "Climate Change Communication on Facebook, Twitter, Sina Weibo, and Other Social Media Platforms", in: *Oxford Research Encyclopedias, Climate Change*, April; at: http://climatescience.oxfordre.com/abstract/10.1093/acrefore/9780190228620.001.0001/acrefore-9780190228620-e-361?rskey=kaowcK&result=1 (5 April 2018).

Tomuschat, Christian, 1994: "Chapter VI. Pacific Settlement of Disputes, Art. 33", in: Simma, Bruno (Ed.): *The Charter of the United Nations. A Commentary* (Oxford: Oxford University Press): 505–514.

Trevors, J. T.; Pollack, Gerald H.; Saier, Jr., Milton H.; Masson, Luke, 2012: "Transformative research: definitions, approaches and consequences", in: *Theory in Bioscience*, June, 131(2): 117–123.

U.S. National Science Board, 2007: *Enhancing Support of Transformative research at the National Science Foundation* (Washington, D.C.: National Science Foundation, 7 May).

UN, 2009: *Climate change and its possible security implications. Report of the Secretary-General.* A/64/350 of 11 September 2009 (New York: United Nations).

UNEP, 2011: *Green Economy Report: Towards a Green Economy: Pathways to Sustainable Development and Poverty Eradication* (Nairobi: UNEP); at: https://sustainabledevelopment.un.org/content/documents/126GER_synthesis_en.pdf.

United States Global Change Research Program, 2009: *Global Climate Change Impacts in the United States* (Cambridge: Cambridge University Press).

UNSC, 2011: "Statement by the President of the Security Council on Maintenance of Peace and Security: Impact of Climate Change", S/PRST/1011/15, 20 July 2011; (27 July 2011).

Vasak, K., 1984: "Pour une troisième génération des droits de l'homme", in: Swinarski, Christophe (Ed.): *Studies in Honour of Jean Pictet* (The Hague: Martinus Nijhoff Publishers): 837.

Waters, Colin Neil; Zalasiewicz, Jan A.; Williams, Mark; Ellis, Michael A.; Snelling, Andrea M., 2014: *A Stratigraphical Basis for the Anthropocene* (London: Geological Society, London, Special Publications 395).

WBGU, 2007: *World in Transition—Climate Change Security Risk* (Berlin: German Advisory Council on Global Change, July).

WBGU, 2011: *World in Transition—A Social Contract for Sustainability* (Berlin: German Advisory Council on Global Change, July).

WCED (World Commission on Environment and Development), 1987: *Our Common Future. The World Commission on Environment and Development* (Oxford–New York: Oxford University Press).

Webel, Charles P.; Barash, David P., [2]2008: *Peace and Conflict Studies* (London: Sage).

Welzer, Harald, 2012: *Climate Wars: What People Will Be Killed For in the 21st Century* (Cambridge: Polity).

Wilson, Edward O., 1998: *Consilience* (New York: Knopf).

Wolfrum, Rüdiger, 1994: "Chapter 1. Purposes and Principles, Art. 1", in: Simma, Bruno (Ed.): *The Charter of the United Nations. A Commentary* (Oxford: Oxford University Press): 49–56.

Zalasiewicz, Jan; Williams, Mark; Smith, Alan et al., 2008: "Are we now living in the Anthropocene?", in: *GSA Today*, 18(2): 4–8.

Zalasiewicz, Jan; Williams, Mark; Steffen, Will; Crutzen, Paul, 2010: "The New World of the Anthropocene", in: *Environment Science & Technology*, 44(7): 2228–2231.

Zalasiewicz, Jan; Waters, Colin N.; Summerhayes, Colin P.; Wolfe, Alexander P.; Barnosky, Anthony D.; Cearreta, Alejandro; Crutzen, Paul; Ellis, Erle; Fairchild, Ian J.; Gałuszka, Agnieszka; Haff, Peter; Hajdas, Irka; Head, Martin J.; Ivar do Sul, Juliana A.; Jeandel, Catherine; Leinfelder, Reinhold; McNeill, John R.; Neal, Cath; … Williams, Mark, 2017a: "The Working Group on the Anthropocene: Summary of evidence and interim recommendations", in: *Anthropocene*, 19 (September): 55–60.

Zalasiewicz, Jan; Waters, Colin N.; Wolfe, Alexander P. et al., 2017b: "Making the Case for a formal Anthropocene Epoch: an analysis of ongoing critiques", in: *Newsletters on Stratigraphy*, 50(2) (March): 205–226.
Zalasiewicz, Jan; Steffen, Will; Leinfelder, Reinhold; Williams, Mark; Waters, Colin, 2017c: "Petrifying Earth Process: The Stratigraphic Imprint of Key Earth System Parameters in the Anthropocene", in: *Theory, Culture & Society*, 34(2–3): 83–104.
Ziegler, Dieter, 2010: "Die Industrialisierung", in: *WBG Weltgeschichte*, vol. V: *Entstehung der Moderne 1700 bis 1914* (Darmstadt: Wissenschaftliche Buchgesellschaft): 41–91.

The International Peace Research Association (IPRA)

Founded in 1964, the International Peace Research Association (IPRA) developed from a conference organised by Quaker International Conferences and Seminars in Clarens, Switzerland, 16–20 August 1963. The participants decided to hold international Conferences on Research on International Peace and Security (COROIPAS), which would be organised by a Continuing Committee in a similar way to the Pugwash Conferences. Under the leadership of John Burton, the Continuing Committee met in London on 1–3 December 1964. At that meeting, it took steps to broaden the original concept of holding research conferences. The decision was made to form a professional association with the principal aim of increasing the amount of research focused on world peace and ensuring its scientific quality.

An Executive Committee including Bert V. A. Röling, Secretary General (The Netherlands), John Burton (United Kingdom), Ljubivoje Acimovic (Yugoslavia), Jerzy Sawicki (Poland), and Johan Galtung (Norway) was appointed. This group was also designated as Nominating Committee for a fifteen-person Advisory Council to be elected at the first general conference of IPRA, to represent various regions, disciplines, and research interests in developing the work of the Association.

Since then, IPRA has held twenty-five biennial general conferences, the venues of which were chosen with a view to reflecting the association's global scope. IPRA, the global network of peace researchers, has just held its 25th General Conference on the occasion of its fiftieth anniversary in Istanbul, Turkey in August 2014, where peace researchers from all parts of the world had the opportunity to exchange actionable knowledge on the conference's broad theme of 'Uniting for sustainable peace and universal values'.

© Springer Nature Switzerland AG 2019
H. G. Brauch et al. (eds.), *Climate Change, Disasters, Sustainability Transition and Peace in the Anthropocene*, The Anthropocene: Politik—Economics—Society—Science 25, https://doi.org/10.1007/978-3-319-97562-7

The 26th IPRA General Conference took place from 27 November to 1 December 2016 in Freetown, Sierra Leone on the theme *Agenda for Peace and Development: Conflict Prevention, Post-conflict Transformation, and the Conflict, Disaster and Development Debate*. The 27th IPRA General Conference will take place in Ahmedabad, India, on 24–27 November 2018, on the theme *Innovation for Sustainable Global Peace*.

IPRA: http://www.iprapeace.org/.
The IPRA Foundation: http://iprafoundation.org/.

IPRA Conferences, Secretary Generals and Presidents 1964–2018

IPRA general conferences	IPRA secretary generals/presidents
1. Groningen, The Netherlands (1965)	1964–1971 Bert V. A. Röling (The Netherlands)
2. Tallberg, Sweden (1967)	1971–1975 Asbjorn Eide (Norway)
3. Karlovy Vary, Czechoslovakia (1969)	1975–1979 Raimo Väyrynen (Finland)
4. Bled, Yugoslavia (1971)	1979–1983 Yoshikazu Sakamoto (Japan)
5. Varanasi, India (1974)	1983–1987 Chadwick Alger (USA)
6. Turku, Finland (1975)	1987–1989 Clovis Brigagão (Brazil)
7. Oaxtepec, Mexico (1977)	1989–1991 Elise Bouding (USA)
8. Königstein, FRG (1979)	1991–1994 Paul Smoker (USA)
9. Orillia, Canada (1981)	1995–1997 Karlheinz Koppe (Germany)
10. Győr, Hungary (1983)	1997–2000 Bjørn Møller (Denmark)
11. Sussex, UK (1986)	2000–2005 Katsuya Kodama (Japan)
12. Rio de Janeiro, Brazil (1988)	2005–2009 Luc Reychler (Belgium)
13. Groningen, the Netherlands (1990)	2009–2012 Jake Lynch (UK/Australia)
14. Kyoto, Japan (1992)	Katsuya Kodama (Japan)
15. Valletta, Malta (1994)	2012–2016 Nesrin Kenar (Turkey)
16. Brisbane, Australia (1996)	Ibrahim Shaw (Sierra Leone/UK)
17. Durban, South Africa (1998)	2016–2018 Úrsula Oswald Spring (Mexico)
18. Tampere, Finland (2000)	Katsuya Kodama (Japan)
19. Suwon, Korea (2002)	**Presidents**
20. Sopron, Hungary (2004)	The first IPRA President was Kevin Clements (New Zealand/USA, 1994–98)
21. Calgary, Canada (2006)	
22. Leuven, Belgium (2008)	His successor was Úrsula Oswald Spring (Mexico, 1998–2000)
23. Sydney, Australia (2010)	
24. Mie, Japan (2012)	
25. Istanbul, Turkey (2014)	
26. Freetown, Sierra Leone (2016)	
27. Ahmedebad, India (2018)	

IPRA's Ecology and Peace Commission

IPRA's *Ecology and Peace Commission* (EPC) addresses the relationship between the Earth and human systems, and their impacts on peace. A special focus is placed on the linkages between problems of sustainable development and sustainable peace. The Ecology and Peace Commission evolved from IPRA's Food Study Group.

The conveners are elected for two years by the participants at IPRA conferences, to prepare the publications of the past conference and the sessions for the next conference. The conveners between the IPRA conferences in Mie (2012) and Istanbul (2014) were:

- Úrsula Oswald Spring (CRIM/UNAM, Cuernavaca, Mexico), Professor/Researcher at the National University of Mexico (UNAM) in the Regional Multidisciplinary Research Center (CRIM), lead author of the Intergovernmental Panel on Climate Change (IPCC);
- Hans Günter Brauch (Free University of Berlin (ret.), Peace Research and European Security Studies [AFES-PRESS], Mosbach, Germany); Chairman, Peace Research and European Security Studies (AFES-PRESS), Mosbach, Germany;
- Keith G. Tidball (Cornell University, Ithaca. NY, USA), Senior Extension Associate in the Department of Natural Resources and New York State Coordinator for the NY Extension Disaster Education Network.

© Springer Nature Switzerland AG 2019   237
H. G. Brauch et al. (eds.), *Climate Change, Disasters, Sustainability Transition and Peace in the Anthropocene*, The Anthropocene: Politik—Economics—Society—Science 25, https://doi.org/10.1007/978-3-319-97562-7

Based on papers presented to the IPRA conference in Mie they published the book:

 Ursula Oswald Spring; Hans Günter Brauch; Keith G. Tidball (Eds.): *Expanding Peace Ecology: Security, Sustainability, Equity and Peace: Perspectives of IPRA's Ecology and Peace Commission.* SpringerBriefs in Environment, Security, Development and Peace, vol. 12. Peace and Security Studies No. 2 (Cham – Heidelberg – New York – Dordrecht – London: Springer-Verlag, 2014).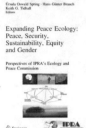

At Istanbul in August 2014 the conveners between the IPRA conferences in Istanbul (2014) and in Freetown (2016) were elected:

- Prof. Dr. Úrsula Oswald Spring (CRIM/UNAM, Cuernavaca, Mexico);
- PD Dr. Hans Günter Brauch (Free University of Berlin (ret.), Peace Research and European Security Studies [AFES-PRESS], Mosbach, Germany);
- Juliet Bennett, Ph.D. candidate (Centre for Peace and Conflict Studies, The University of Sydney, Australia).

Based on papers presented to the IPRA conference in Istanbul (August 2014) they published these two peer-reviewed books:

 Hans Günter Brauch, Úrsula Oswald Spring, Juliet Bennett, Serena Eréndira Serrano Oswald (Eds.): *Addressing Global Environmental Challenges from a Peace Ecology Perspective.* APESS No. 4 (Cham: Springer International Publishing, 2016).

Úrsula Oswald Spring, Hans Günter Brauch, Serena Eréndira Serrano Oswald, Juliet Bennett (Eds.): *Regional Ecological Challenges for Peace in Africa, the Middle East, Latin America and Asia Pacific.* APESS No. 4 (Cham: Springer International Publishing, 2016).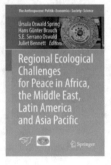

In November 2016 these four conveners were elected in Freetown for the period 2016–2018:

- PD Dr. Hans Günter Brauch (Free University of Berlin (ret.), Peace Research and European Security Studies [AFES-PRESS], Mosbach, Germany);
- Juliet Bennett, Ph.D. candidate (Centre for Peace and Conflict Studies, The University of Sydney, Australia);

- Prof. Dr. Andrew E. Collins, Northumbria University, Newcastle upon Tyne, UK;
- Rod Mena, Ph.D. candidate, Erasmus University, International Institute of Social Studies (ISS), The Hague, The Netherlands.

For the 2018 IPRA Conference in November 2018 in Ahmedabad, the conveners of the new *Ecology, Conflict Risks, Forced Migration and Peace* (ECR–FMP) Commission suggest devoting several sessions to the following themes:

A. **Ecology and Peace:**

  i. Conceptual Approaches to Peace Linkages, Peace with Nature and Peace Ecology
  ii. Societal Impacts of Global Environmental Change on Conflicts and Sustainable Peace
  iii. Sustainable Development Goals
  iv. The 2030 Agenda: Linking Sustainable Development and Sustainable Peace
  v. Decarbonisation of the Economy and Sustainability Transition in the Anthropocene.

B. **Disaster and Conflict Risks:**

  i. Complex Emergencies
  ii. Linking Early Warning of Natural Hazards and Conflicts
  iii. Integrated Disaster and Conflict Risk Reduction
  iv. High-intensity Conflict, Environmental Hazards and Disasters
  v. Post-conflict Peace-building, Environmental Hazards and Disasters.

C. **Forced Migration:**

  i. Theoretical approaches to migration
  ii. War-induced forced migration
  iii. Environmental forced migration

It is again planned to publish selected papers as a special issue of a peer-reviewed scientific journal or as a book in this series. The sessions at the IPRA conference in 2018 will be prepared and coordinated by:

- Prof. Dr. Andrew Collins, University of Northumbria, Newcastle upon Tyne, UK;
- Dr. Lydia Gitau, University of Sydney, Australia;
- Mr. Rodrigo Mena, Rotterdam University, The Netherlands;
- PD Dr. Hans Guenter Brauch (Free University of Berlin (ret.), Peace Research and European Security Studies [AFES-PRESS], Mosbach, Germany).

About the Editors

Hans Günter Brauch (Germany), Dr., until 2012 Adj. Prof. (Privatdozent) at the Faculty of Political and Social Sciences, Free University of Berlin; since 1987 chairman of *Peace Research and European Security Studies* (AFES-PRESS). He is editor of: *Hexagon Book Series on Human and Environmental Security and Peace*; *SpringerBriefs in Environment, Security, Development and Peace* (ESDP); *SpringerBriefs on Pioneers in Science and Practice*; *Pioneers in Arts, Humanities, Science, Engineering, Practice*; and of *The Anthropocene: Politik—Economics—Society—Science* with Springer International Publishing, He was guest professor of international relations at the universities of Frankfurt a Main, Leipzig, Greifswald, and Erfurt, a research associate at Heidelberg and Stuttgart, and a research fellow at Harvard and Stanford Universities. In autumn and winter 2013–2014 he was a guest professor at Chulalongkorn University in Bangkok. He has been a member of the IPRA Council (1996–2000 and 2016–2018), of its Executive Committee (2016–2018), Chairman of IPRA's Subcommission on Statutes (2016–2018) and Co-convenor of IPRA's *Ecology and Peace Commission* (EPC) from 2012–2018, as well as a co-editor of its four peer-reviewed book publications (see below).

He has published on security, armaments, climate, energy, migration and Mediterranean issues in English and German, and has been translated into Spanish, Greek, French, Danish, Finnish, Russian, Japanese, Portuguese, Serbo-Croatian, and Turkish. His recent books in English and Spanish include: (co-edited with Liotta, Marquina, Rogers, Selim), *Security and Environment in the Mediterranean. Conceptualising Security and Environmental Conflicts*, 2003; (co-edited with Oswald Spring, Mesjasz, Grin, Dunay, Chadha Behera, Chourou, Kameri-Mbote, Liotta), *Globalization and Environmental Challenges: Reconceptualizing Security in the 21st Century, 2008;* (co-edited with Oswald Spring, Grin, Mesjasz, Kameri-Mbote, Chadha Behera, Chourou, Krummenacher), *Facing Global Environmental Change: Environmental, Human, Energy, Food, Health and Water*

© Springer Nature Switzerland AG 2019   241
H. G. Brauch et al. (eds.), *Climate Change, Disasters, Sustainability Transition and Peace in the Anthropocene*, The Anthropocene: Politik—Economics—Society—Science 25, https://doi.org/10.1007/978-3-319-97562-7

Security Concepts, 2009; (co-edited with Oswald Spring), *Reconceptualizar la Seguridad en el Siglo XXI*, 2009; (co-edited with Oswald Spring, Mesjasz, Grin, Kameri-Mbote, Chourou, Dunay, Birkmann), *Coping with Global Environmental Change, Disasters and Security—Threats, Challenges, Vulnerabilities and Risks*, 2011; (co-edited with Scheffran, Brzoska, Link, Schilling), *Climate Change, Human Security and Violent Conflict*, 2012; (co-edited with Oswald Spring, Grin, Scheffran), *Handbook on Sustainability Transition and Sustainable Peace*, 2016; (co-edited with Oswald Spring, Bennett and Serrano Oswald), *Addressing Global Environmental Challenges from a Peace Ecology Perspective*, 2016; and (co-edited with Oswald Spring, Serrano Oswald and Bennett), *Regional Ecological Challenges for Peace in Africa, the Middle East, Latin America and Asia Pacific*, 2016.

Address: PD Dr. Hans Günter Brauch, Alte Bergsteige 47, 74821 Mosbach, Germany.
Email: brauch@afes-press.de.
Website: http://www.afes-press.de and http://www.afes-press-books.de/.

Úrsula Oswald Spring (Mexico), full-time Professor/ Researcher at the National University of Mexico (UNAM) in the Regional Multidisciplinary Research Center (CRIM), has been national coordinator of water research for the National Council of Science and Technology (RETAC-CONACYT), first Chair of Social Vulnerability at the United National University Institute for Environment and Human Security (UNU-EHS); founding Secretary General of El Colegio de Tlaxcala; General Attorney of Ecology in the state of Morelos (1992–1994), National Delegate of the Federal General Attorney of the Environment (1994–1995); and Minister of Ecological Development in the state of Morelos (1994–1998). She was President of the International Peace Research Association (IPRA, 1998–2000), and General Secretary of the Latin American Council for Peace Research (2002–2006).

She studied medicine, clinical psychology, anthropology, ecology, and classical and modern languages. She obtained her Ph.D. from the University of Zürich in 1978. For her scientific work she received the Price Sor Juana Inés de la Cruz (2005), the Environmental Merit award in Tlaxcala, Mexico (2005, 2006), and the UN Development Prize. She was recognised as Women Academic at UNAM (1990 and 2000) and Woman of the Year (2000). She works on non-violence and sustainable agriculture with groups of peasants and women and is President of the Advisory Council of the Peasant University.

She has written 46 books and more than 328 scientific articles and book chapters on sustainability, water, gender, development, poverty, drug consumption, brain damage due to under-nourishment, peasantry, social vulnerability, genetic modified organisms, bioethics, and human, gender, and environmental security, peace and conflict resolution, democracy, and conflict negotiation.

Address: Prof. Dr. Úrsula Oswald Spring, CRIM-UNAM, Av. Universidad s/n, Circuito 2, Col. Chamilpa, Cuernavaca, CP 62210, Mor., Mexico.
Email: uoswald@gmail.com.
Website: http://www.afes-press.de/html/download_oswald.html.

Andrew E. Collins (United Kingdom) is Professor of Disaster and Development and Leader of the Disaster and Development Network (DDN), Northumbria University, UK. His research, based on a range of countries, informs theoretical, methodological and policy aspects of disaster risk reduction and response, health ecology and sustainable development. This addresses crisis and well-being in rapid- and slow-onset environmental, social and political change using people-centric approaches and multi-sector partnering. He entered academia following voluntary skill sharing in places affected by war and extreme poverty. He gained a Ph.D. from King's College, London in 1996 and in 2000 established the world's first and currently ongoing integrated disaster management and sustainable development postgraduate programme at Northumbria.

He is an elected Board Director of the International Society for Integrated Disaster Risk Management (IDRiM), Chair of the Global Alliance of Disaster Research Institutes (GADRI) and Co-Chair of the UK Alliance of Disaster Research Institutes (UKADR), and has a number of United Nations roles. He recently completed a four-year period as elected Chair of the Enhanced Learning and Research for Humanitarian Assistance (ELRHA) Steering Group supported by the UK government, the Welcome Trust, and humanitarian agencies.

Previous books include *Disaster and Development* (Routledge, 2009) and *Hazards, Risks and Disasters in Society* (Elsevier, 2015).

Address: Professor Andrew Collins, B308, Department of Geography and Environmental Sciences/Disaster and Development Network (DDN), Northumbria University, Newcastle upon Tyne, United Kingdom, NE1 8ST.
Email: andrew.collins@northumbria.ac.uk.
Website: https://www.northumbria.ac.uk/about-us/our-staff/c/andrew-collins/.

Serena Eréndira Serrano Oswald (Mexico) is associate professor at the Regional Multidisciplinary Research Center, National Autonomous University of Mexico (CRIM-UNAM). She holds a Ph.D. in Social Anthropology (UNAM), an M.Sc. in Social Psychology (LSE, London), a Master's in Family Therapy (MFT) in Systemic Family Therapy (CRISOL), and a BA (Hons) in Political Studies and History (SOAS, London). She has a Post-doctorate in Sociology and Gender (UNAM), a professional diploma in translation and interpreting (Institute of Linguists), and eight specialised postgraduate trainings in person-centered therapy, psychopathology, Gestalt therapy, couples therapy, adolescents and infants, thanatology, addictions, and teaching and supervision (at CRISOL and the Gestalt Institute). She has worked on 19 research projects, has 14 edited books and 70 articles and book chapters, has taught over 70 courses, and has participated 270 times in congresses. Certified by the National Council of Researchers (SNI I), she presides over the Latin American and Caribbean Regional Development Association (LARSA-RSAI), and was president of the Mexican Regional Development Association (AMECIDER) from 2013 to 2016. She has been Secretary General of the Latin American Council for Peace Research (CLAIP) since January 2017 and is a member of eight national and eight international professional and research associations. Among her recent book publications she has co-edited: *Nuevos Riesgos Socio-ambientales para la Paz y los Derechos Humanos en América Latina* (2018, CRIM-UNAM); *Risks, Violence, Security and Peace in Latin America* (2018, Springer); and *Ciudad, género, cultura y educación en las regiones* (IIEc-UNAM/AMECIDER, 2018).

Address: Dr. Serena Eréndira Serrano Oswald, CRIM-UNAM, Av. Universidad s/n, Circuito 2, Col. Chamilpa, Cuernavaca, Morelos, Mexico CP 62210.
Email: sesohi@gmail.com.

About the Other Contributors

Hans Günter Brauch (Germany), Dr., Adj. Prof. (Privatdozent) at the Faculty of Political and Social Sciences, Free University of Berlin (ret.); since 1987 chairman of *Peace Research and European Security Studies* (AFES-PRESS); see above under editors.

Andrew E. Collins (United Kingdom) is Professor of Disaster and Development and Leader of the Disaster and Development Network (DDN), Northumbria University, UK; see above under editors.

Samantha Melis (The Netherlands) is a Ph.D. candidate at the International Institute of Social Studies (ISS), The Hague, part of Erasmus University (EUR). She is currently involved in the project "When disasters meet conflict", funded by the Netherlands Organisation for Scientific Research (NWO). Within this project, she focuses on the response to socio-natural disasters in post-conflict scenarios. She received an MA in Conflict Studies and Human Rights from Utrecht University in 2009 and an M.Sc. in Cultures and Development Studies from KU Leuven in 2011. She has worked for an international NGO in Burundi and has gained extensive experience in monitoring and evaluation and conducting research in conflict-affected countries. The outputs of her research have been presented in articles, at high-level conferences, and in blogs.

Address: Ms Samantha Melis, International Institute of Social Studies (ISS), Kortenaerkade 12, 2518 AX The Hague, The Netherlands.
Email: melis@iss.nl.
Website: https://www.eur.nl/people/samantha-melis.

Rodrigo Mena (Chile) is a socio-environmental researcher, currently involved in the project "When disasters meet conflict" at the International Institute of Social Studies (ISS) at Erasmus University Rotterdam. His main focus is on humanitarian aid response to disasters occurring in high-intensity conflict-affected scenarios. For more than a decade he has worked in academia, the public and private sectors,

© Springer Nature Switzerland AG 2019
H. G. Brauch et al. (eds.), *Climate Change, Disasters, Sustainability Transition and Peace in the Anthropocene*, The Anthropocene: Politik—Economics—Society—Science 25, https://doi.org/10.1007/978-3-319-97562-7

government, local and international NGOs and the UN. As well as fieldwork in hostile, complex and remote areas, he has also led humanitarian aid groups and conducted research following a number of disasters. As the author of articles and book chapters, his research focuses on resilience and vulnerability to socio-environmental disasters and conflicts, as well as on applied fieldwork research. From 2017 he has co-chaired the Ecology and Peace Commission (EPC) of the International Peace Research Association (IPRA).

Address: Mr. Rodrigo Mena F, International Institute of Social Studies (ISS), Kortenaerkade 12, 2518 AX, The Hague, The Netherlands.
Email: mena@iss.nl.
Website: https://www.eur.nl/people/rod-mena-fluhmann.

Mokua Ombati (Kenya) is a doctoral fellow affiliated to the Anthropology and Human Ecology Department, Moi University, Kenya. His academic and research interests are focused on African indigenous knowledge systems; gender, children and youth issues; and peace, security and non-violence. This article is an improved version of a research manuscript prepared for the 26th *International Peace Research Association* (IPRA) general conference, held in Freetown, Sierra Leone from 27 November to 1 December 2016. I am overly grateful to the *International Peace Research Association Foundation* (IPRAF) for a partial scholarship to facilitate my participation at the conference. I am also immensely indebted to the co-editors and anonymous reviewers who spent much of their unpaid time and effort to read and make specific contributions, which helped shape and develop the manuscript into a publishable book chapter.

Address: Mokua Ombati, Anthropology and Human Ecology Department, School of Arts and Social Sciences, Moi University Main Campus. P.O. Box 3900-30100, Eldoret, Kenya.
Email: keombe@gmail.com.

Úrsula Oswald Spring (Mexico), full-time Professor/Researcher at the National University of Mexico (UNAM) in the Regional Multidisciplinary Research Center (CRIM), Co-Secretary General of IPRA (2016–2018); see above under editors.

Serena Eréndira Serrano Oswald (Mexico) is research professor at the Regional Multidisciplinary Research Centre, National Autonomous University of Mexico (CRIM-UNAM), Secretary General of the Latin American Council for Peace Research (CLAIP) (2017–2018); see above under editors.

Printed in the United States
By Bookmasters